DIGITAL TWIN FOR HEALTHCARE

DIGITAL TWIN FOR HEALTHCARE

Design, Challenges, and Solutions

Edited by

ABDULMOTALEB EL SADDIK
School of Electrical Engineering and Computer Science at the University of Ottawa
Ottawa, ON, Canada

ACADEMIC PRESS
An imprint of Elsevier

ISBN: 978-0-323-99163-6

For information on all Academic Press publications
visit our website at https://www.elsevier.com/books-and-journals

Publisher: Mara E. Conner
Acquisitions Editor: Sonnini R. Yura
Editorial Project Manager: Howi M. De Ramos
Production Project Manager: Maria Bernard
Cover Designer: Vicky Pearson

Typeset by VTeX

Working together
to grow libraries in
developing countries

www.elsevier.com • www.bookaid.org

Contents

4. Challenges of Digital Twin in healthcare

Hazim Dahir, Jeff Luna, Ahmed Khattab, Kaouther Abrougui, and Raj Kumar

5. Intelligent digital twin reference architecture models for medical and healthcare industry

Zhi Wang and Abdulmotaleb El Saddik

6. Artificial intelligence models in digital twins for health and well-being

Rahatara Ferdousi, Fedwa Laamarti, and Abdulmotaleb El Saddik

7. COVIDMe: a digital twin for COVID-19 self-assessment and detection

Roberto Martinez-Velazquez, Fernando Ceballos, Alejandro Sanchez, Abdulmotaleb El Saddik, and Emil Petriu

8. Improving human living environment and human health through environmental digital twins technology

Zhihan Lv and Dongliang Chen

9. Role of smart technologies in detecting cognitive impairment and enhancing assisted living

Devvrat Bhardwaj, Jeffrey Jutai, and Pascal Fallavollita

10. Digital twins and cybersecurity in healthcare systems

Issam Al-Dalati

11. Potential applications of digital twin in medical care

Kaouther Abrougui, Hazim Dahir, Ahmed Khattab, Jeff Luna, Raj Kumar, and
Rashika Verma

16. Digital twins for allergies

Kamran Gholizadeh HamlAbadi, Monireh Vahdati, Ali Mohammad Saghiri, and
Kimia Gholizadeh

Contributors

Kaouther Abrougui   Voltron Data Inc., Mountain View, CA, United States

Issam Al-Dalati   Cyber Security Architect, Ottawa, ON, Canada

Moayad Aloqaily   Machine Learning Department, Mohamed Bin Zayed University of Artificial Intelligence (MBZUAI), Abu Dhabi, United Arab Emirates

Devvrat Bhardwaj   School of Electrical Engineering and Computer Science, Faculty of Engineering, University of Ottawa, Ottawa, ON, Canada

Ouns Bouachir   Zayed University, Dubai, United Arab Emirates

Fernando Ceballos   Department of Information Technology, University of Colima, Colima, Mexico

Dongliang Chen   College of Computer Science and Technology, Qingdao University, Qingdao, China

Hazim Dahir   Cisco Systems, Research Triangle Park, NC, United States

Abdulmotaleb El Saddik   University of Ottawa, Ottawa, ON, Canada
Mohamed bin Zayed University of Artificial Intelligence, Abu Dhabi, United Arab Emirates

Mohd Faisal   University of Ottawa, Ottawa, ON, Canada

Pascal Fallavollita   School of Electrical Engineering and Computer Science, Faculty of Engineering, University of Ottawa, Ottawa, ON, Canada
Interdisciplinary School of Health Sciences, Faculty of Health Sciences, University of Ottawa, Ottawa, ON, Canada

Rahatara Ferdousi   University of Ottawa, Ottawa, ON, Canada

Kimia Gholizadeh   Department of Computer and Electrical Engineering, Mazandaran University of Science and Technology, Babol, Iran

Kamran Gholizadeh HamlAbadi   Young Researchers and Elite Club, Qazvin Branch, Islamic Azad University, Qazvin, Iran
Faculty of Computer and IT Engineering, Qazvin Branch, Islamic Azad University, Qazvin, Iran

Jeffrey Jutai   Interdisciplinary School of Health Sciences, Faculty of Health Sciences, University of Ottawa, Ottawa, ON, Canada

Fakhri Karray   Machine Learning Department, Mohamed Bin Zayed University of Artificial Intelligence (MBZUAI), Abu Dhabi, United Arab Emirates

Ahmed Khattab   Cisco Systems, San Jose, CA, United States

Raj Kumar   Cisco Systems, San Jose, CA, United States

Fedwa Laamarti   University of Ottawa, Ottawa, ON, Canada
Mohamed bin Zayed University of Artificial Intelligence, Abu Dhabi, United Arab Emirates

Jeff Luna   Cisco Systems, San Jose, CA, United States

Zhihan Lv   Faculty of Arts, Uppsala University, Visby, Sweden

Roberto Martinez-Velazquez   University of Ottawa, Ottawa, ON, Canada
Mohamed bin Zayed University of Artificial Intelligence, Abu Dhabi, United Arab Emirates

Roberto Alejandro Martinez Velazquez   University of Ottawa, Ottawa, ON, Canada

Emil Petriu   University of Ottawa, Ottawa, ON, Canada

Ali Mohammad Saghiri   Computer Engineering Department, AmirKabir University of Technology, Tehran, Iran

Alejandro Sanchez   Department of Information Technology, University of Colima, Colima, Mexico

GilAnthony Ungab M.D.   Lucia Health Guidelines, San Francisco, CA, United States

Monireh Vahdati   Young Researchers and Elite Club, Qazvin Branch, Islamic Azad University, Qazvin, Iran
Faculty of Computer and IT Engineering, Qazvin Branch, Islamic Azad University, Qazvin, Iran

Rashika Verma   Rutgers New Jersey Medical School, Newark, NJ, United States

Zhi Wang   University of Ottawa, Ottawa, ON, Canada

1

Introduction

Abdulmotaleb El Saddik[a,b]

[a]University of Ottawa, Ottawa, ON, Canada, [b]Mohamed bin Zayed
University of Artificial Intelligence, Abu Dhabi, United Arab Emirates

1.1 History of digital twin

The future of humankind depends on developing systems that can address the computational demands of expanding digitized data and related advanced hard/software solutions. For the next few years, new societal trends and challenges will be increasingly prevalent in the fields of health and wellness, security and safety, manufacturing and construction, transport and energy, and mobility and communications. The convergence of technologies and scientific knowledge will boost citizens' wellness and enhance their quality of life. Hence, a promising solution to accomplish this convergence is through the development of the digital twin. According to Market Research Future, it is expected that the global digital twin market will reach $35,462.4 million by 2025.[1]

Gartner identified digital twins as one of the Top 10 Strategic Technology Trends since 2017. The use of the digital twin became popular during the digitization of machinery and production systems in the manufacturing industry which was introduced in the early 2000s. For example, General Electric (GE) builds digital twins for its machines, which are cloud-hosted software models that use information collected from sensors and processed using artificial intelligence, physics-based models, and data analytics. The DT was defined by Grieves and Vickers as "a set of virtual information constructs that fully describes a potential or actual physical manufactured product from the micro atomic level to the macro geometrical level. At its optimum, any information that could be obtained from inspecting a physical manufactured product can be obtained from its Digital Twin". This definition speaks to the prevalence of the manufacturing

[1]https://www.marketresearchfuture.com/reports/digital-twin-market-4504.

1

industry in the usage of the DT. However, the potential of this technology in other domains became slowly more apparent. In 2018, the DT has thus been redefined as a digital replica of a living or nonliving physical entity [1]. Given this redefinition, the benefits of DT spread to multiple other domains, where the DT of human beings promises to play a crucial role in the increase of humans' well-being. This redefinition has been matched by Gartner in their 2019 classifying the digital twin as one of the top 10 strategies for the second year in a row where digitals twins are referred to as "digital representations of people, processes and things" [3] people being one of the focuses of DT works.

Indeed, Digital Twin is by no means a new concept. Simulation, physics-engines and data-driven modeling, virtual prototypes or digital threads, device shadows were all concepts that could easily been understood as digital twins. In general, a key challenge hindering the widespread adoption of the Digital Twin concept is having an accepted shared definition and common language amongst experts and users. (See Table 1.1.)

Considering the speed, direction, and major advances that modern technology is able to achieve, the digital twin will disrupt industries beyond manufacturing. Therefore, it is critical to redefine the digital twin. Instead, a digital twin should be defined as a **digital replica of a living or non-living entity synchronized at a specified frequency and fidelity,** in which the virtual entity will exist simultaneously with the physical entity through the seamless **bi-directional** transmission of data, which works in a closed loop process. In this regard, the digital twin will facilitate the means to monitor, understand, and optimize the functions of the physical entity be it living or non-living in a closed loop manner.

Unlike the current trend of the digital twin, this definition suggests that a digital twin can be implemented on a living entity. Once implemented, the digital twin will revolutionize many domains. Let us consider the study of biology as an example: if a tree were to have a digital twin, a great amount of critical data could be collected. Scientists would be able to digitally monitor and examine the external components of a tree as well as the internal components. They could measure the amount of oxygen released by a tree, the amount of water consumed, and the amount of sunlight received. They would be able to determine the age of the tree and follow its growth from seedling to adult. Moreover, a digital twin could help scientists to monitor and combat parasites and other harmful pests or diseases before they spread to other parts of the tree.

A digital twin for humans would also involve the collection and processing of a large amount of data. It could collect physical, physiological, and context data. Physical and physiological data could help to predict illnesses or disease. A stroke could be predicted before it occurs providing the possibility to prevent the stroke. Machine and deep learning

TABLE 1.1 Digital Twin definitions across industry and academia.

Source	Definition
General Electric	A digital twin is a living model that drives a business outcome [4].
IBM	A digital twin is a virtual representation of a physical object or system across its lifecycle, using real-time data to enable understanding, learning and reasoning [5].
Microsoft	A digital twin is a virtual model of a process, product, production asset or service. Sensor-enabled and IoT- connected machines and devices, combined with machine learning and advanced analytics, can be used to view the device's state in real-time. When combined with both 2D and 3D design information, a digital twin can visualize the physical world and provide a method to simulate electronic, mechanical, and combined system outcomes [6].
Siemens	A digital twin is a virtual representation of a physical product or process, used to understand and predict the physical counterpart's performance characteristics [7].
NASA	A digital twin integrates ultra-high fidelity simulation with the vehicle's on-board integrated vehicle health management system, maintenance history and all available historical and fleet data to mirror the life of its flying twin and enable unprecedented levels of safety and reliability [8].
Deloitte	A digital twin is a near-real-time digital image of a physical object or process that helps optimize business performance [9].
Michael Grieves	The digital twin is a set of virtual information constructs that fully describes a potential or actual physical manufactured product from the micro (atomic level) to the macro (geometrical level) [10].
Abdulmotaleb El Saddik	A digital Twin is a virtual representation of any living or nonliving entity [1].
Digital Twin Consortium	A digital twin is a virtual representation of real-world entities and processes, synchronized at a specified frequency and fidelity [11].
Gartner	A digital twin is a digital representation of a real-world entity or system. The implementation of a digital twin is an encapsulated software object or model that mirrors a unique physical object, process, organization, person or other abstraction [12].

techniques could be used to analyze data to detect patterns and predict potential health problems. Context data, such as information about environment, age, light, emotions, and preferences could also be collected to fully understand and characterize the comprehensive condition of a user. In this way, the digital twin would be able to assist the user in analyzing and interpreting significant amounts of data.

It is a trend that promises to deliver great value in and across many application domains such as healthcare, manufacturing, and supply chains. However, Managers, CEOs and politicians need to understand many of its underlying technology as well as evaluate the benefits and drawbacks of applying digital twins. Municipalities, and Companies must overcome many organizational and technological barriers to achieve a feasible solution that brings the value digital twin technology can offer.

According to the digital twin consortium, the following advantages can be identified:

- Digital twin systems transform business by accelerating holistic understanding, optimal decision-making, and effective action.
- Digital twins use real-time and historical data to represent the past and present and simulate predicted futures.
- Digital twins are motivated by outcomes, tailored to use cases, powered by integration, built on data, guided by domain knowledge, and implemented in cyber-physical systems.

1.2 Elements of changes

Many technologies have developed over the years, and some of them have witnessed very fast progress recently. These technologies have had a big impact in many areas, and their potential has not even been explored in all possible ways. In this chapter, we will be introducing these technologies, defining them, and presenting the possible convergence of these technologies. We will explain how each of them is a big player in itself, and the potential that lays in bringing them together in an impactful emerging technology that is the Digital Twin.

1.2.1 What has changed regarding content?

Many people will define multimedia as things in the middle, such as simple images, video, or audio. Those are indeed part of multimedia. However, this definition is far from completed. First, there are many other multimedia types, such as olfactory media, which are the media involving the sense of smell, and gustatory media which involve the sense of taste. These media, even though not developed nearly as much as the audio/video systems, are already gaining interest in the research community and have the potential to play very useful roles in multimedia systems. Second, the multimedia field is not concerned only with the media files but also and mainly with communicating those media bits between different parties. Indeed, the field of multimedia communications is at the heart of the current development and the technologies involved in this field play a major role in most cutting-edge applications involving the internet of things and network communications.

We are all good at using and handling types of media such as text, audio, and video because we are used to them. We use them on social media every day, like Twitter, Instagram, YouTube, etc. However, we are much less used to media types like haptics, smell, and taste. So let us start first with an overview of the existing types of media and on the creation process of these media and how they make their way to you. Let us consider the world in which we live in today. There are many media sources around us, such as landscapes than we can photograph, scenes that we can record, different weather conditions that we can register. And with the internet of things, there is now a huge amount of different sensory data measured by more and more sensors everywhere. There is the internet as a source of media, like social media which has high amount of content uploaded by users every day. This data is captured, and usually transmitted to the cloud where it is stored. Then this data needs to be represented in a better way, so we need synchronization: in order for this data to be synchronized properly. Subsequently, we want to present this media to users/consumers in the best possible way, and on the most appropriate device depending on the use of these multimedia. Therefore, we consider the whole process of multimedia capturing, transmitting, representation, and presentation as multimedia communications research field.

One of the emerging media types is haptics, which refers to the science of touch. Haptics is composed of mainly two types of information. The first one is the tactile, where we feel the texture of things or a bump for example, and the second one is kinesthetics, where we feel the force feedback, like when we press a soft ball, or hold a dumbbell, etc. We all make use of some form of haptics in our daily life. For example, whenever you have your cellphone on vibrate mode and you receive a call, you feel the phone vibration which is a unidimensional force feedback, a form of haptic feedback. Now this is a trivial kind of haptic feedback that we came to take for granted as we became accustomed to it. However, think about the doors that haptic feedback can open in our daily lives. Haptic can have a bigger impact such as the possibility of hand shake over the internet during a video conference for example, which would add much more interactivity to the traditional audio-video conferencing.

New types of media can change the way we go about online shopping. We can feel the fabric of a new outfit over the internet, smell a flower or perfume remotely, or even tasting food online before ordering it. Touch, smell and taste are all media types that will soon emerge to serve the ongoing change in the way we communicate. Furthermore, multisensory media or hybrid media that go beyond the audio-visual, can deliver an even more immersive interaction between multiple parties or between a stakeholder and the digital world.

1.2.2 Content and the significance of velocity, scope, and impact

The data we use to which we refer as multimedia content is captured from the world around us. Two main types of sensors are used for that.

Hard sensors and soft sensors. Hard sensors are physical devices that we can find everywhere, such as at home like a temperature sensor, or a security camera, or in the car such as gas information sensor etc. With the Internet of Things, the number of these hard sensors is exploding, as everything around us is being equipped with sensing capability. These sensors capture data in different formats and at a different frequency, which constitute challenges when we want to aggregate all of this data, to understand it and give the full picture. The other main source of multimedia content is soft sensors. These refer to data entered by humans, such as social media data, medical data records, financial data, etc. The soft sensors provide a huge amount of data. Take for example Fakebook, with about 2.4 billion monthly active users, or Instagram which facilitates better inclusion of a diversified multimedia content of images and video, has 1 billion active users so far.

Many think that the use of this multimedia content is for entertainment only, or even mainly for entertainment. This is a misconception about multimedia. In fact. This data can serve many purposes. For example, on an individual level, information extracted from this data can assist with recommendations for the person about a sport or a set of physical activities that are suitable specifically for them, or it can assist them with shopping by making recommendations about e-commerce platforms. In can assist companies by giving a better understanding of customer profiles to help with customer retention or give a better view on the stock market. How can we obtain such information from the collected data, and how can we handle the wide variety of data formats of the collected multimedia content?

1.2.3 Making sense of the data

Traditional approaches, methods and algorithms cannot achieve this requirement and handle such data discrepancy and data volumes. The need for Artificial Intelligence is clear. A subset of Artificial Intelligence is machine learning, for which many algorithms have been developed in recent years. And a subset of machine learning is deep learning. Deep learning is mainly used for machine vision, such as image classification for which deep learning is known to perform very well.

Indeed, if we visit an ecommerce website like eBay or Amazon, after just a few interactions with the website, we quickly notice the recommendations that start being displayed as we navigate through the website. Hence, items that are likely to be of interest to us are displayed on top or suggested to us. The reason for this is that a lot of soft-sensory information about us has been recorded by the system and processed, to recommend items that we are likely to purchase as per the system's findings. This process of gathering costumer information and processing it to extract customer behavior and purchase tendencies is called profiling. Machine learning has made profiling much more effective. It is being used by many companies for customer retention.

Given the gathered and constructed content, many multimedia applications become possible, like video on demand, interactive TV, such as provided by video streaming services. Multimedia intelligence plays an important role here again, where movies or series recommendations are provided based on history of movies consumption of each costumer. The same tendency has been used for electronic commerce, video on demand, and hopefully it will also be used in the future for distance learning, providing customers with the best learning experience through intelligent multimedia systems.

1.2.4 Touching, smelling and tasting data

Here, we do not mean the type of basic functions such as pressing the play or stop buttons to control a video, or even the rewind or fast-forward buttons, there is no interaction there. The environment can be interactive with the use of virtual environments and their immersive nature, where the virtual world responds to the users' actions, speech, or even thoughts through computer-brain interfaces. Content will be even more interactive, especially with the integration of the sence of touch (haptic), and in the future smell and taste. With the integration of these modalities, even regular movies can be turned into interactive ones in the future.

To achieve this, smart actuators are incorporated within the system. Actuators are physical things that generate motion, force, heat, or flow in the physical environment based on a specific command. Suppose for example that you are watching an interactive movie, and in one of the scenes, the chair you are sitting on starts vibrating in sync with the action in the scene. Similarly, the temperature in the room is controlled by the system connected to the movie, and the room may become colder or warmer depending on the scenes being watched.

Multimedia interaction can happen through multiple modalities. A modality is a channel by which information is communicated between the computer and the human. Today, this information is communicated mostly as text, image, or video. However, other modalities as we mentioned can be used to convey information, such as by means of heat, chair vibrations, air flow, etc.

1.2.5 Everyone and everything are getting connected

Data is transmitted all the time and by virtually everyone with access to internet. A high number of images and videos are uploaded to the cloud and downloaded from the cloud daily by users, companies, organizations, etc., for a variety of purposes including entertainment, advertising, and learning. An image or video transmission time can vary depending on the size of the media. However, the transmission time also depends to a large extent on the type of network in use. If a 4G network is used it takes longer to transmit the media, comparing to transmission over a 5G network. This will be detailed in the chapter on communication networks. Nonetheless, media also have their own network requirements, depending on their size but also on the nature of the media. For example, if an image is to be transmitted, the delay in transmission may not be pleasant, but it will not affect the viewing of the image once received. On the contrary, if we are sending a video, which is a succession of compressed images, the delays in transmission of those images will be introduced lags in the video display when the video is being streamed to the user. This will greatly affect the quality of experience the user will have. In the case of haptic media transmission, the problem is even more noticeable. Indeed, haptic media consist of blocks of information small in size, but that need to be transmitted at high speed. This is due to the nature of applications where haptics plays an important role in the interaction, as we will detail in the chapter on Multimodal Interaction. Consequently, the characteristics of the media in question are determining factors in which type of network is needed for a successful and timely multimedia transmission.

1.2.6 Big brother is watching

Is the multi-sensory data secure? The short answer is no, it should never be assumed that the use of multimedia is secure in and by itself. The long answer is we can make multimedia use more secure by taking the necessary measures at multiple levels.

Security at the content level: This is concerned with making sure only the authorized parties are granted access to specific data, and that this access is granted with the proper rights such as read only access or read and write access, etc. Besides, in multimedia content security, we are also concerned with copyright protection, which is achieved through watermarking as we will see in the chapter on multimedia cybersecurity.

Security at the device level: The smartphone which started as voice communication tool, has become now one of the best multimedia tools in 2020. This multimedia device has access to internet most of the time and is used around the clock, which makes it very important to ensure that these devices are secured. This has evolved from simple passwords to being now achieved through biometrics such as fingerprint mapping or face recognition. Even though leaps have been made in the technology of biometrics, we will talk in the chapter on cybersecurity about the flaws to be aware of as far as the security of multimedia devices is concerned.

Security at the networking level: As we have discussed, high amounts of multimedia content are transmitted over the internet every day. However, internet is a public network that can be accessed by everyone, including hackers and intruders. Consequently, it is necessary to ensure multimedia security at the network level as well. To achieve this, we make use of Virtual Private Networks (VPN), or Hypertext Transfer Protocol Secure (https), or other types of secure protocols.

Furthermore, another important aspect to take into consideration in multimedia application is privacy protection. Suppose you are using one of the social media platforms such as Instagram, and you post a picture with some text and hashtags. Then you go to another social media network like Facebook for instance, and surprisingly you find posts related to those same hashtags popping up in this other social media platform. This is because those platforms are interconnected, which means that your data in one platform is shared on the other platform. To ensure that only authorized user information is being shared, privacy protection measures have to be put in place.

1.3 The convergence of technologies

What is a Digital Twin (DT)? DTs are "digital replications of living as well as nonliving entities that enable data to be seamlessly transmitted between the physical and virtual worlds" [1]. This means that any interaction happening in the real world will be transmitted to the digital world, and any interaction in the virtual will also be transmitted to the real world. What makes the DT unique is the convergence of five cutting edge technologies.

Let us summarize the process we have described previously. In a nutshell, we collect the data, hard and soft, we transmit this data, we make this data intelligent, we facilitate people's interaction with this data, and we secure this data. So, the DT is the convergence of these five technologies mentioned earlier.

A literature review conducted in [2] in 2018 discusses the need to identify the key technologies involved in the DT. We analyze this matter in the current section of this book. "Digital twins facilitate the means to monitor, understand, and optimize the functions of the physical entity and provides continuous feedback to improve quality of life and well-being" [1]. To achieve this, multiple steps have to be undertaken. Let us check this process more closely.

To constantly monitor the physical entity and create a digital replica of it, we need the internet of things (IOT) technology, as shown in Fig. 1.1. Advances in IOT facilitate continuous data collection by means of connected sensors attached to the real twin. Continuous tracking also involves data collected from social networks. Other sources of data include health records, financial systems, etc.

Given the continuous data collection over time, as well as the variety of sensors types collecting various kind of data about the real twin, huge

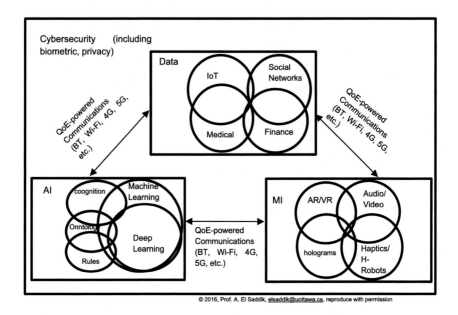

FIGURE 1.1   Convergence of the Digital Twin key technologies.

amounts of data are gathered. Hence, there is a need for the use of big data technologies. Once large amounts of data are collected, and continue to be collected as we monitor the real twin, we need data analytics to extract useful information from this data. Traditional algorithms are not very useful here given the nature of the data, which consists of huge amounts of heterogeneous and unstructured data. We need Artificial Intelligence algorithms to make sense of this data, be it machine learning, deep learning, ontologies, rules, etc.

As soon as the data is analyzed, the next focus becomes how to efficiently deliver this information to the real twin, and allow them to interact with their DT. Multimodal interaction can enable this kind of communication, interfacing the real and digital twin. Multimodal interaction can happen through a physical representation of the DT using soft robotics, haptics, or holograms, or also through virtual representation of the DT such as an avatar or simply through traditional audio video interaction.

These three components, namely the data collection, the artificial intelligence, the multimodal interaction must be interconnected. This connection happens through communication networks, from its simplest forms such as over Bluetooth, which is now used commonly every moment of the day, to the very advanced wide area network 5G. Often, this connection should happen over a performing network. Indeed, some data is time sensitive and requires a high-speed network to be transmitted to the cloud and the response communicated back to the real twin in a matter of milliseconds.

Importantly, this communication should happen in a secured manner, ensuring the protection of privacy and identity of the real twin, and security of the data at all times. This is achieved through cybersecurity including latest advances in biometrics, etc.

Therefore, the DT concept is the convergence of five existing technologies that, when brought together in a unique combination, can improve the quality of life of the real twin. Fig. 1.1 illustrates this convergence of DT key technologies.

1.4 DT characteristics

The characteristics of the DT are the following:

- **Unique identifier.** In order to be able to interact with the DT anywhere, anytime, we certainly need to identify it uniquely.
- **Sensors and actuators.** Real twins could be equipped with sensors so that digital twins could replicate their senses—sight, hearing, taste, smell, and touch—using the appropriate actuators, depending on application needs.
- **AI.** Digital twins are equipped with a controller embedded with ontologies, machine learning, and deep learning techniques in order to make fast and intelligent decisions on behalf of their real twin. Furthermore, the AI engine of the DT allows making recommendations to the real twin, upon which more data continues to be collected on the real twin and fed back to the AI engine so that it can recommend a continuously improved interaction model to ensure the real twin satisfaction.
- **Communication.** Digital twins should be able to interact in near real time with the environment, real twins, and/or other digital twin. Communication, including the sense of touch (haptics), must occur within 1 ms and thus must follow 5G and Tactile Internet standards.
- **Multimodal communication.** Digital twins can have a representation as a 3D avatar, hologram, or even a humanoid social robot. However, it could also just be software components without a tangible representation, depending on the application. The interaction with the DT is quality of experience-based, which involves the degree of acceptance of the interaction with the data, and the satisfaction of the user with this interaction. This entails not only the traditional quality of service, such as delay jitter, but also how the data is represented, whether it makes sense to the user, and how the user perceives the DT feedback.
- **Privacy and security.** Digitals twins should be authentic and should be able to protect the identity and privacy of their real twin. This requires the use of advanced cryptography algorithms and biometrics

techniques (ECG-biometrics, haptic biometrics, and so on) as well as the resolution of regulatory issues.

- **Feedback loop:** the major characteristic of digital Twins is the quality of experience feedback loop. In which the interaction with the analyzed data is further transmitted to the user and new sensory data is generated, analyzed, transmitted, interacted with and looped back to the user.

1.5 Identify opportunities

Having the virtual replica of the physical entities (human/products, etc) and gathering historical data about them allow running data analytics to predict performance and provide output optimization. Besides, the DT provides the possibility to run simulations by use of the gathered data and predict performance in multiple possible scenarios.

All of these benefits have inspired us to adopt the DT concept for humans, and use this emerging technology to improve the state of health and well-being of individuals. Given the advances in personal devices for health data collection, and AI for advanced analytics, the DT concept opens doors for many applications for us humans as we will discuss throughout this book. Examples of these applications are presented in the next section.

In healthcare, the idea of DT applications in this domain is just starting to be explored. The concept is highly promising and can help improve individuals' health and well-being. When a person visits the doctor or caregiver, they don't have time to go through the health history, and even if they did, they don't know what has been happening with the person since the last visit. So they ask questions and perform some tests in order to try and guess where the health problem comes from. However, the DT tracks the person continuously and captures their physiological data. With the use of the right AI algorithms, the DT not only has all the data to infer the source of the health issue, but can even predict it way in advance, before it happens. For example, DT can detect the person's stress level using sensory technology, can detect if they are not sleeping well, if they are dehydrated. So it will predict that some health conditions can result from to those types of behaviors, and as a consequence the DT will recommend to the person a healthy sleep time, wake up time, it will remind them to drink more if they have not done so, to get up if they are sitting too long. For example, you have been reading this book for a little while, did you stand up and walk or stretch your body? In such case, you DT will be recommending to you to stand up because sitting for too long is unhealthy.

However, the DT applications in this domain are different in many aspects than their equivalent in the other industries. In both cases, we want to monitor constantly to build a digital replica of the physical entity. However, one of the goals of DT in manufacturing is to control the physical entity, while in the DT for health, we the goal is to allow the real twin to control their DT. Another difference lies in the fact that in the DT for health, a very important benefit is a deeper and better understanding of the human body functions and what factors influence them, through constant tracking. However, in the DT for manufacturing, these are things that are already known and the focus is rather put on profit improvement and cost reduction.

In the rest of this book, we will be addressing the DT with a focus on humans. The reason behind this is that in our era, humans are the center of all technological development, which should consider humans well-being. So we will be discussing the DT from a human perspective, and looking at the interaction between the real and digital twins for healthcare and well-being.

References

[1] A. El Saddik, Digital twins: the convergence of multimedia technologies, IEEE Multimed. 25 (2) (Apr. 2018) 87–92.

[2] W. Kritzinger, M. Karner, G. Traar, J. Henjes, W. Sihn, Digital twin in manufacturing: a categorical literature review and classification, IFAC-PapersOnLine 51 (11) (2018) 1016–1022.

[3] M. Kerremans, B. Burke, D. Cearley, A. Velosa, Top 10 Strategic Technology Trends for 2019: Digital Twins, 2019.

[4] GE Digital, Minds + machines 2016, meet the digital twin, https://www.youtube.com/watch?v=2dCz3oL2rTw, 2016. (Accessed 14 September 2021).

[5] IBM, Watson Internet of things, https://www.ibm.com/internet-of-things/trending/digital-twin. (Accessed 14 September 2021).

[6] Microsoft, The promise of a digital twin strategy, https://info.microsoft.com/rs/157-GQE-382/images/Microsoft%27sDigitalTwin%27How-To%27Whitepaper.pdf. (Accessed 14 September 2021).

[7] Siemens, Digital twin, https://www.plm.automation.siemens.com/global/en/our-story/glossary/digital-twin/24465. (Accessed 14 September 2021).

[8] NASA, Apollo 13, https://www.nasa.gov/mission_pages/apollo/missions/apollo13.html, 2009. (Accessed 14 September 2021).

[9] A. Parrott, L. Warshaw, Industry 4.0 and the Digital Twin, Deloitte Univ Press, 2017, https://dupress.deloitte.com/dup-us-en/focus/industry-4-0/digital-twin-technology-smart-factory.html. (Accessed 14 September 2021).

[10] M. Grieves, J. Vickers, Digital twin: mitigating unpredictable, undesirable emergent behavior in complex systems BT – transdisciplinary perspectives on complex systems, in: New Findings and Approaches, 2017, pp. 85–113.

[11] Digital Twin Consortium, Definitions, https://www.digitaltwinconsortium.org/initiatives/the-definition-of-a-digital-twin.htm. (Accessed 14 September 2021).

[12] Gartner, Gartner glossary, https://www.gartner.com/en/information-technology/glossary/digital-twin. (Accessed 14 September 2021).

Underactuated digital twin's robotic hands with tactile sensing capabilities for well-being

Mohd Faisal[a],
Roberto Alejandro Martinez Velazquez[a],
Fedwa Laamarti[a,b], and Abdulmotaleb El Saddik[a,b]

[a]University of Ottawa, Ottawa, ON, Canada, [b]Mohamed bin Zayed University of Artificial Intelligence, Abu Dhabi, United Arab Emirates

2.1 Introduction and background

The year 2020 has shown our globalized society how vulnerable we are. The new coronavirus SARS-CoV-2 that causes the disease COVID-19 has taken the lives of our loved ones in addition to wearing down the world economy. Different recommendations were issued by health authorities aiming to reduce the infection rate in the general population. Physical distancing was quickly adopted world-wide. The approach to physical distancing is to limit physical interactions among individuals outside the household, especially those deemed nonessential. However, physical distancing also comes with unwanted consequences. Humans are social beings and physical touch plays an important role in our mental well-being Conversely, many interactions that used to routinely happen in the real world, have now converted to the virtual world instead. However, the research on interaction between virtual world and real world started a few years back and may provide some answers to this issue.

David Gelernter introduced the concept of Mirror Worlds (MW) as a "software models of some chunk of reality" [1].

15

In Gelernter original idea, one had to create a Computer Assisted Model (CAD) of the real world and that model would receive vast amounts of information from the real object to a point where the "mirror world" would be able to reproduce real world in detail. The mirror worlds were intended to help people better understand and explore reality through a virtual interface. One could have a mirror world of a university with models of students, buildings, bank accounts, faculties and every real-world object could have a counterpart in the "mirror world". The idea was then picked-up by Michael Grives, who took it further by introducing the concept of Digital Twin (DT) [2]. A DT is actionable, which means that the model can simulate the effects of external forces applied to elements in the model itself.

Alam et al. [3] took Grieves definition of DT to extend it and introduce a DT architecture reference model that augments the original DT idea by establishing the existing relations between the different Real Twins (RTs) and pushing their data to the cloud. This means that the DT is not only a copy of a RT and its internal processes but holds also a copy of its interactions with other RTs. However, El Saddik raises the question: How can a DT interact directly with the real world on behalf of its RT? The answer that El Saddik suggests combines Artificial Intelligence (AI), sensors, and actuators. This will be useful in situations where the DT is required to take action in the real world on behalf of the RT.

A DT can interact with the real world directly by rendering a representation of the RT. The range of options to represent a RT include but are not limited to a robot, hologram, an avatar or simply a software entity sending messages to other systems. A robot can use haptics, speech, or sound, and more. However, a software entity could be limited to using network messages. The DT as suggested by El Saddik is capable of modeling relations and interacting directly with the real world. El Saddik et al. introduced an ecosystem to implement DTs that lay-out all the necessary tools and technologies to implement a DT for health and well-being [4].

We propose the concept of Robo Twin, which is conceived as a mediator of physical interaction. It aims at providing a form of direct interaction between individuals, that feels as close as possible to an actual physical contact with others. We believe that to achieve such an immersive experience, we need to provide the Robo Twin with a "body" that feels as close as possible to the body of a human and the capacity to capture touch-related information. In other words, the Robo Twin should be able to convey the sense of touch as accurately as possible, such as in a situation of handshake interaction for example. As a first step towards this concept, we survey the research works related to the design and implementation of anthropomorphic robot hands. Our objective is gathering information that will support researchers' decisions in the use and improvement of existing designs and implementations of a DT's robotic physical representation, with a focus on the robotic hand. In the following sections, we will first review the existing humanoid robots designs in Section 2.2, then discuss the additive manufacturing of robotic hands in Section 2.3. We will then review various underactuated robotic hand designs in Section 2.4, while Sections 2.5 and 2.6 review the use of temperature sensors and pressure sensors in DT robotic hands. In Section 2.7, we discuss the findings of this survey. Finally, we use Section 2.8 for concluding and suggesting future research directions.

2.2 Humanoid robots

Humanoid robots are complex anthropomorphic artificial machines. The growing interest in humanoid robots accompanied by the latest and ever-increasing technological advancements in the field of robotics, locomotion, and AI, achieved by engineers, has speeded up their development over the past decade. Moreover, because of their human-like shape, these robots can use the same equipment and environment as humans, hence, making them more compatible to be used as a building platform for the physical implementation of the Digital Twin.

Today's humanoid robots come in different shapes and sizes and are extensively being used for research and space exploration, personal assistance and care-giving, education and entertainment, search and rescue operations, manufacturing and maintenance, public relations, and most importantly healthcare sector. Some of the several potential prototypes of humanoid robots that can serve the purpose are discussed below.

ASIMO (Advanced Step in Innovative Mobility) [5] is one of the humanoid systems developed by Honda Motor Co. in the year 2000. Considered to be the world's most advanced humanoid robot, ASIMO uses Honda's new "i-WALK" technology to achieve a remarkable human-like ability to walk, climb stairs, hop on legs continuously, run backwards, and even walk over uneven terrain with high stability. However, the price for this advanced humanoid system is tagged at over 1 million USD, making it prohibitively expensive.

HRP-1, developed by Honda R & D, mainly for the maintenance task at the industrial plants and to provide security services for home and office costs about 400K USD. The more advanced and developed humanoid robot HRP-4 [6], is a life-size and has achieved a light-weighted body compared to its predecessors HRP-2 and HRP-3. Although, this humanoid robot can coexist with humans in their living spaces, thereby improving their quality of life, it excessively priced costing approximately 50K USD. HOAP-1, a miniature humanoid robot [7] from Fujitsu, designed as an ideal tool for research and development in robotics technology also costs about 50k USD.

TABLE 2.1   Functional characteristics of humanoid robots.

References	Humanoid robot	Height (h) (m)	Weight (w) (Kg)	BMI (Kg/m²)	Degrees of Freedom (DOF)	Price (USD)
[11]	ASIMO	1.3	50	29.58	57 [7/arm(wrist); 13/hand] **	>1 million
[6]	HRP-4	1.51	39	17.10	34 [7/arm; 2/hand] **	250K
[7]	HOAP-1	0.48	6	26.04	20	50K
[8]	NAO	0.58	4.5	13.5	25 [5/arm; 1/hand] **	8K–16K
[9]	*HBS-1	1.2	5	3.48	51[4/arm; 15/hand] **	10K
[10]	*Inmoov	NA	NA	NA	NA	2K

*BMI: Body Mass Index; NA: Not Available; *Manufactured using 3D printing technology, **More detailed information of DOF is not provided in the cited references.*

The work in [8] details another humanoid system – NAO Robot, developed by a French robotics company 'SoftBank Robotics', in the year 2008, with a motive to make the humanoid system affordable to more people. Because of its versatile nature, moderate size and human-like behavior, the NAO robot is being used as an optimized teaching-aid tool for in-depth research on human-machine interaction, autonomous navigation, and cognitive computing, in numerous academic institutions worldwide. Besides its use in academia, the NAO robot is also employed in the healthcare industry to improve the standard of care provided and for efficient administration. This robot is commercially available over a price range of 8k to 16K USD.

In [9] a child-sized 3D printed humanoid robot named HBS-1 is presented by the University of Texas, designed primarily for children's education and rehabilitation research, costing approximately 10K USD.

Inmoov [10], the first Open-Source 3D printed life-size humanoid robot, designed by French designer Gael Langevin, is another example of an economical humanoid robot, costing around 2K USD.

From the above literature, it is evident that to have an economical humanoid robot, additive manufacturing technology should be incorporated in the design and fabrication of the humanoid robots. Table 2.1 illustrates some of the functional characteristics of the existing humanoid robots.

2.3 Additive manufacturing of robotic hands

Additive manufacturing is a relatively new and progressively developing branch of technology, generally known as 3D printing. With the advent of this technology, prototyping and manufacturing of desired elements with the consecutive application of layers have seen a significant boost lately. This technology intends to provide relatively simple and cost-effective production of complex 3D intelligent structures such as robotic hands which needs distributed sensors equipped within the structural design besides reducing the material waste, energy consumption and prototyping time.

3D printing can be defined as a technology that creates three-dimensional objects with one ultra-fine layer on the other. Each layer bonds to its preceding layer of melted or partially melted building materials. The additive technologies employ various methods of fabrication. Of the many, Fused Deposition Modelling or Fused Filament Fabrication (FDM/FFF) is the most widely used method as it offers a low cost of production. This technique involves the extrusion of thermoplastic material like Polylactic Acid (PLA), Acrylonitrile Butadiene Styrene (ABS), Polyethylene terephthalate Glycol (PET-G), Nylon (PA-12), High Impact Polystyrene (HIPS), and Acrylonitrile Styrene Acrylate (ASA), through a heated nozzle. The dimensions of the type of polymers used in this process have been standardized as either 1.75 mm or 3 mm, with 1.75 mm being more popular. We discuss in the next section 3D printed robotic hands that are already being used for several applications ranging from pick and place operation to realistic human-robot handshake.

The authors in the paper [12] present a 3D printable, six-axis anthropomorphic manipulator with only rotational joints as a robotic arm. The project of this manipulator is available under the Creative Commons Attribution-ShareAlike 4.0 International (CC BY-SA 4.0) license. The design of this manipulator is modified as per the needs of the study and is printed using the FDM technology.

This 3D printed robotic arm is capable of handling a maximum load of 0.75 kg and consists of a total of 54 printable parts that can be printed on any FDM printer with a printing area of $20 \times 20 \times 20$ cm. The maximum range of the robotic arm is about 62 cm. The motion of this 3D printed robotic arm is controlled by seven bipolar stepper motors and the whole manipulator is mounted on the support structure for its stability. The authors were able to demonstrate a connection between Robotics Engineering and Additive Manufacturing technology by developing a cost-effective anthropomorphic robotic arm but with only rotational joints. However, for Robo-twin, a more human-like robotic hand with additional degrees of freedom is needed to make the overall human-machine interaction as close to real as possible.

Inspired from human hand anatomy, the researchers in [13], introduce a customized 3D printed robotic hand design to produce a realistic human-robot handshake. With a motive to imitate a human-like hand behavior, the robotic hand is mounted on a Kuka LWR 4+ robotic arm, a collaborative serial manipulator with seven degrees of freedom. The prototype mainly focuses on achieving realistic palm compliance and finger grasping by implementing a position-controlled feedback loop. Most of the finger and palm parts which are complex in design and are not exposed to excessive forces were 3D printed using ABS plastic as a building material. However, parts like the frame of the hand on which the motors are mounted and the palm links, which are subjected to excess pressure from the human user were made of aluminium for durability and robustness of the prototype.

One of the major limitations of this prototype that was studied from the results of the experiments is the lack of the fourth finger and better actuation mechanism for the thumb. It was also mentioned that the emulation of the human skin was not accurate. Although, the researchers were able to achieve palm compliance and finger grasping with a 3.7 rating out of 5 for comprehensive haptic rendering, the overall realism of natural human handshake was drastically affected.

In [14], the authors detail the development of a 3D printed soft capacitive sensor, with a capacitance-to-digital converter chip on a PCB, fully embedded into the 3D printed robotic hand to provide pressure sensing and signal processing abilities. This integrated capability is achieved using innovative design and multimaterial additive manufacturing technology. Firstly, a five-finger 3D printed robotic hand with embedded actuators for the motion of fingers, is 3D printed using PLA as a building material. Then a desktop fused deposition modeling (FDM) printer with an extra printing nozzle for paste/ink printing is used to print the conductive and dielectric layers of capacitive sensors, and the conductive tracks on the robotic hand to obtain soft capacitive pressure sensing phalanges.

Different combinations of flexible dielectric and conductive materials were tested. The best performing pair of silver paint as a conductive material and soft rubber (Ecoflex 00-30) as a flexible dielectric is used in the fabrication of the soft capacitive sensor because of their capability of reliably sensing pressures as low as 1 kP a while exhibiting good response in terms of sensitivity with an average of 0.00218 kPa^{-1} and a linear response in the entire tested range, compared to other tested pairs.

Another reported advantage of this customized 3D printed robotic hand is that in some way the design of the phalanx resembles the human distal phalanx which consists of the bone with soft tissue and skin around. In this design this pattern can be represented with a rigid base made from black PLA as the bone, the pair of flexible dielectric and conductive material mimicking the soft tissue, and the top layer of the TPU 3D printing filament with a Young's Modulus of 12 MPa reflecting the elastic properties of the human skin which has a Young's Modulus of 5–20 MPa. The researchers were able to illustrate that the multilayer additive manufacturing technique can be used to create robotic hands with integrated sensor providing more functionalities than conventional rigid robotic hands. However, this design is not compatible to integrate any other haptic modalities or sensing capability except pressure.

The research group at Nazarbayev University (NU) in [15] presents a preliminary prototype of a new semianthropomorphic multigrasp robotic hand to serve as an end effector in industrial and service robots. To maintain low cost and weight, and to have a relatively easy assembly, additive manufacturing techniques were widely used. Most of the hand's structure is 3D printed to keep its overall weight less than 1 Kg, else the heavy weight of the hand will limit the payload capabilities of the base manipulator. The NU hand consists of 10 joints which are actuated by five tendons connected to four servo motors. Also, the rubber-like paddings on fingers and palm which are integrated into the design use 3D printing techniques. These paddings feel similar to the soft tissues present on human fingers and palms thereby helping in the conformal grasping of objects.

Besides, in [16] a 5-fingered anthropomorphic robotic hand is used to provide force feedback to the user through pneumatic haptic muscles on the glove. This robotic hand design is open-source 3D printable and is mostly used for research on prosthetic hands. The backside of the hand is printed in a black PLA material while the front palm and fingers are printed in gold Ninjaflex, a flexible filament produced by Ninjatek which gives the palm a softer texture thereby allowing for some compliance while grasping. The fingertips are embedded with soft force sensors for developing an intuitive teleoperating system based on IMU's where the user is able to move freely while the hand is still in operation.

Both of the prototypes in [15], [16] are suitable to be used, but they are restricted to the use of one specific sensor and only offer the design of the wrist and not the complete hand i.e., the forearm, the bicep and the shoulder part.

In the papers [17], [18], the Inmoov (Open-Source 3D printed life-size humanoid robot) design of both hand and arm is used. The researchers in [17] modified the Inmoov design using CREO Parametric 3.0T M to incorporate both temperature and pressure sensors inside the fingertips. These sensors and some other parts of the hand are covered in a layer of silicone to provide better gripping capability and a more human-like appearance. All the hand and arm parts are 3D printed on a Fortus 3D printer, using Acrylonitrile Butadiene Styrene (ABS) material, with an overall volume of 78 in^3. The printed hand is about the same size as of a male human's hand with forearm being a little longer and wider to accommodate the standard sized servo motors. In [18], a wireless-controlled 3D printed robotic hand system consisting of a gesture control glove and a 3D printed Inmoov forearm is demonstrated. The forearm with a total of 46 individual parts was 3D printed with white biodegradable polylactic acid (PLA) material, for its low printing temperature and smooth printing finish. The forearm replaces the use of human hand to do dangerous tasks remotely, with each finger consisting of three rotational joints.

To be useful for daily tasks, the hand of the Robo-Twin requires the use of various sensors (temperature and pressure) to have more than one haptic modality. Hence, the 3D printed design of the robotic hand should be able to incorporate this very vital need. With most of the available hand designs being restricted to only one specific sensing functionality, the open source Inmoov design of the robotic hand seems to be the best available option as the design can be easily customized.

Another aspect is reducing the cost of an anthropomorphic robotic hand, using underactuated mechanism techniques, i.e., using fewer actuators than the degrees of freedom. It not only makes the system less complex but also reduces its cost to a greater extent. However, the human hand can perform complex tasks while retrieving information from the surrounding environment. This is the grasping organ of the human body which uses approximately 29 bones and 34 muscles to move all the fingers. Due to such a high degree of freedom and massive neural connections the human hand can produce all kinds of finger movements like pinch, grasp, touch, squeeze, throw, and hold while demonstrating immense flexibility. Therefore, it is very important to select the appropriate underactuated mechanism to reflect, as closely as possible, the enormous flexibility of the human hand while reducing the cost and complexity of the system. In the following section, we survey underactuated designs available from literature.

2.4 Underactuated designs

A highly underactuated design of a robotic hand is discussed in [13], wherein only two electric motors (namely finger motor and palm motor) are used to control the entire motion of the hand. The mechanical design of the robotic hand consists of three fingers and one thumb. The motion of all fingers is obtained in a similar manner as tendons in the human hand. The finger motor provides a maximum of 111N of force. This force is connected to a parallel slider bar, used for equally distributing the force to the finger cables attached to it. This force distribution technique is used to achieve a homogeneous grasp as each finger should apply an almost equal amount of force on the opponent's hand while making a handshake. The slider bar is actuated by the finger motor to move towards it along the guiding rods while pulling in the attached finger cables thereby making the hand grasp. The palm bars are attached to a simple pin joint which provides a systematic palm motion with two degrees of freedom. To synchronize the system, a gear wheel is used to interlock the palm bars. However, it was observed that the palm mechanism is not capable of transmitting a minimum required force of 50N to the palm (compared to its calculated force of 17.3N). To overcome this, rubber pads of dimension 15 mm × 25 mm × 10 mm with a shore hardness of 40A are placed between the palm bars. These rubber pads absorb the additional compressive forces and exhibit a similar elastic hysteresis behavior as observed in the human palm.

Although his highly underactuated design of the robotic hand is able to provide a similar force distribution contributing to human-like haptic feedback and making the grasp of the handshake fully self-adaptive, a lack in the grasping strength and a longer duration in making a handshake with a human subject was observed.

The paper [19], illustrates an improvised version of the underactuated mechanism of the robotic hand in [13]. In the previous prototype, the motors were used beyond their load limit leading to overheating. Therefore, the actuators were replaced to provide the required forces and operated within their respective limits. The motors are mounted on either side of the hand to keep the center of mass of the hand relatively closer to its geometric center, whereas in [13] the motors were located on the same side of the hand and the parallel slider bar mechanism prevented the shaft from being doubly supported. Also, some of the structural components are made up of aluminium to avoid wear and tear of the plastic printed components.

The fingers are actuated in a similar manner as described in [13], except that the current prototype's thumb is redesigned to provide 2 DOF rather than 1 DOF. It is not actuated by the motor, instead passively driven by the compression of the palm. This is done by connecting its string to the opposite finger than to the slider bar. As a result, the compressive force applied by the thumb on the human hand is depended on how strongly the palm is compressed.

The lever design proposed in [13] to achieve the palm motion is also redesigned. The new design uses two bars and a cylindrical guide instead of three bars connecting the levers to the middle finger, resulting in the linear motion of the middle finger by always having it on the symmetric axis between the two levers. Also, the disk and bar mechanism in [13] to transfer the middle finger motion to the motor has a tendency to reach a singularity, this is avoided by replacing it with a simple cable and pulley system.

The major drawback of this underactuated design is the inability of the thumb to produce an adequate natural grasping effect because of its passively driven motion.

In [20], the robot Vizzy, a four-fingered robot, is used to carry out a study of the handgrip strength for comfortable handshakes. The thumb and the index finger are controlled by a single motor while the middle finger and the ring finger are connected to another motor. A pulley is attached to the motor of a finger, and a fishing line string runs through it to the last joint in the finger. Hence, the motion of the motor causes movement in all three joints of the finger. Although the robot Vizzy follows an underactuated mechanism, a detailed description of such is not mentioned.

A Tendon-Pin Underactuated (TP-UA) mechanism of Columbia Hand is discussed in [21]. The Columbia Hand consists of 3 fingers having 3 degrees of freedom each, and is actuated with only two motors, one for closing the fingers and the other for rotating the thumb around its base. In order to compensate the missing DOF's, a number of sensors has been integrated in the hand design that can evaluate the joint angles and tactile contacts on each part of the hand.

In TP-UA, every part of the robotic finger is connected with pin joints and a tendon with the response being studied in three different stages while grasping an object, i.e., the Initial stage, Pre-Shaping stage, and the Closing stage. In the Initial stage, the finger is in straight position by means of return springs and the tendons running through every segment of the finger with the other end connected to the actuator. During the Pre-Shaping stage, the finger will start closing because of the tensile strength applied by the actuator on the tendon, simultaneously rotating the joints in a coupled relationship. Finally, in the Closing stage, the middle phalanx is blocked, and the tendon is still being pulled down by the actuator while the distal phalanx continues to bend. The joint angle between the two is decoupled by the object.

One of the advantages of using this underactuated mechanism is that it can complete the grasp task with the use of a single actuator with minimum number of lifting mechanisms as used in other underactuated mechanisms.

In [15], an underactuated design of NU hand is detailed. This hand consists of 10 joints, actuated by 5 tendons connected to 4 servo motors. The forward and backward motion of the thumb is achieved by connecting it directly to the micro servo motor. The second joint of the thumb is controlled by a digital brushless servo motor (Futaba BLS153). The motion of index finger is realized by connecting it to a standard digital servo motor (HITEC HS322). Finally, the motion of the rest of the three fingers is controlled by one servo motor (HITEC HS322) via differential coupling, i.e., by distributing the applied force between the three fingers using elastic elements. The elastic elements used in the design act in series with tendons and are 3D printed using Object Connex 260 printer. The curling motion of the fingers is achieved by means of tendons while the torsion springs are used to release the fingers back to their normal state.

A 10 degree of freedom NU Hand is actuated with only four servo motors. Hence, the highly underactuated design allows the fingers to perform conformal grasping without the need of any addition actuator.

The 3D printed, 5-fingered anthropomorphic robotic hand in [16] employs four linear actuators, namely Actuonix PQ12-R, to achieve the motion of its fingers, one for each first, second and third finger and the remaining one for both fourth and fifth finger. Although the design of the robotic hand is underactuated, the mechanism used to achieve the under actuation is not discussed.

Both [17], [18] use an underactuated design to control the motion of each finger phalanxes using only one actuator per finger, i.e., the distal, middle, and proximal phalanxes. These are controlled using only one HS-35HD Hi-tech servo motor, placed inside the forearm. The plastic servo horns that come with the servo motors are replaced by 3D printed servo pulleys through which a fine braided fishing line, acting like the tendons in the human body, runs to the tip of each finger passing through the holes incorporated in it. At the servo end, the string is wound around the servo pulley and tied at one of its edges such that when the servo is operated, it rotates, and the string is pulled further to its edge thereby causing the fingertip to move in towards the palm. Springs presented in each of the three-finger joints add to the smooth and more controlled motion of the finger.

Therefore, the tension in the string causes the distal phalanx to bend inwards. As the tension keeps on increasing due to the torque provided by the servo motor, the middle phalanx followed by the proximal phalanx curls in too. Hence, the curling motion of the finger which constitute 3 degrees of freedom is achieved by making use of only one actuator. Fingers are returned to their normal resting position by just rotating the servos in the opposite direction by the same amount/angle.

However, to achieve better grasping motion, two servos are used to control the motion of the thumb. The extra servo is used to directly drive the thumb along an axis parallel to the palm of the hand and its closing motion is achieved in the same manner as the rest of the four fingers. This underactuated mechanism is used in the Inmoov hand design of the robotic hand.

Apart from just demonstrating high flexibility in producing various finger movements, the human hand carries out a variety of functions with sensory and motor being among the most important ones. Both sensory and motor functions of the human hand make it an important organ as they help in collecting information and carrying out some daily tasks. Therefore, the DT anthropomorphic robotic hand must be able to carry out some of the sensory and motor functions of the human hand.

With temperature and pressure sensing being important senses to humans, the following sections discuss some of the available temperature and pressure sensors that have/can be used in the anthropomorphic DT robotic hand, thus making it capable of sensing the physical temperature and pressure of the contact subject or the surrounding environment.

2.5 Temperature sensors

The research in [17] uses the temperature sensor PT502J2, a 10K thermistor from US Sensors in their robotic hand for temperature measurement.

This small size, low cost, highly accurate and stable device is designed especially for temperature sensing and control applications where it is necessary to eliminate the expensive individual circuit calibration to make the overall process cost-effective. Operating in the range of $-80°C$ to $+150°C$ with an accuracy of $\pm0.2°C$, it provides fast thermal response time with a thermal constant of maximum 1 sec and a dissipation constant of 1 mW/°. These sensors were covered with a 0.05–0.07-inch-thick layer of silicone elastomer for better-grasping capability. Although silicone is known to be a heat resistant material, the overall impact on the performance characteristics of the thermistors was found to be less effective. For example, the time constant for the chosen thermistors with and without silicone covering was observed to be 16.5 seconds to 15 seconds respectively with a reduction of 2°C in its steady-state temperature value.

Initial testing of the temperature sensor indicated that the hand is capable of identifying hot objects. However, during the sensing test, it was found that the cooling rate of the temperature sensors was rather slower.

The design and fabrication of a liquid metal-based wearable tactile sensor capable of measuring contact forces and temperature independently is presented in [22]. This sensor is designed to be worn on an index finger and thumb of the human hand for the measurement of tactile information while grasping an object. A proper balance between temperature and force sensing is achieved by employing the Wheatstone bridge circuit.

Among a variety of available liquid metals, Galinstan (gallium-indium-tin) is used in the fabrication process because of its wide temperature range (low melting point of $-20°C$ and a very high boiling point of 2300°C), high surface tension (718×10^{-3} Nm^{-1}), l ow viscosity (2.4×10^{-3} Pa.s), high electrical and thermal conductivity (3.46×10^{6} Sm^{-1} and 16.5 Wm^{-1}K^{-1}). Galinstan was then filled in fingerprint-shaped fluidic channels and oval-shaped bulges enclosed in an elastomer structure made up of Polydimethylsiloxane (PDMS) providing excellent flexibility, easy operation, and better skin contact.

The temperature sensing sensitivities are calibrated to 0.41%°C^{-1} from 20°C to 50°C and to 0.21%°C^{-1} from 50°C to 80°C. Grasping experiment of water-filled beaker demonstrated that the developed sensor is capable of accurately measuring contact temperatures and can be reliably used in tactile sensing applications.

In [23] a small passive EPCOS fast response 10K NTC thermistor is used to estimate the surrounding air temperature before and during the contact with the object. It is then incorporated within a handheld portable device with tactile sensing modalities to have a data-driven thermal recognition of contact with people and object.

The paper [15] highlights the use of Melexis MLX90614, as a contact-less infrared temperature sensor, housed inside the sensor module placed on the NU hand to provide autonomous intelligent object manipulation capability to the robotic hand. This noncontact temperature sensor prevents the NU hand from avoiding contact with hot objects.

However, the papers [23], [15] does not detail the thermal characteristics of the temperature sensors used while performing grasping and sensing tests.

[24] details the manufacturing of a printed skin alike temperature sensor with Poly (3,4-Ethylene DioxyThio-phene) PolyStyrene Sulfonate (PEDOT: PSS) and Graphene oxide composite as a temperature-sensitive layer and silver as contact electrodes. To illustrate the temperature sensing test of this printed temperature sensor, it was then placed on the distal phalanx of the thumb of a 3D printed robotic hand and then placed over a heating arrangement. Upon reaching a set threshold value, the finger is moved away from the hot object, actuated with a linear servo motor placed inside the palm region.

To demonstrate its reliability for temperature measurement applications, the performance characteristics are compared with RS PRO Thermistors DO-35 100KΩ, a commercially available thermistor. Both these temperature sensors were placed on the hot plate at room temperature while gradually increasing the temperature of the hot plate to 100°C. The printed temperature sensor showed high sensitivity towards temperature changes with around 80% change in resistance at 100°C with a response time of 18 s in comparison to 90% change in resistance at 100°C with a response time of about 65 sec as demonstrated by the commercial thermistors.

Also, the recovery time of these sensors is evaluated by allowing them to cool down to room temperature after removing from the hot plate. From the recovery paths plotted for both, the printed sensor recovered completely after 32 seconds compared to 120 seconds taken by the commercial thermistors.

Therefore, it can be concluded that the printed skin like temperature sensor showed a faster response than the used commercial thermistors and is a promising solution to be used in eskin applications in humanoid robots. However, this printed sensor demonstrated more response time than the PT502J2, a 10K thermistor from US Sensors used in [17].

The paper [25] presents the design of a low-cost robotic finger with soft fingertip, having position, temperature, normal and shear force sensors, entirely embedded inside the design to be used in the new version of humanoid robot ARMAR-6. The sensory system employs four 3D shear force sensor which consists of a 3D Hall effect based digital sensor (MLX90393, Melexis) with their corresponding magnets cast in silicone rubber and two normal force sensors. Although both pressure and hall effect sensors can measure the temperature of the object or material it is in contact with, pressure sensors provide more accurate and reliable thermal information.

To assess the response of the sensor to thermal flux, four different materials, i.e., wood, PVC, steel and aluminium were placed inside the freezer at −20°C. After the immediate removal of these materials outside the freezer, their temperature was measured by placing one of the pressure sensors on the finger in direct contact with the material while the temperature was allowed to cool down to room temperature. The Wood showed minimal impact on the temperature measured by the sensor because of its low thermal conductivity while steel demonstrated the highest impact. Hence, the fingertip can differentiate various strengths of thermal flux within seconds and identify high-temperature surfaces.

The construction of a vision based soft somatosensory system for tactile modalities is discussed in [26]. In this, a transparent coating of Thermochromic Liquid Crystal (TLC) Ink is used on the finger-shaped sensing surface. TLC is sensitive to temperature changes as the distance between its molecules vary once the temperature starts changing within its working range. As the distance changes it reflects different wavelengths of light. On the color spectrum, it varies from red to purple of the visible spectrum of light when the temperature goes from lower limit to higher limit of the operating temperature range of TLC ink.

To evaluate its accuracy in measuring temperature, the sensor is placed inside the oven and heated from 25°C to 31°C. The images of color variation during this process as well as the corresponding real-time temperature are recorded using a camera and a reference thermometer respectively. The captured images are then converted from RGB to HSV color space to avoid lighting interference while measuring the TLC color. It is observed that the hue values of the reflected light increase while its wavelength decreases. Therefore, a linear relationship between the hue values of the reflected light from the TLC ink and the estimated temperature is established.

From the test results obtained, it is concluded that the temperature sensing range of this sensor is from 25°C to 31°C with a temperature estimation resolution of 0.4°C i.e., 6.7% of its sensing range. However, to employ this sensor in robotics applications such as anthropomorphic robotic arm, its small sensing range becomes one of its major drawbacks.

The temperature sensors and their functional properties discussed above are summarized in Table 2.2. The pie-chart in Fig. 2.1 indicates that the Force Sensitive Resistor (FSR) is the most widely used method in determining the contact pressure or forces in any application involving a robotic hand.

2.6 Pressure sensors

The research work in [17] uses A101 Flexi Force sensor to measure the applied pressure because of its small size, low cost, high durability and easy implementation. Pressure sensors were placed on the fingertips for achieving maximum sensing capabilities, inspired by the work presented

TABLE 2.2 Functional properties of studied temperature sensors

Reference	Type of Temperature Sensor	Working Principle	Operating Range	Accuracy of measurement	Thermal Constant	Dissipation Constant	Time Constant	Temperature Sensing Sensitivities	Response & Recovery Time
[17]	PT502J2 10K Thermistor	Thermistor	-80°C to +150°C	±0.2°C	1 s max	1 mW/°C	15 s	N/A	N/A
[22]	Wearable Tactile Sensor	Galinstan Liquid Metal-based	20°C to 2300°C	N/A	N/A	N/A	N/A	$0.41\%°C^{-1}$ from 20°C to 50°C and to $0.21\%°C^{-1}$ from 50°C to 80°C	N/A
[23]	Passive EPCOS 10K NTC	Thermistor	N/A	N/A	N/A	N/A	N/A	N/A	N/A
[15]	Melexis MLX90614	Infrared	-40°C to +125°C	0.5° C	N/A	N/A	N/A	N/A	N/A
[24]	Printed Temperature Sensor	PEDOT: PSS & Graphene Oxide Composite	25°C to 100°C	N/A	N/A	N/A	N/A	$1.09\%°C^{-1}$	18 s & 32 s
[26]	Vision based soft somatosensory system	Variation in temperature properties of Thermochromic Liquid Crystal	25°C to 31°C	0.4°C	N/A	N/A	N/A	N/A	N/A

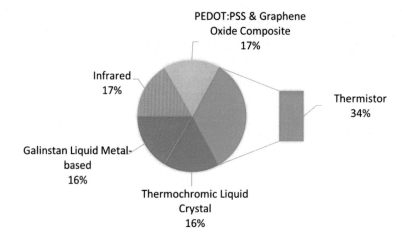

FIGURE 2.1   Breakdown of Temperature Sensing technologies used in the literature.

by Mirkovic and Popovic in [27]. Therefore, the sensors are incorporated into the distal phalanxes of each finger.

However, using one sensor per fingertip imposes a difficulty to measure the applied force on any part of the fingertip rather than the sensing area of A101 Flexi Force sensor which is only 0.15 inches. To overcome this problem, a cantilever beam is designed and placed hanging over the sensor's surface. This cantilever beam concentrates the applied force anywhere on the fingertip to the sensor's center, thereby increasing the sensing area of the sensor. The results obtained from the grasping test performed by the hand demonstrated that it could grasp various objects of different shapes and sizes and the corresponding pressure sensor readings were plotted for each fingertip. The sensors responded readily to the change in the pressure values.

Research in [20] uses tactile sensors made up of soft elastomer body consisting of a 3-axis Hall-effect sensor placed below a small permanent magnet inside it. On the application of external force on the elastomer, the relative position of the permanent magnet changes. This variation in the magnetic field is detected by the hall-effect sensor and is proportional to the applied external force. 3-axis Hall-effect sensor detects the magnetic field variations in 3 axes, allowing the tactile sensor module to measure the magnitude and direction of force in 3 dimensions. These sensors have the capacity to detect minimum forces in the order of 10 mN. Experiments were conducted to analyze the qualitative feedback from human subjects while they make a handshake with robot Vizzy (a four fingered humanoid robot). A total of 15 tactile sensors are placed on the robot's hand with 3 on thumb and 4 on each of the other fingers. The given placement of sensors yields the force distribution on several contact points rather than just fingertips. As a result, many of the human subjects reported experiencing more comfortable handshake than expected. However, they reported that the length of the hand is slightly larger than normal and recommended to have same tactile sensing feeling in the palm as in the fingers.

In paper [21] a highly under actuated design of Columbia Hand with integrated joint angle and force sensors is presented. In this, a Force Sensitive Resistor (FSR) is used as a force/pressure sensor because of its low cost and relatively high sensitivity compared to capacitive tactile sensors and piezoelectric polymer films. Although FSR is a nonlinear device which is highly sensitive to changes at low forces and much less sensitive to changes at high forces, a voltage divider with nonlinear transfer characteristics is employed to provide greater values of output at smaller values of sensor's resistance.

In this, FSR's are only used to determine whether the Columbia Hand is in contact with an object or not by setting a reference voltage. No more details on the pressure/force sensing capacity of the hand are provided in the paper.

The development of a complex 3D printed robotic hand with an embedded soft capacitance touch/pressure sensor having dimensions 19.6 mm wide, 2.6 mm thick, and 28 mm high is presented in [14]. The phalanxes are designed with gaps to house the electronics related to and the pressure sensor. A total of five different capacitive pressure sensing phalanxes were fabricated with a combination of conductive and dielectric materials, namely Ecoflex-silver (Eco-Ag), Ecoflex-graphite (Eco-Grp), Ecoflex-conductive PLA (Eco-PLA), thermoplastic polyurethane-silver (TPU-Ag), and thermoplastic polyurethane-graphite (TPU-Grp). All these five devices are tested not only for their sensing capabilities under a constant pressure of 20 KP a for 8 mins but also for hysteresis and dynamic cycle response with increasing pressure from 0 KP a to 50 KPa.

The pair of silver paint as a conductive material and soft rubber (Ecoflex 00-30) as a flexible dielectric is used in the fabrication of the soft capacitive sensor because of their capability of reliably sensing pressures as low as 1 kP a while exhibiting good response in terms of sensitivity with an average of 0.00218 kPa^{-1} and a linear response in the entire tested range, compared to other four pairs. Although the authors have demonstrated a simple and cost-effective fabrication method for a tactile sensing system, this 3D printed hand is only capable of sensing pressure. Moreover, the design is not compatible to adapt the addition of extra sensors for measurement of other physical parameters such as temperature in our case.

A variety of methods is described in the literature of [28] for the measurement of contact forces between the human fingertip and an object. A strain gauge as a force sensor attached to the object, allows accurate measurement with high resolution. The only limitation is that for every different object a custom-made device has to be developed and calibrated for conducting experiments. The process becomes laborious with the increase in number of objects.

A force sensitive resistor can be placed on the fingertip to measure the contact forces. Although FSR provides advantages of low cost, small thickness, and flexibility, the major limitation is the lack of sense of touch as the sensor comes in between the human fingertip and the object. Moreover, its nonlinearity, drift, saturation and hysteresis makes it difficult to be used in custom-made solutions.

Another proposed method is to examine the color changes of the fingernail while making contact with different contact forces. This way the user will not lose the haptic sense as the finger is in direct contact with the object. Nevertheless, the limitations are prominent, as the result changes drastically from one person to other and the calibration process is grueling.

Lately, some researchers have proposed to measure the change in the width of the finger caused by the normal deformation force of the finger pad while contacting an object. This change can be measured by a sensor placed at the side of the fingertip, thereby, eliminating the use of sensor between the fingertip and an object. This will provide the whole haptics information to the user grasping the object.

In [29], tactile data is obtained using JACO arm (a product of Kinova Robotics) equipped with Tekscan tactile sensor. This sensor consists of 18 segments to be connected to fingers and different regions of the palm. The Tekscan Grip system for R & D is used to obtain data related to pressure distribution while grasping an object over a 10 second time interval. Grip Research 6.70 is the software that converts the output voltage of the resistive sensor elements into relative pressure values in pounds per square inch (psi). Though a 95.21% of classification accuracy with a processing time of 2.6166 s is achieved in identifying different objects using the acquired pressure data, this sensor is designed only for thumb, ring finger and little finger of the human palm.

The mathematical analysis and fabrication of a cost-effective, flexible, aluminum-coated, polyimide paper based, touch mode capacitive pressure sensor is described in [30]. This fabricated pressure sensor is tested in all four regions of its operations demonstrating highest sensitivity in touch mode operation from 10 to 40 kPa of exerted pressure compared to the normal mode (0 to 8 kPa), transition mode (8 to 10 kPa) and saturation mode (after 40 kPa) operations with a response time of 15.85 ms.

The MEMS-based capacitive pressure sensors offer various advantages over Piezoresistive pressure sensors such as high stability, low power consumption, high pressure sensitivity and repeatability, and low temperature drift. However, they require large diaphragm area, and has low dynamic range and exhibits nonlinear response. The touch based capacitive sensors not only overcome these drawbacks but also, they offer less separation gap between the electrodes at high-pressure values. Because of their linear response, touch-based capacitive sensors can be efficiently used in making cost-effective skin-like material for prosthetic hands.

As mentioned earlier, the liquid metal-based wearable tactile sensor in [22] is capable of measuring both the contact forces and the temperature independently. The Wheatstone bridge circuit helps in decoupling the force and temperature signals. The force sensing sensitivities are calibrated to 0.32 N^{-1}. Grasping experiment of water-filled beaker demonstrated that the developed sensor is capable of accurately measuring contact forces and can be reliably used for multimode tactile sensing in real applications.

A single 2.5 cm square of a stretchable fabric-based force sensor is used in [23], the raw ADC output of which is converted in Newtons assuming uniform pressure distribution over the fabric. However, authors didn't discuss the functional aspect of this fabric-based pressure sensor in detail.

The design of a low-cost robotic finger with soft fingertip, and position, temperature, normal and shear force sensors, entirely embedded inside the design to be used in the hand of a new version of humanoid robot ARMAR-6 is discussed in [25]. The sensory system employs four 3D shear force sensor which consists of a 3D Hall effect based digital sensor (MLX90393, Melexis) with their corresponding magnets cast in silicone rubber and two normal force sensors. The experiment results demonstrated that the pressure sensors showed higher sensitivity to pressure than the Hall-Effect sensors, but they get saturated at high pressure values greater than 126 kPa, while the Hall-Effect sensors continued to respond. Therefore, this combination of sensors for pressure measurement not only provides higher sensitivity but also wider range of measurement, thereby utilizing the benefits of both the sensors. The placement of these sensors allows forces over the whole area of the fingertip to be determined.

However, this combination imposes a complexity in maintaining a balance between signal clarity and mechanical benefits during grasp action as the tactile sensor signals are influenced by material creep induced by using a flexible material with a less ShA hardness.

The functional properties of the above mentioned pressure senors are summarized in Table 2.3. Pie-chart in Fig. 2.2, indicates that the Force Sensitive Resistor (FSR) is the most widely used technology in determining the contact pressure or forces in any application involving a robotic hand in the cited references.

2.7 Discussion

From our review of the literature, some relevant points are discussed in this section.

TABLE 2.3 Functional properties of various Pressure sensors

Reference	Type of Pressure/Force Sensor	Working Principle	Force/Pressure Resolution or Range	Force/Pressure Sensitivities	Curve Characteristics	Response Time	Saturation level
[17]	A101 Flexi Force	Force Sensitive Resistor	N/A	N/A	N/A	N/A	N/A
[20]	3-axix Hall Effect Sensor	Magnetic Field variation	10 mN	N/A	N/A	N/A	N/A
[21]	Force Sensor	Force Sensitive Resistor	N/A	N/A	NonLinear Response	N/A	N/A
[14]	Soft Capacitance Touch Sensor (Silver paint and Ecoflex 00-30 based)	Capacitance Variation	1 kPa	0.00218 kPa^{-1}	Linear Response	Few 100 ms	N/A
[29]	Tekscan Tactile sensor	Force Sensitive Resistor	N/A	N/A	N/A	N/A	N/A
[30]	Flexible touch-based pressure sensor	Capacitance Variation	(10 to 40) kPa	N/A	Linear Response	15.85 ms	N/A
[22]	Wearable Tactile Sensor	Galinstan Liquid Metal-based	0 to 13.5 N (0 to 54 kPa)	0.32 N^{-1}	Linear Response	N/A	N/A
[25]	3D Shear force sensor (MLX90393, Melexis)	Hall Effect	50 grams	N/A	N/A	N/A	>44N (126 kPa)
[25]	Normal force sensor	MEMS Barometer	0.5 grams	N/A	N/A	16.5 s Barometer	0.1N-44 N (100 kPa– 126 kPa)

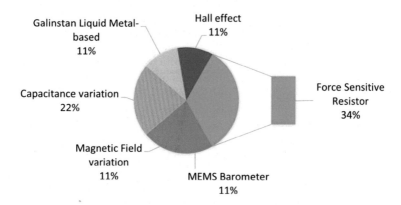

FIGURE 2.2   Breakdown of Pressure Sensing technologies used in the literature.

1. 3D printing technology has been found by researchers to be a promising solution in reducing the cost of DT representation, making it affordable to the general public.
2. The Robo-Twin hand requires different types of sensors, including temperature and pressure, to be useful in assisting in day-to-day tasks. However, most of the hand designs available in literature are restricted to only one specific type of sensor. We found the Inmoov design of the robotic hand to be the best available option as of now, as the design can be easily customized to incorporate both temperature and pressure sensors.
3. In one of research works studying the design a robotic hand that is able to produce a realistic human robot handshake, the experiments showed that the designed palm compliance proved to be quite realistic. However, it was found that the imitation of the human soft skin was not accurate. More research is needed to provide insights on this aspect for a more realistic robotic hand skin,
4. There is a scarcity in research analyzing the sensors placement on the robotic hands. Be it temperature sensors or pressure sensors, we did not find in the papers surveyed much information on the choice of the placement of sensors. We believe however that this has to potential to make a difference in the accuracy of the measured and reported temperature and/or pressure, depending on the application. In the example of the handshake, different parts of the palm and the back of the hand are pressed to different extents, and different parts are more/less sensitive to temperature. Hence, the choice of the sensory placement is of high importance in such application.
5. One of the reported issues, in the surveyed papers integrating temperature sensors in their design, in the issue of the cooling down rate of the temperature sensors. Indeed, even though the testing found that

the hand was capable of identifying hot objects, the cool down time afterwards was slow. This remains an open problem that needs more research and analysis to suggest suitable solutions for it.

6. This paper gives an insight into how difficult it is to provide a set of human skills to a humanoid robot to carry out the complex task of making a handshake. The following two scientific questions were addressed in order to illustrate tactile sensing problems in the context of handshaking between a person and a humanoid robot:

    a. Does the hand-shaped end-effector of our humanoid robot provide a comfortable sensation in terms of force and touch interaction during the handshake? and

    b. Is the robot capable of detecting if it is grasping a human hand or another inanimate object during handshake interactions?

Besides the previous points, the following research questions need to be considered, to propose ideas and solutions for robotic hands with haptic modalities:

1. Sensory data: Once the data is collected from multiple sensors and multiple types of sensors, how will the synchronization of this data will be achieved? And subsequently, how will this synchronized data be structured? Are any of the existing sensory meta languages satisfactory to be used to structure this synchronized data, or is the development of a new language necessary to this effect?

2. Once the captured sensory data is synchronized and structured, will it need to be compressed for effective storage and transmission? If yes, how will this be achieved?

3. Following the transmission of DT robotic hand's sensory data over the network to achieve a virtual handshake, what is the best way to reproduce this handshake remotely, and how accurate will the handshake replication be?

Further research is warranted to answer part of these questions and provide more insights towards the design and development of the DT robotic representation.

2.8 Conclusion

In this chapter, we reviewed the existing literature in the field of Digital Twin robotic representation with a focus on Digital Twin robotic hand. A limited number of research works were proposed in this domain and much more work still needs to be achieved. Overall, we found that the DT robotic hand needs to incorporate multimodal haptic sensing. It should provide a good accuracy in terms of temperature sensing and pressure sensing, while keeping the cost low enough for use by the public. Additionally, it should provide human-like characteristics, which can be challenging, such as the emulation of human skin to provide a realistic handshake. Many areas need to be explored and analyzed such as optimal sensors' locations on the robotic hand, which will need conducting many studies and experimenting with different potential locations to select the best points in terms of accuracy of sensing, while still providing cost effectiveness. More research is needed to consolidate the existing works, and to move further towards building the DT robotic representation.

References

[1] D. Gelernter, Mirror Worlds, Oxford University Press, 1991, https://doi.org/10.1093/oso/9780195068122.001.0001.

[2] M. Grieves, J. Vickers, Digital twin: mitigating unpredictable, undesirable emergent behavior in complex systems, in: Transdisciplinary Perspectives on Complex Systems, Springer International Publishing, Cham, 2017, pp. 85–113, https://doi.org/10.1007/978-3-319-38756-7_4.

[3] K.M. Alam, A. El Saddik, C2PS: a digital twin architecture reference model for the cloud-based cyber-physical systems, IEEE Access 5 (2017) 2050–2062, https://doi.org/10.1109/ACCESS.2017.2657006.

[4] A. El Saddik, H.F. Badawi, R. Velazquez, F. Laamarti, R. G´amez Diaz, N. Bagaria, J.S. Arteaga-Falconi, Dtwins: a digital twins ecosystem for health and well-being, IEEE COMSOC MMTC Commun. Front. 14 (2) (2019) 39–43.

[5] Honda Debuts New Humanoid Robot "ASIMO", Ind. Robot 28 (2) (Apr 2001), https://doi.org/10.1108/ir.2001.04928bab.002.

[6] Topio – Robotic Infrastructure Services Provider — Robot Center, https://www.robotcenter.co.uk/products/humanoid-robot-hrp-4?variant=187380732.

[7] Fujitsu Introduces Miniature Humanoid Robot, HOAP-1, https://pr.fujitsu.com/en/news/2001/09/10.html.

[8] D. Gouaillier, V. Hugel, P. Blazevic, C. Kilner, J. Monceaux, P. Lafourcade, B. Marnier, J. Serre, B. Maison-nier, Mechatronic design of NAO humanoid, in: Proceedings – IEEE International Conference on Robotics and Automation, 2009, pp. 769–774, https://doi.org/10.1109/ROBOT.2009.5152516.

[9] L. Wu, M. Larkin, A. Potnuru, Y. Tadesse, HBS-1: a modular child-size 3D printed humanoid, Robotics 5 (1) (2016), https://doi.org/10.3390/robotics5010001.

[10] InMoov – open-source 3D printed life-size robot, https://inmoov.fr/.

[11] ASIMO by Honda, Honda – the world's most advanced humanoid robot, https://asimo.honda.com/asimo-specs/.

[12] R. Siemasz, K. Tomczuk, Z. Malecha, 3D printed robotic arm with elements of artificial intelligence, Proc. Comput. Sci. 176 (2020) 3741–3750, https://doi.org/10.1016/j.procs.2020.09.013.

[13] M. Arns, T. Laliberte, C. Gosselin, Design, control and experimental validation of a haptic robotic hand performing human-robot handshake with human-like agility, in: IEEE International Conference on Intelligent Robots and Systems 2017-Septe, 2017, pp. 4626–4633, https://doi.org/10.1109/IROS.2017.8206333.

[14] M. Ntagios, H. Nassar, A. Pullanchiyodan, W.T. Navaraj, R. Dahiya, Robotic hands with intrinsic tactile sensing via 3D printed soft pressure sensors, Adv. Intell. Syst. 2 (6) (2020) 1900080, https://doi.org/10.1002/aisy.201900080.

[15] Z. Kappassov, Y. Khassanov, A. Saudabayev, A. Shintemirov, H.A. Varol, Semi-anthropomorphic 3D printed multigrasp hand for industrial and service robots, in: 2013 IEEE International Conference on Mechatronics and Automation, IEEE ICMA 2013 (c), 2013, pp. 1697–1702, https://doi.org/10.1109/ICMA.2013.6618171.

[16] S. Li, R. Rameshwar, A.M. Votta, C.D. Onal, Intuitive control of a robotic arm and hand system with pneumatic haptic feedback, IEEE Robot. Autom. Lett. 4 (4) (2019) 4424–4430, https://doi.org/10.1109/LRA.2019.2937483.

[17] D. Lanigan, Y. Tadesse, Low Cost Robotic Hand that Senses Heat and Pressure, 2017.

[18] F. Salman, Y. Cui, Z. Imran, F. Liu, L. Wang, W. Wu, A wireless-controlled 3D printed robotic hand motion system with flex force sensors, Sens. Actuators A, Phys. 309 (2020) 112004, https://doi.org/10.1016/j.sna.2020.112004.

[19] J. Beaudoin, T. Laliberte, C. Gosselin, Haptic interface for handshake emulation, IEEE Robot. Autom. Lett. 4 (4) (2019) 4124–4130, https://doi.org/10.1109/lra.2019.2931221.

[20] J. Avelino, T. Paulino, C. Cardoso, R. Nunes, P. Moreno, A. Bernardino, Towards natural handshakes for social robots: human-aware hand grasps using tactile sensors, Paladyn 9 (1) (2018) 221–234, https://doi.org/10.1515/pjbr-2018-0017.

[21] L. Wang, J. DelPreto, S. Bhattacharyya, J. Weisz, P.K. Allen, A highly-underactuated robotic hand with force and joint angle sensors, in: IEEE International Conference on Intelligent Robots and Systems, 2011, pp. 1380–1385, https://doi.org/10.1109/IROS.2011.6048748.

[22] Y. Wang, Y. Lu, D. Mei, L. Zhu, Liquid metal-based wearable tactile sensor for both temperature and contact force sensing, IEEE Sens. J. 21 (2) (2021) 1694–1703, https://doi.org/10.1109/JSEN.2020.3015949.

[23] T. Bhattacharjee, J. Wade, Y. Chitalia, C.C. Kemp, Data-driven thermal recognition of contact with people and objects, in: IEEE Haptics Symposium, HAPTICS 2016-April, 2016, pp. 297–304, https://doi.org/10.1109/HAPTICS.2016.7463193.

[24] M. Soni, M. Bhattacharjee, M. Ntagios, R. Dahiya, Printed temperature sensor based on PEDOT: PSS-graphene oxide composite, IEEE Sens. J. 20 (14) (2020) 7525–7531, https://doi.org/10.1109/JSEN.2020.2969667.

[25] P. Weiner, C. Neef, T. Asfour, A multimodal embedded sensor system for scalable robotic and prosthetic fingers, in: 2018 IEEE-RAS 18th International Conference on Humanoid Robots (Humanoids), IEEE, 2018, pp. 286–292, https://doi.org/10.1109/HUMANOIDS.2018.8624955.

[26] C. Yu, L. Lindenroth, J. Hu, J. Back, G. Abrahams, H. Liu, A vision-based soft somatosensory system for distributed pressure and temperature sensing, IEEE Robot. Autom. Lett. 5 (2) (2020) 3323–3329, https://doi.org/10.1109/LRA.2020.2974649.

[27] B. Mirkovic, D. Popovic, Prosthetic hand sensor placement: analysis of touch perception during the grasp, Serb. J. Electr. Eng. 11 (1) (2014) 1–10, https://doi.org/10.2298/sjee131004001m.

[28] A.M. Almassri, W.Z. Wan Hasan, S.A. Ahmad, A.J. Ishak, A.M. Ghazali, D.N. Talib, C. Wada, Pressure sensor: state of the art, design, and application for robotic hand, J. Sens. 2015 (2015), https://doi.org/10.1155/2015/846487.

[29] A. Khasnobish, M. Pal, D. Sardar, D.N. Tibarewala, A. Konar, Vibrotactile feedback for conveying object shape information as perceived by artificial sensing of robotic arm, Cogn. Neurodyn. 10 (4) (2016) 327–338, https://doi.org/10.1007/s11571-016-9386-0.

[30] R.B. Mishra, S.M. Khan, S.F. Shaikh, A.M. Hussain, M.M. Hussain, Low-cost foil/paper based touch mode pressure sensing element as artificial skin module for prosthetic hand, in: 2020 3rd IEEE International Conference on Soft Robotics, RoboSoft 2020, 2020, pp. 194–200, https://doi.org/10.1109/RoboSoft48309.2020.9116035.

Digital twin for healthcare immersive services: fundamentals, architectures, and open issues

Moayad Aloqaily[a], Ouns Bouachir[b], and Fakhri Karray[a]

[a]Machine Learning Department, Mohamed Bin Zayed University of Artificial Intelligence (MBZUAI), Abu Dhabi, United Arab Emirates, [b]Zayed University, Dubai, United Arab Emirates

3.1 Introduction

The emerging era of digitalization everything has revolutionized the way we understand systems, process data, and make decisions. In addition, the advances in big data processing and communications approaches have introduced new methods to efficiently exploit the data and deduce insights from them [1]. Due to these advancements, we are witnessing the birth of newel technological concepts such as digital twin (DT). By creating the digital replication of any physical asset or function, called the DT, using the gathered information, it is possible to simulate and analyze the captured behavior in a comprehensive way to build valuable insights about the asset characteristics. Thus, DT can be used to test new decisions and adjust them based on the predicted results before applying them in real life scenarios [2]. DT has proved its benefits in various sectors including engineering, automotive and manufactures, etc.

On the other hand, the immersive technology is another digital advance that allows to immerse the users in a mirrored environment where reality is combined with virtual objects for distinct purposes, informative and entertaining [3]. This technology is evolving very fast and is being developed for several other applications. In just a few years, different immersive models have been introduced, mainly, Virtual Reality (VR), Augmented Reality (AR), Mixed Reality (MR), and Extended Reality (XR) and they have all successfully integrated with several industrial sectors.

Combining the DT concept with the immersive technologies of its different types leverages the benefits of one another by providing more comprehensive manner to interact with the DT and by bringing various advantages. In this chapter, we mainly focus on the integration if these technologies in the medical domain. Indeed, by immersing the healthcare professionals in a virtual world where they can manipulate and interact with the digital information of any patient, the doctors can rely on advanced data analytics, ML and AI tools to build valuable insights about health status [4]. Thus, new therapies and medicines can be tested in the digital environment allowing to provide exhaustive details about the expected patient reactions. Subsequently, modifications, adjustments, and decisions on the needed doses can be validated on virtual twin of the patient before adopting to real cases [5]. Furthermore, creating 3D models based on real data gathered from individuals and analyzed by the advanced DT mechanisms revolutionize the training strategies and the telemedicine approaches. Distant doctors can visualize and interact with 3D models of their patients as if they were just in front of them. The created 3D models use accurate medical information and real-time data collected from the patients allowing the professionals to better understand the symptoms and identify various issues. Using the same 3D model, the doctors can smoothly manipulate telesurgeries using their precise hand techniques that are captured by the immersive technology trackers in real-time to provide guidance to the robot or professionals who are locally manipulating the operation. As the 3D model is based on the DT of the patient, real-time information about the case progress can be shown to the surgeon during the operation. Also, the virtual twin of patients can be highly useful in leveraging the traditional training strategies by allowing professionals to test and practise their techniques on the DT of individuals as it can simulate the reactions of the human bodies. In other words, combining the DT technology with immersive healthcare is a tremendous and cost-effective solution that can revolutionize the medical sector by providing more customized therapies that are tested and validated before their application, thus reducing risks [6].

Adopting new technologies in the medical sector is a strenuous decision. The medical domain is a delicate area as it deals with human health and lives. The used technologies and approaches must meet several strict requirements to gain the trust of the experts and patients including the highest level of efficiency and accuracy, and the lowest possible risk. Indeed, to make crucial decisions related to patient health, the professionals need to rely on approaches and tools that guarantee precise information and accurate analyses with the lowest delay as errors are not tolerated in this case. Thus, designing new applications for healthcare systems that use XR and DT must meet very strict requirements in terms of precision, accuracy and delay.

This chapter introduces the concept of immersive healthcare services and their relationship to DT technology. Several use cases are discussed to show this mutual association where each technology takes advantage and the benefits of the other. Additionally, various requirements need to be taken into consideration to build a robust platform able to meet the strict demands of the medical sector using different advanced approaches including data sensing, transmission, analysis, and manipulation with the help of ML and AI-based models. The topology of the designed system and the methods adopted to manage the various tasks are highly essential to provide an efficient performance able to guarantee the needs and to acquire the trust of the patients and the professionals. This chapter also proposes several advanced paradigms that have proved their efficiency in such systems including edge intelligence and federated learning.

The remainder of this chapter is presented as follows: Section 3.2 introduces the fundamentals of the DT and the various immersive technologies: VR, AR, MR, and XR, then it presents the advantages and several use cases of the alliance of these two digital technologies in healthcare services, called XR-DT for healthcare. Section 3.3 provides a discussion on the requirements and detailed steps needed by the XR-DT platform to guarantee the required outcomes while section 3.4 proposes emerging paradigms that can enable the deployment of such a platform. Section 3.5 presents the various open issues that should be taken into consideration to enable the adoption of the XR-DT in the healthcare sector. Section 3.6 summarizes the learned lessons and finally section 3.7 concludes this chapter.

3.2 Fundamentals of DT and XR

This section presents the fundamentals of the DT and the various immersive technologies. Then, it discusses the advantages of the alliance between these two technologies in healthcare systems.

3.2.1 Digital twin (DT)

DT is a digital replication of a physical asset that reproduces real characteristics and behaviors in the digital environment. It provides an efficient way to reach a high level of physical system awareness that is highly essential to drive useful insights and make accurate decisions to enable performance optimization [7,8]. To this end, DT relies on the collection of important amounts of real-time data describing the asset including any useful information related to its shape or its behavior. This data enables the creation of a virtual DT of the physical assets that can reproduce all its actions and demonstrate various precise details allowing to create better system understanding. Moreover, the collected data goes through different filtering and processing steps to extract some important insights and situational awareness that can be useful in analyzing and comparing the newly gathered information. This comparison and the gained awareness of the system are very useful in predicting future behaviors thus making accurate decisions that can improve the system performance and avoid problems. In other words, DT is based on analyzing the past information and observing the current actions to predict future behaviors (Fig. 3.1).

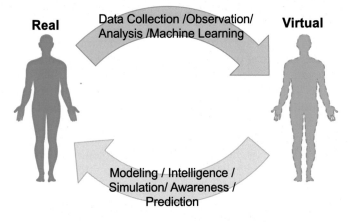

FIGURE 3.1   Digital Twin Model.

The DT of any physical asset or system integrates modeling, simulation, and data intelligence. It leverages the traditional simulation concepts by providing an advanced platform able to test and evaluate the impact of various actions and situations. Using the acquired understanding of the system performance, the influence of any new event can be studied with exhaustive details that allow to adjust and validate the needed actions that guarantee the best results before making decisions. Various advanced approaches constitute the base of DT starting from IoT and communications to cyber-physical system components as shown in Fig. 3.2. Indeed, IoT enables the collection of various types of data showing the real-time progress of any physical system. The transportation of this huge amount of data relies on advanced communication technologies such as 5G and beyond 5G (B5G) networks to guarantee the minimum delay and the highest accuracy. Subsequently, different data manipulation mechanisms are highly needed to build and update the virtual DT of any system. These mechanisms include data cleaning, filtering, validation, analysis, and ML-based

FIGURE 3.2 DT various components.

approaches that allow understanding, modeling, simulation, and prediction of future behavior.

DT offers a smart way to use the collected data efficiently and mirror the real world in various domains such as automotive, training, and healthcare. It provides a wide range of advantages including operation optimization, anomalies detection and prediction, and enabling system autonomy [4].

3.2.2 Immersive services

The impressive advances in technology are bringing features from science fiction to the real world. Nowadays, the users can be immersed in different environments depending on the used application such as communication, entertainment, learning, and training. Immersive technologies are taking more and more importance in various activities and they are progressing quickly offering various types of services: VR, AR, MR, and XR (Fig. 3.3).

3.2.2.1 Virtual reality (VR)

Virtual Reality is a computer-generated virtual environment using 360 videos and images allowing to completely immerse the user in a 3D world that is totally disconnected from his real one. The generated environment is perceived through a dedicated headset called VR Headset or helmet. This headset allows the users to be immersed in a virtual environment rather than viewing a 2D scene in front of them. The users have the feeling to be inside the scene and are able to interact with it. For instance, whenever they turn their heads the visualized graphics react accordingly.

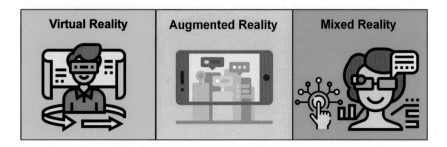

FIGURE 3.3 The different immersive technologies.

VR is used in various domains such as entertainment, education, architecture, tourism, and medicine. By creating illusions of reality, VR allows the students to better understand mechanisms and expand their horizons. Also, it enables engineers to visualize their designs before creating physical prototypes, what can facilitate the realization of new creative ideas. In the medical domain, VR is successfully used in training new doctors and in psychological and physical therapies by allowing to immerse the patients inside a well-chosen environment that can help to reduce the pain and speed up the rehabilitation process [9].

3.2.2.2 Augmented reality (AR)

Augmented Reality (AR) is another immersive technology that is different from VR, but sharing its same main target which is offering a virtual immersive experience. AR overlays the real environment and objects with digital images generated by the used device. Unlike VR that removes the users from the real surrounding world to completely immerse them in virtual scenes, AR uses the real environment and objects as a framework where digital images are placed. Thus, the users are not disconnected from their real world. In other words, AR allows to digitally augment the real world to enhance the user experience such as adding information related to some objects or locations, or also, by inserting funny icons. Wearing a dedicated headset is not strictly necessary with AR. The mobile phone or google glass can be used in this case such as with Google Maps AR, various games (e.g. Pokémon Go) or the funny filters used in social media applications.

The combination real and virtual objects in AR is visualized in 2D images that can be shown on the mobile phone screen. Just by using the camera of the mobile phone, AR detects the captured environment and injects virtual objects for various purposes (e.g. informative, guidance, or entertainment). Using advanced AI, ML and image analysis mechanisms, the captured images are analyzed and real objects and/or locations are recognized, subsequently, a predefined related digital image is placed in

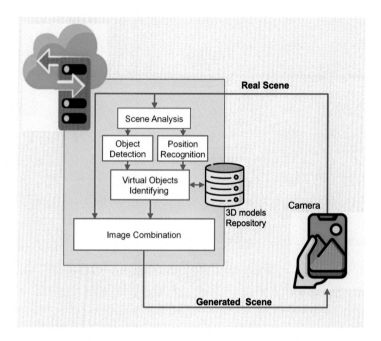

FIGURE 3.4   AR concept.

the captured scene as defined by the AR application (Fig. 3.4). AR is very useful in various sectors including industry, education, culture, tourism and medical domain [10]. It helps the doctors to visualize important information about the patients in a smart efficient way to help making the most appropriate diagnosis and following up with situation during surgeries [11,12]. The market of AR in healthcare is expected to reach USD 77.0 billion by 2025[1] representing a grow up at CAGR at 38.1%.

3.2.2.3 Mixed reality (MR)

Mixed reality (MR) is a hybrid technology that combines AR and VR to provide an interactive virtual experience over the real world. It merges the real and virtual worlds to create a new environment where physical and 3D digital objects coexist and interact in real-time. With MR, the users are not totally removed from their real environment, like with VR. The created virtual scenes take into consideration the real view of the users and changes with their locations using a different type of headset (i.e. a headset that does not completely blank the real world). Unlike AR, with which the users are able to see the virtual images in a 2D flat screen with no possible

[1]https://www.marketsandmarkets.com/Market-Reports/augmented-reality-virtual-reality-market-1185.html, Accessed: October 2021.

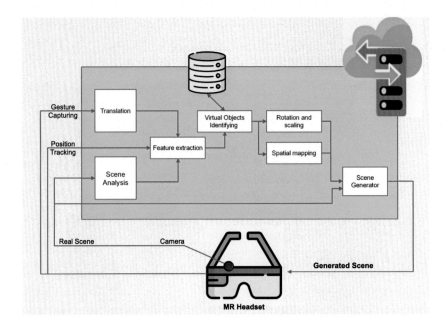

FIGURE 3.5   MR concept.

interaction, MR enables the manipulation of the digital images that are overlaid into the real world. In other words, the users are able to see 3D holograms of the digital objects superimposed in their real environment in front of them and they can manipulate them by changing their locations, sizes, and shapes; rotate and move them, etc. Microsoft HoloLens[2] is one of the current projects that integrate MR. Table 3.1 provides a comparison between MR, AR and VR.

MR leverages AR tools by adding advanced tracking strategies to keep monitoring the user's real views, position, motion, etc. This real-time data including the exact position of the user's head should be processed and analyzed to generate new scenes where the user is able to see the real world with the overlaid digital images that appear as one single environment (Fig. 3.5). Various precise calculations and critical decisions should be taken to provide the highest level of credibility. For instance, distances between the user and the real virtual objects should be determined to define the closest one to the user so as to place it in front of the other.

MR brings several benefits to various domains including education, entertainment, engineering, and healthcare. In the education sector, MR enables the students to be immersed in an exciting learning experience with interactive 3D projections and simulations to better gain a deeper un-

[2]https://www.microsoft.com/en-us/hololens, Accessed: October 2021.

TABLE 3.1 Comparison between VR, MR and XR.

	VR	AR	MR
Main Task	Completely Immerse the user in a digital world	Augment the real world by overlying virtual objects into the real 2D scene	Combination of VR and AR.
Generated Scene	3D computer generated environment and hidden real world	Virtual objects are layered on the top of the real environment in a 2D scene	A new world is created where physical and virtual objects coexist
Presence of the user	Completely immersed in the digital world. No existence of the real world	Presence in the real world which is augmented by digital objects	Presence in an environment containing real and 3D virtual objects
Interaction with the digital objects	Possible interaction with the digital objects using hand controllers. The responses are seen in the virtual world	No possible interaction	High level interaction with the 3D objects thanks to various types of sensors. The responses are in the MR environment
Equipments	VR headset	Smartphone or AR glasses	MR headset
Application	Gaming, training, simulations, learning, medical	Gaming (Pokemon Go), learning, engineering, medical	Various advanced sectors that require AI-based training

derstanding of the studied concepts and systems. As entertainment, MR allows the spectators to have their favorite celebrities performing their shows wherever the users are, instead of watching them on a flat 2D screen. In healthcare, MR is considered a promising technology that enhances the learning and training strategies of the new doctors and revolutionizes traditional imaging and spectroscopy [13–15]. Indeed, MR allows healthcare experts to visualize 3D holograms of the patient's organs showing details about what is happening inside. Experts are able to interact with these digital objects by moving or rotating them, getting inside the organs, etc. It facilitates and speeds up the understanding of the patient's situation allowing the experts to make an accurate diagnosis or decisions when monitoring a surgery. MR is a promising technology that can leverage telemedicine by allowing remote surgeries and consultations.

3.2.2.4 Extended reality (XR)

Extended Reality (XR) is the term used to refer to VR, AR, and MR (Fig. 3.6). It is an umbrella covering these various technologies. XR creates a smart real reality supported by the digital world and real-life interactions. It combines real and virtual worlds in a single environment

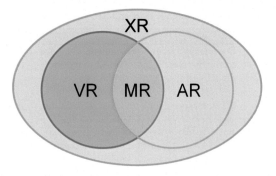

FIGURE 3.6   Relationship between VR, AR, MR and XR.

that highly depends on human-machine interactions in different levels as shown in Fig. 3.7. This technology will have a big impact on the traditional working, learning, entertaining, and consuming ways. It creates a new reality where digital and physical worlds are brought together in different levels to offer better user experience and to create significant opportunities in various sectors: entertainment culture, travel, marketing, shopping, city operations [16], engineering, education, healthcare, and the list is so long.

COVID-19 has imposed a new lifestyle based on social distancing and big restrictions in movements that have changed the collaboration, learning, and consuming ways increasing the interest in advanced immersive technologies. The global market of XR is expected to reach the size of USD 136.8 billions in 2024[3] and USD 1,246.57 billions in 2035.[4] Indeed, with the outstanding advances, XR devices are very accessible as the needed headset should include some sensors, cameras, gaze trackers, accelerometers, and microphones to be interfered with the user to the digital world. Other important requirements are discussed in section 3.3.

3.2.3 Immersive DT in healthcare: a use case

Nowadays, IoT devices generate a huge amount of data, more important than whatever was expected requiring more complex and creative mechanisms to manage and extract valuable information as none has time and capacity to sort all the gathered data. DT and XR are different technologies that provide distinct services. However, they can be complementary offering a new way of data exploitation that will revolutionize several sectors [17].

The alliance between XR and DT, called XR-DT, provides a platform for simulating, planning, testing and evaluating physical systems. In this

---

[3]https://e.huawei.com/ae/eblog/industries/insights/2020/immersive-city-experiences.
[4]https://www.emergenresearch.com/industry-report/extended-reality-market.

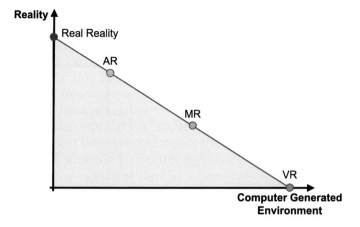

FIGURE 3.7   Relationship reality, virtual world in VR, AR and XR.

platform, the collected data should go through a heavy processing step following DT mechanisms to create the DT of a physical asset, while XR is used to show a 3D hologram of that twin with its real-time updates. In other words, the user will be able to visualize and interact with a 3D virtual representation of the created DT showing all the real-time updates and history. For instance, automotive engineers can see a hologram of the virtual twin of the designed car in front of them, so that they can interact easily with it to test various options. This combination of DT with immersive technology enables deeper understanding and speeds up the creation of valuable insights. In addition, it facilitates the interaction with the created DT simplifying various tasks such as simulation manipulation. The immersive technology in these systems has a heavier task compared to other traditional applications such as learning or entertainment as it represents the interface between the user and the DT. Indeed, XR must visualize the continuous real-time updates affecting the DT and at the same time, respond to the dynamic user interaction.

XR-DT brings several benefits to various sectors including healthcare, engineering, automotive, constructions, etc. In healthcare, this technology leverages the traditional medical concepts to enhance various areas including training, telemedicine and personalized healthcare. Indeed, by collecting various types of clinical information about patients, a DT of their bodies and organs can be created simulating and visualizing precise information about their medical situations [4]. Healthcare experts are able then, to interact with the 3D virtual duplication of the patient body or organ to create a deeper understanding of the situation and to make an accurate diagnosis that can be used in various activities including (1) training professionals and testing drugs, (2) personalized healthcare and (3) telesurgeries.

3.2.3.1 Testing drugs and training professionals

The immersive DT of a patient allows understanding the human body reactions under distinct circumstances, providing a tremendous replacement to the traditional learning strategies that rely on huge amount of texts and heavy training materials tough to digest and comprehend [18]. This solution solves an important issue related to the cost and availability of medical machines and patients for training. Healthcare professionals use immersive DT of the medical devices for training and the virtual DT of patients to simulate and plan their treatments and test the new drugs on the digital replica before applying them on humans. This technology allows professionals to understand the body responses so as to avoid mistakes that can be done in the real world, and improves the quality of the medical treatment as in the virtual environment, it is safe to fail and learn from mistakes.

3.2.3.2 Personalized healthcare

By collecting real-time data about the patient's health, DT can provide a scan of the whole body and/or its organs. Professionals are able then, to get a precise report about the patient's situation and can interact with his organs to get the needed information and build useful insights. Using the gained understanding of the one health situation, the doctors can test and plan a personalized treatment that better fits each patient, specifying customized therapy including the selected medicines or the exact doses needed for each case. Moreover, customized diagnoses and treatments can also be provided using genetic information collected from different people allowing the DT system to map between the various cases to better understand their medical and genetic characteristics to predict future diseases for each person. Thus, customized actions and treatments will be provided for each patient to enhance the general health.

Immersive DT enables moving the medical domain from a general care to a personalized care and from treating sicknesses to preventing diseases.

3.2.3.3 Telesurgeries

Immersing medical professionals in a world where they are able to interact and visualize the DT of their patients enables them to train before surgeries and to remotely manipulate robots during operations. Indeed, by visualizing and interacting with the 3D hologram of the patient organ DT, the professionals can monitor the progress of the surgery since they are able to see the real time responses of the body, thus they can remotely manipulate robots performing a surgery or guide and cooperate with other professionals participating in the same telesurgery. Also, immersive DT allows visualizing the history of a surgery that is very useful in learning, and mistake understanding.

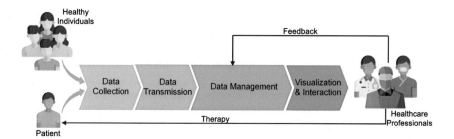

FIGURE 3.8   XR-DT for healthcare system process.

Various other applications can be designed for immersive DT in healthcare that will revolutionize the sector.

3.3 XR-DT-based system for healthcare requirements

The medical domain is a very critical area since it deals with human health and life. It is mandatory that the used technologies are accurate, fast and precise. Adopting new solutions has always been taken seriously as it has to go through various validations processes to guarantee its efficiency. Providing a technology that can help in testing new strategies and treatments can revolutionize the traditional medical area. XR-DT is a tremendous solution to reduce risks by providing an alternative virtual platform for simulation and testing. As it combines two recent technologies, several requirements should be taken into consideration to meet the heavy demands of the healthcare systems and to design a robust, fast and accurate platform.

The involvement of immersive DT in hospitals necessitates the highlight of various features that should be taken into consideration. To understand these features, it is important to go through the different steps of generating XR-DT service. Different devices are needed at the two sides of the end users: Individuals (healthy and patients) and the medical professionals including different type of sensors and XR headsets. As presented in Fig. 3.8, the process can be divided into four main steps: (1) Data collection, (2) Data transmission, (3) Data management and (4) Visualization and interaction.

3.3.1 Data collection

To offer the highest level of accuracy, XR-DT relies on the collection of different types of information related to patient health. This big data is used in building the patient's DT and in understanding his reactions. To this end, various types of information are needed to provide a precise update on the one health situation that can be divided into four main categories: clinical data, real-time health data, third-party data, and genome data.

- *Clinical data:* This type of information provides details about the medical situation of the patient and its progress that can be part of the hospital information system including medical notes, examination results, medication history, medical images, laboratory reports, etc. The collection of this type of information is challenging due to its sensitivity and the privacy of the hospital databases. Various countries are adopting the strategies of building a centralized database to collect the medical information of the different hospitals and pharmacies helping to bring all the needed information in one place.
  In addition, XR technologies require, in some applications such as telesurgeries, the use of 360 and Ultra-High Definition (UHD) cameras to take high resolution images and videos.
- *Real-time health data:* This category is related to IoT information that can be collected using IoT devices, wearable or nano IoT sensors, and that provides real-time reports about the patient's health condition. It includes heart rate, blood pressure, blood glucose, body fat, temperature, etc. The continuous generation of this category creates some challenges due to the size of the collected information. Specific mechanisms to transfer, process, and store this big data are needed to achieve the highest level of accuracy and efficiency with the lowest delay. Various types of sensors that can be portable, wearable, or embedded, can be used to monitor the patient's condition. Indeed, with the advances in nanotechnology and robotics, nano-sensors can be implemented inside the human body to collect accurate real-time data and to inject medicine doses like the glucose monitor that is inserted under the skin [19]. These devices should be used for long periods of time; thus, some specific requirements have to be provided such as designing ergonomic lightweight tools to enable the users to continue their daily activities in a normal way.
- *Third-party data:* This category can be collected from any entity that may have a relationship with the patient's health such as medical insurance, gym, dietitian customized meal companies, etc. This information helps to further understand and monitor the progress of the medical condition as it presents complementary data about the patient behavior and lifestyle. Sensitivity and privacy are the main challenges in this category.
- *Genome data:* With the spread of the bio-bank concept, genetic data is more and more involved in research and medical diagnosis revealing impressive information about the anatomy and physiology of the human body. Several genetic characteristics can be compared and analyzed providing a medical profile of any person including possible

future diseases and personalized behavior and reactions. Genetic information is essential to build the DT of patients as it helps health professionals to make an accurate diagnosis and to understand and predict the patient reactions to any tested therapy. Some samples (e.g. tissues or blood) collected from the biggest possible number of people create a genetic database that can provide precise details about the public health and help in developing new medicines.

The gathering of a big amount of heterogeneous data introduces several challenges to the XR-DT platform as it has to be forwarded, manipulated, and exploited in an efficient and fast way to reach the needed level of accuracy and to help in enhancing the healthcare system.

3.3.2 Data transmission

The advent of innovative technologies that rely on data collection overburdens the communication infrastructure. Various applications with different characteristics and demanding requirements are designed to offer a high level of quality of experience (QoE) overwhelming the communication network. In the healthcare domain, the proposed solution requirements are mainly related to system efficiency in terms of accuracy and speed. The collected information is essential to make a precise diagnosis about the patient's condition in time as errors and delays are not tolerated in such critical areas in which human lives are at risk. It is the task of the communication technology to transport this data with the highest precision and the lowest delay. As discussed in section 3.3.1, various types of data have to be transported from where it is generated to where it is going to be processed and stored and then, to where it will be visualized and consumed. Each type has specific demands that have to be guaranteed by the network anywhere and anytime such as specific bandwidth and latency. For instance, transmitting a big amount of real-time data is one of the demanding tasks mainly in dynamic environments where the patient is mobile and the load of the network keeps changing with the crowd in the visited areas. Also, to create the XR-based DT of an organ and the interaction with it, UHD (Ultra-High Definition) images and 4k/8k videos with a big number of frames per second (i.e. few Terabits of data) should be forwarded to the visualization location in time. All the needed demands should be guaranteed by the network to provide the needed QoE and to reach a high level of efficiency mainly in emergency situations (e.g. telesurgeries).

With that being said, 5G and B5G (beyond 5G) are promising technologies able to guarantee the needed bandwidth independent of the environment (i.e. crowd) and of what is happening around. Indeed, 5G provides extremely low latency that is 200 times faster than 4G (i.e. it reduces the latency from 200 ms in 4G to 1 ms in 5G) making it very suitable for massive IoT and real-time interaction and immersive services such as telesurgery that has stringent latency requirements. The main concern of 5G is to satisfy the exponential increase in the number of connected devices and their demanding requirements imposed by the various applications [20]. It is based on the use of smaller cells with high-frequency mm-wave band ranging from 2.4 to 7.2 GHz used currently, and up to 300 GHz in future deployment to support hundreds of times more data and capacity compared to 4G. 5G relies on various emerging technologies to provide the most efficient services. For instance, advanced access technologies are used in 5G such as Beam Division Multiple Access (BDMA) and Filter Bank Multi-Carrier (FBMC). Also, to significantly increase the network capacity, massive Multiple-Input Multiple-Output (MIMO) is deployed with software-defined networking (SDN) mechanisms to achieve high flexibility level in the network through network slicing. Network slicing refers to the capability to have independent logical networks on the same physical network, where each slice is customized based on the end-to-end QoS requirements. Network slicing is important to satisfy the needs of XR-DT mainly with mobile patients and healthcare professionals in a dynamic environment and variable requirements. With B5G, a new level of intelligence will be introduced to leverage the traditional network slicing approach through the integration of AI and ML mechanisms to reconfigure the network slice parameters in accordance with the dynamic situation of the network. In other words, it will provide an autonomous network platform, able to adapt itself according to the variation of the network conditions in a smart and efficient way offering more suitable solution to the complex and irregular data of IoT and immersive services and their requirements.

3.3.3 Data management

The efficient management of heterogeneous massive data is so complex and has to be processed over several steps using emerging technologies. In XR-DT platform, data management is divided into two main steps: DT and then XR. The gathered information should go through various processing mechanisms to provide an accurate service that meets the requirements of health care systems, mainly the highest precision and the lowest delay. For instance, the DT of the patients allows healthcare professionals to test customized therapies and to analyze patients' bodies' reactions before the application in real life. In this case, the simulation results should provide an accurate and precise analysis of the impact of the tested treatment taking into consideration all the possible information that can be related to the anatomy and physiology of the studied patient or any similar cases. Indeed, the main target of this platform is to increase the efficiency of the healthcare system by reducing error risks.

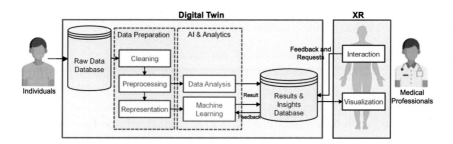

FIGURE 3.9   Digital twin mechanism in XR_DT based healthcare system.

3.3.3.1 DT mechanisms in healthcare

Various types of data (presented in section 3.3.1) are heavily processed and analyzed to precisely understand the characteristics of the human body and create its DT (Fig. 3.9). The created DT can be the replication of the human body, organ (e.g. heart, liver, etc.), tiny body component (e.g. cells, molecular level, etc.), or body function/system (e.g. respiratory, nervous, digestive, etc.). DTs belonging to different individuals (healthy and patients) can be aggregated for clinical matching purposes to trace and track various medical characteristics that lead to discover relations between distinct data elements over time and to build timely and valuable insights about the human body. Using the gained knowledge, health professionals can test or simulate the impact of new therapies on the created DT to provide an exhaustive prediction about the body's reactions and make more informed and proactive decisions [21].

The gathered data has to go through a preprocessing phase to remove redundancy and faulty information and extract important parameters [22]. This phase including data cleansing, filtering, and classification mechanisms, allows for the reduction of the size of the raw data and reducing the system complexity to reach the needed level of efficiency. Then, the selected data goes through data analysis mechanisms to provide a detailed understanding of the medical condition and through ML-based processing to build classifiers and predictive models allowing to detect anomalies and predict the body reactions. Due to the various types of gathered information (images, text, videos, etc.), multimodal ML approaches are highly needed to provide a smart way to analyze this heterogeneous data and extract valuable information. Different ML models can be used in this system including Convolutional Neural Network (CNN), Long-Short-Term Memory (LSTM) model, Multi-Layer Perception (MLP), Logistic Regression (LR), and Support Vector Classification (SVC) [4]. The result of these procedures is used as continuous feedback for further corrections and model optimizations. Also, feedback from medical professionals is highly essential in this system as they provide advice and a different vision of the treatment based on their formal training and experience.

3.3.3.2 *Data management in XR for healthcare*

XR represents the interface between the service consumer and DT mechanisms. It must perform several tasks to combine the DT outcomes with the real world in a comprehensible way and to offer a simple tool to interact with the created virtual twin [23]. At this level, two categories of data are used by the XR platform: DT information resulting from the various examination and processing performed by the DT management approach (section 3.3.3.1), and the real-world data gathered from the user surrounding environment including real scenes and captured gestures (detailed in section 3.3.4).

The combination of the real scenes with digital information requires a high level of intelligence to manipulate the large and complex information generated by the DT mechanisms [24]. AI-and ML-based approaches are highly needed to investigate and understand the real and digital data and make decisions on the relation between them and thus, their combination. The information generated by DT is highly complex and requires dedicated tools to build an understanding that allows the design of 3D content reflecting the various outcomes. An autonomous AI-based platform should be designed to receive DT data, user gestures and requests, and images and videos from their surrounding real world to generate XR-based virtual scene with a high level of interaction. As presented in Fig. 3.10, the received information should go through different processing steps, mainly, data analysis, 3D construction, and data linking.

Data analysis and 3D construction

Each type of the received data must go through a dedicated processing phase to extract valuable information that can be translated into an element of the XR scene:

- *User tracking data:* This information is generated by several monitoring devices and sensors provided with the XR headsets to enable the interactivity feature. It is highly important to decide on the 3D model that should be visualized in front of the user to reflect the conclusions and observations of the DT such as the full human body, an organ, a part of the organ, some molecules, or a selected mechanism or function. These preferences are mainly received from the users and sent to the DT module as a request for further data filtering, extraction, and observations. On the other hand, the tracking information provides real-time bio-feedback about the user during the immersive experience such as eye or head movement tracking. The analysis of this information creates comprehensive visibility about the users' positions, vision directions, and the exact location where their eyes are landed and allow to make

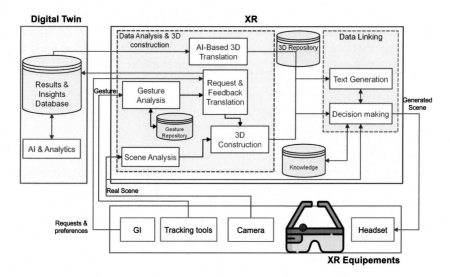

FIGURE 3.10   XR data management.

some predictions about future requests. These analyses are then shared with the 3D reconstruction module to update the immersive experience in real-time to guarantee customized and natural scenes for every user.

- *Digital twin data:* Based on the user interaction with the 3D objects, DT data is filtered, analyzed, and calculated. The dedicated information is then extracted as a response to the received requests. This interaction allows the user to choose to freely perform some modifications on the object or express some preferences such as visualizing a given behavior of a selected organ, or part of the organ, zoom in/out, make some operations to simulate a therapy, etc. Using an AI-based translation module, the received data from the DT module is then analyzed and examined to create customized 3D content. A repository is created containing XR standard models that can be used, modified, and optimized based on the received DT data such as the 3D representation of the heart, liver, molecule, etc.
- *Camera real scenes:* The images and videos captured by the camera embedded in the headset represent the real side of the XR output. The examination of these videos must go through various steps including scene recognition, object detection, and spatial and temporal investigations to provide valuable insights about the user's real environment useful in interpreting and recognizing its characteristics and dimensions (e.g. office, lab, etc.). Thus, the video shots should be detected and their various key frames have to be extracted and separated to simplify their examination using deep learning models such as CNN as used in [25,26] and video segmentation mechanisms, such as Spatial and tem-

poral CNN-based Markov Random Field as proposed in [25]. Then, the object detection and localization module should determine the different objects that are present in the scene with their exact locations such as desk, chair, etc to be used in the 3D reconstruction. The location of the physical objects in relation to the user is highly important to decide on the 3D element placement in the generated scene what allows to precisely interpret the user gestures (mainly, gestures in relation with the virtual objects) and create a natural realistic scene.

Data linking

This module receives the various needed elements and then generates the new XR scene. In other words, it allows combining the heterogeneous selected data (i.e. DT information, text showing details, 2D and 3D elements). Decisions related to the 3D environment reconstruction are then made to localize the virtual objects in the created 3D space and sent to the XR headset. A real-time scene update is mandatory for all users, mainly, in case several professionals collaboratively manipulate a 3D model at the same time, all the modifications made by any user should be visible to all the others in time. Everyone should have a customized view based on his location. Also, all the details should be represented fairly to help in training doctors. For instance, in dentistry, the dentist has to operate inside the mouth of the patient with limited movements and visibility of the teeth. XR-DT platform should put the trainees in a realistic environment by respecting the real dentists view in the provided scenes.

3.3.4 Visualization and interaction

Most of this step is handled at the end-user side. At this stage, the user is the medical professional who manipulates the DT of a patient. The XR technology allows immersing the users in an interactive virtual environment where they are able to visualize and manipulate the digital information in a comprehensive, natural and simple way [27]. In other words, XR represents an interface between the users and the complex digital information provided by the DT. Various mechanisms and tools are designed to capture the user preferences and requests in a very natural manner as per the XR's main objective.

To interact with the DT objects and data, the users are able to use dedicated tracking devices, and/or an application interface that can be installed on their smartphones, tablets, or computers.

3.3.4.1 *Application graphical interface (GI)*

Using a GI, that can be downloaded on a smartphone, tablet, or computer, the user can easily communicate with the XR-DT platform and set the various parameters needed in the DT immersive experience. The GI provides a familiar management tool that allows the medical professional to choose various options regarding the visualized elements (e.g. choosing a specific patient) and makes several modifications such as filtering the requested data, testing some elements, comparing with other models, etc. Distinct applications can be designed using XR-DT offering a wide range of customized options to the user that may depend on the application target (e.g. training, testing, etc.) and the medical discipline (e.g. cardiology, dentistry, neurology, etc.)

3.3.4.2 Tracking devices

XR relies on various advanced cameras and sensors to track the user behavior in a very smooth and natural manner. The captured behavior is then analyzed and translated into requests and preferences that are used to update the generated virtual environment in real-time.

The tracking sensors and cameras are embedded in the headset and/or handheld tools collecting bio-feedback information and monitoring the user motion.

- *Biofeedback*: Using eye-tracking technology and neural interface, various indications can be communicated to the XR-platform revealing the user preferences and behavior. For instance, analyzing the eye movement provides a clue about the part of the visualized scene the user is focusing on, the decisions made by the user, and predictions on the future ones. Also, a neural interface can be used to understand the user gestures such as pinching. Bio-data tracking is a complex technology that requires important training to be able to understand user behavior and gestures.
- *Handheld trackers*: Providing a familiar tool to interact with the virtual object, the handheld controllers allow to detect the user's hand gestures and translate them into requests. They track the hand motion, rotation, and position and contain buttons to monitor distinct events and interactions inside the virtual world (e.g. select, grab, drag, etc.). Some handheld devices are able to provide haptic feedback to simplify the user experience. Indeed, the utilization of handheld trackers can be very complex to the users as they may require some level of training to properly use this technology, mainly in the medical domain, putting more pressure on the users and increasing their cognitive load which stands in contradiction with the natural principle of the XR immersive experience.
- *Camera-based gesture tracking*: The most natural way to track the users' gestures is to let them move in a normal manner while detecting their positions, rotations, and motions. Using a camera embedded in the headset, video, and images of the hand movement and head position are captured. The motions of the fingers and the palm are recognized and mirrored in the virtual environment. Various predefined gestures can be recognized by the system including dragging, pinching, twist-

ing, tapping, selecting, zooming, and objects rotating and placing, etc. Images of the hand and fingers' motions are detected and shared with the module responsible for the analysis to recognize these various gestures in real-time. More customized gestures and complex interactions can be designed to provide more freedom to the user to perform various tasks in a normal manner. For instance, hand techniques used by the medical professional to operate a surgery or any other type of therapy (e.g. dentistry) can be detected by these camera-based-trackers to allow them to practise their techniques in a normal manner. The leap motion sensor[5] is one example of these trackers. It captures several images during the hand movement used to recognize its poses and gestures. For the highest level of movement freedom, these trackers are embedded in the XR headset to provide the inside-out tracking [28] as in Microsoft HoloLens[6] and Samsung HMD Odyssey[7] headsets. Instead of surrounding the user with various types of sensors to track his movements and position, inside-out trackers are placed on the user (i.e. on the headset) to capture his gestures, body and head positions, movements, and the virtual environment is updated accordingly in real-time to create a customized XR scene.

The progress in the tracking tools aims to reach the highest level of natural interactivity and put the users in XR realistic environments increasing the efficiency of the XR-DT in the healthcare domain. Indeed, medical professionals are able to practise and manipulate surgeries using all their delicate and critical gestures in a very high level of accuracy.

3.4 XR-DT for healthcare architecture: emerging paradigms

XR-DT is an intricate technology that should meet the highly demanding healthcare requirements. As presented in section 3.3, various complex tasks should be provided to gather the needed information, process the collected data using different ML approaches, and offer an interactive visualization of the DT. The fulfillment of these tasks requires a resourceful strategy using advanced paradigms to reach the highest level of efficiency and accuracy with the lowest delay. Answers for several questions need to be considered like where is the most efficient place to process this data? How to process and access it? Where to store it? This section provides tremendous paradigms as answers for all these questions as they provide the efficiency and accuracy required by the XR-DT for healthcare.

---

[5]https://www.ultraleap.com/product/leap-motion-controller/.
[6]https://www.microsoft.com/en-us/hololens.
[7]https://www.samsung.com/us/support/computing/hmd/hmd-odyssey/hmd-odyssey-mixed-reality/.

TABLE 3.2   Comparison centralized and distributed computing approaches.

	Centralized approach	Distributed approach
Geographic distribution	Centralized	Distributed
Intelligence location	At the center of the network	At the edge
	Far from the end-devices	Close to the end-devices
Transmission distance	Far	Near
	Through multiple hops	Single/few hops
Computing capability	Few high performance servers	Multiple lower performance servers
Storage location	Centralized	Distributed
Latency	High	Low
Advantage	High capacity permanent storage	More suitable for mobile end-devices

3.4.1 Cloud/edge-based hybrid computing architecture

The efficiency of XR-DT for healthcare systems highly depends on the accuracy of the processing models and the variety of the gathered data. Indeed, the participation of the biggest number of healthy individuals and patients allows creating more accurate understanding of human health and discovering new valuable insights. The increase in the number of data producers (i.e. participating individuals) and service consumers (i.e. health professionals and individuals) enlarges the geographic distribution of the service raising the load of the servers in charge of the various processing tasks.

Relying on a centralized cloud server to manipulate the collected data decreases the efficiency of the system due to the processing complexity that increases with the size of the collected data. Also, the centralized approach requires gathering the generated data in a focal point (i.e. the cloud server) causing important accuracy issues and transmission latency due the distinct geographic locations of the data producers in relation to the cloud server. On the other hand, the distributed computing approach, known as edge computing, has proved more efficiency in dealing with the manipulation of impressive amounts of data with the lowest delay [29]. Indeed, having various cooperative edge servers distributed in various geographic locations brings the intelligence close to where the data is generated and the service is consumed minimizing the transmission latency making this approach more suitable for delay sensitive domains like healthcare. Table 3.2 presents a comparison between the centralized and distributed computing approaches.

The edge-based distributed computing architecture enriches the provided service by providing a local speedy high level of intelligence enabling the processing and the manipulation of the data close to where it is generated [30]. This feature is very suitable with the highly interactive and the delay sensitive applications, such as XR-DT responses. Indeed, to guarantee a natural immersive experience, XR-DT must process the captured requests, analyze the requested situation using DT mechanisms, and provide the updated virtual scene to the user with the lowest delay and highest accuracy. Thus, relying on edge computing meets the strict real-time requirement of the XR-DT and enhances the service quality. On the other hand, edge computing has several drawbacks due to the limited heterogeneous capabilities of the used devices in comparison with cloud servers.

The distributed edge servers are able to operate in a cooperative manner in which various collaborating tasks can be allocated to the edge nodes based on their locations and capabilities to provide a progress toward the global application target. For instance, some edge servers can focus on the preprocessing phase of the gathered data, locally inside a given geographic area and share it with other servers with higher capabilities where it is going to be analyzed with data received from different areas. Due to their heterogeneous capabilities, the cooperative edge servers perform distributed tasks as a single logical device. Several tasks of the XR-DT for healthcare systems are so complex requiring important computing capacity to process the huge amount of the gathered data, thus the involvement of the performance cloud servers is a wise solution to handle these tasks and work in collaboration with the distributed edge. To this end, a hybrid architecture based on cloud and edge is more suitable for XR-DT applications to provide the best service quality and efficiency as healthcare is a critical domain in which efficiency and precision are highly needed. Fig. 3.11 presents the adopted computing architecture for XR-DT-based healthcare applications. It is composed of two main layers: (1) end user layer and (2) data management layer.

- **End-user layer**: This layer groups the various types of users who may generate and/or consume the services. Healthy individuals and patients are the data generators as they share their medical data with the XR-DT platform for further processing while the medical professionals are the consumers who, using the XR interface, can request access to the processed information and the provided services. The individuals are also able to use specific types of XR-DT-based applications allowing them to receive advises about a customized healthy lifestyle.
- **Data management layer**: It contains cooperative cloud and edge servers where distributed edge nodes provide their services close to the end-users. The cloud servers play the role of stabilizers by taking charge of the most complex tasks that need very high computing capabilities. It also provides permanent storage to a part of the processed data including the created insights. For instance, analyzing and translating the captured user gestures into requests are performed in the nearest edge cloud while performing comparison tasks between DTs of various in-

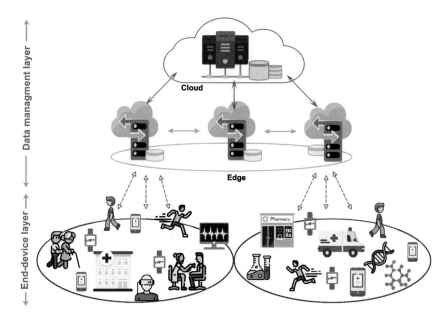

FIGURE 3.11   Hybrid Cloud/Edge architecture for XR-DT based applications.

dividuals is a heavy task that can be performed in the cloud where the global data can be stored.

Relying on 5G network with the Multi-Access Edge Computing (MEC) technology provides a tremendous solution to simplify the deployment of the distributed edge nodes while increasing their capabilities providing a ground for fast and easy adoption of intelligent XR-DT-based technologies and applications in healthcare.

3.4.2 Distributed cooperative data processing: federated learning

Using distributed computing architecture enables the adoption of an advanced smart learning paradigm, which is Federated learning [31]. Federated learning provides a distributed processing approach in which various participating devices cooperate to generate a global model using huge amounts of collected data. In the traditional centralized approach, a central entity is in charge of processing the gathered data from the various participating devices using ML tools to generate a global model, while in federated learning, this task is performed by a group of cooperative nodes. Each participating device processes the locally gathered raw data to generate a local model that is shared with a central entity selected to perform a different task. In essence, the central entity builds a global model after per-

TABLE 3.3 Comparison of traditional centralized learning approach and federated learning.

	Centralized learning	Federated learning
Data distribution	Centralized Processing storage	Distributed processing and storage
Privacy of the raw data	The raw data is shared with a central entity	The raw data is kept with the generated device locally
Data training	Centralized using all gathered data	Cooperative distributed locally gathered data only
	Big amount of data	Smaller amount of data
Failure	Single point of failure	Failure of one device does not have a big impact
Advantages	Less impact of end-device failure	Cooperative processing
		Faster convergence
		More accuracy
Challenge	Heterogeneous massive amount of data	Heterogeneous devices' capabilities

forming several aggregation and averaging mechanisms and subsequently shares the obtained model with all the participating devices to update and adjust their local results for the next round.

The main advantages of federated learning are the privacy, scalability, model simplification and fast convergence [32]:

- **Privacy**: Each participating device should locally process its data and share only the obtained model with the other nodes. The raw data is kept locally with the device, so no other entity can access it, increasing its privacy.
- **Scalability**: The distributed feature of the federated learning allows the system to adapt easily to any variation in the number of the participating devices.
- **Model simplification**: By allowing various cooperative devices to conduct several parallel trainings using small amounts of data, federated learning simplifies the traditional centralized approach in which one single entity should process the impressive amount of data each time.
- **Fast convergence**: Using simpler models, the cooperative participating devices in federated learning perform faster several iterations as they learn from other devices' experiences resulting in faster building of a robust global model.

Table 3.3 presents a comparison between the traditional centralized learning approach and federated learning.

In the XR-DT-based system, various individuals in different locations generate health data that should be processed to create the DT of a person with several demanding requirements including efficiency and privacy. Data privacy is an important requirement in such systems as the collected data contains sensitive personal information about each participant. Thus, federated learning is very suitable for the XR-DT-based applications in healthcare, mainly to the DT processing. Indeed, the data can be processed locally by the user devices or the edge nodes to share the updated parameters only with the system platform to generate the global model. The adoption of federated learning in XR-DT-based applications brings several advantages such as privacy and efficiency, and fast system convergence.

3.4.3 Dynamic data storage

To reach the highest level of accuracy, the DT mechanism needs to process huge amounts of data gathered from various individuals (healthy and patients) to make analysis and comparisons resulting in building novel insights and conclusions. The continuous data gathering and DT conclusion generation accumulate more and more data in the system raising a challenging issue related to data storage. Different questions in relation to this challenge need resourceful answers and strategies: Where to store the accumulated data? How to reduce its size? and how to access it quickly?

Various strategies using advanced techniques can be used to enable storing the huge amount of data. Relying on distributed databases is an efficient solution that allows the distribution of data over various storage locations. This solution provides better results in terms of scalability and accessibility compared to the traditional centralized storage where all the data is placed at one central database. However, various advanced features are needed to handle the multiple databases to localize the needed data with the lowest delay.

The adopted mechanisms must provide support to the distributed data storage like federated learning, in which the private data is stored locally close to where it is generated, while only the updated parameters are stored in other servers mainly in the cloud. Also, several techniques can be implemented to reduce the size of the stored data such as data cleansing, filtering, averaging and aggregation that allow optimizing the size of the stored information. On the other hand, dedicated AI-based approaches can be adopted to enable the dynamic storage and manipulate the cache close to where the service is requested to enhance the data availability, to speed up the information accessibility, and to support the interactivity feature. These tools should provide a prediction about the user requests and interests using ML mechanisms such as the LSTM model allowing to identify the frequently requested data in each location to update the cache accordingly in real-time.

FIGURE 3.12   XR-DT-based healthcare application open issues.

3.5 Open issues

Adopting XR-DT for healthcare is a tremendous approach to revolutionize the traditional medical domain. It combines demanding novel technologies: XR and DT with the critical healthcare domain. Indeed, healthcare is a delicate domain with various strict requirements in terms of efficiency, precision, delay and privacy making professionals very cautious about adopting the new technologies. Thus, several challenges and issues in XR-DT should be addressed to enable the implementation of the new customized preventive care approach.

Four main open issues can be considered (Fig. 3.12): Privacy and security, Trust, Building dedicated models and standardization.

3.5.1 Privacy and security

The most important requirements in any system that needs to handle sensitive personal data related to various participants are privacy and data security. In XR-DT-based healthcare applications, massive amounts of sensitive personal data are gathered and stored making this system a target for cyber-attacks. Strict mechanisms to stop and eliminate any risk of unauthorized access, abuse, modification or disclosure should be implemented at the various stages from the data collection to the DT visualization. Divers mechanisms and devices are considered by this feature (e.g. IoT sensors used to generate data)increasing the complexity of this challenge.

On the other hand, dealing with private sensitive data raises regulatory issues. Collecting, processing and exploiting this personal information should be done based on a legal policy under the approval of the various users (data producers and health professionals). Standard privacy regulations such as General Data Protection Regulation (GDRP)[8] in Europe or California Consumer Privacy Act (CCPA)[9] in California or any regulation at relevant national protection laws should be considered when designing the XR-DT-based healthcare systems.

3.5.2 Trust

The success of XR-DT-based systems for healthcare relies on several features including the number and variety of participants sharing their information and the efficiency of the adopted models. To persuade people and professionals to participate in this system, efforts should be taken to encourage them to trust this new technology which is a challenging task mainly in the medical domain in which adopting new technology may take time. The new technologies always have a gap as they rely on smart devices and data transmission that may crush or get disconnected at any time for any reason. Also, the adoption of ML and AI-based models raises several questions about trust and accuracy of resultant predictions and decisions. People are still discussing trusting a machine that can make crucial decisions.

The concept of XR-DT requires the contribution of qualified and ethical medical professionals to get their accurate feedback about the decision making, verification and validation processes to keep improving the system performance and increasing the trust level. In essence, building trust at the various stages of the XR-DT-based systems is important to assure confidence around the concept and to encourage professionals and users to adopt it. Thus, a strategy based on setting standards, continuous technology improvement and raising awareness are all required to reach the needed trust. This is not an easy task as it needs effort and time.

3.5.3 Dedicated models and approaches

In relation to trust, various dedicated AI and ML models should be designed exclusively for XR-DT systems. This platform could make crucial decisions about human lives increasing the responsibility of building accurate models taking into consideration various challenges such as the diversity of the used devices and the heterogeneity of the collected data. Medical experts play an important role in monitoring and adjusting the outcomes of the proposed technology. Their contribution and feedback are highly essential to improve the designed models. To simplify this role,

---

[8]https://www.fendahl.com/gdpr-data-protection-statement.
[9]https://oag.ca.gov/privacy/ccpa.

various paradigms can be adopted such as explainable AI to enable professionals to easily interpret and interact with the analysis and conclusions made by the various models.

3.5.4 Standardization

Setting global standards is a highly important factor to spread the XR-DT technology and encourage users to trust it faster. Standardization enables to organize the implementation of this technology mainly between various stockholders by providing a ground for fast and easy adoption of intelligent XR-DT-models to design and build various innovative applications. Also, it may affect different features of this technology including privacy, security, models, interactions, data transmission, etc.

3.6 Learned lessons

Several lessons can be extracted from this chapter:

- VR is a computer-generated virtual environment using 360 videos and images allowing to completely immerse the user in a 3D world that is totally disconnected from his real one, while AR overlays the real environment and objects with digital images generated by the user device. It uses the real environment and objects as a framework where digital images are placed. MR is a hybrid technology that combines AR and VR to provide an interactive virtual experience over the real world. It merges the real and virtual worlds to create a new environment where physical and 3D digital objects coexist and interact in real-time. XR is the umbrella covering VR, AR and MR.
- Developing an XR-DT-based healthcare system is a tremendous and cost-effective solution, on the other hand, it can revolutionize the medical sector by providing more customized therapies that are tested and validated, thus reducing risks.
- XR-DT healthcare systems rely on the combination of the power of understanding data using intelligent models, to provide a platform that can provide an exhaustive understanding of individual health.
- Information collection is essential to make a precise diagnosis about the patient's condition in real-time as errors and delays are not tolerated in such critical environments in which human lives are at risk. The gathered data has to go through a preprocessing phase to remove redundancy and faulty information and extract important parameters that include data cleansing, filtering, and classification mechanisms. Then, the selected data goes through data analysis mechanisms to provide a detailed understanding of the medical condition and through ML-based processing to build classifiers and predictive models allowing to detect of anomalies and prediction of body reactions.

- To meet the healthcare requirements in terms of efficiency, precision, delay and privacy, XR-DT has to face several challenges including: privacy and security, trust, building dedicated models and standardization.

3.7 Conclusion

Integrating DT and XR in one platform takes complements one another. XR provides an advanced and a new means to visualize and interact with complex data, while DT leverages the XR 3D models to facilitate the way we interact with virtual twins which all rely on real-time data. Adopting such advanced XR-DT for healthcare systems will revolutionize the medical sector in terms of how we interact and process health data. It enables the radical transformation from traditional healthcare operations to a customized one that enables preventive medicine.To this end, the XR-DT healthcare systems rely on the combination of the power of data with the use of advanced AI and ML models, with the knowledge of competent medical experts to provide a platform that can provide an exhaustive understanding of individual health. This platform is useful in testing new treatments, training professionals, and providing customized reports about health conditions, possible future diseases, and tailored therapies. Indeed, gathering data from multiple healthy individuals and patients builds DT banks where various DTs can be analyzed and compared helping in discovering new insights and building knowledge to predict sicknesses and to enable preventive medicine.

Immersing healthcare professionals digitally enable better understanding of complex data. The natural interaction mechanism adopted by the XR-DT platform is very useful to immerse the professionals in a training and test experience, in which they are able to use their natural hand techniques. This feature is highly important in managing telesurgeries. Moreover, this chapter has introduced the advantages of the XR-DT system in the healthcare domain and provided several uses cases. It detailed the various steps and approaches needed to provide the required output and proposed emerging paradigms to build a robust platform.

References

[1] Xingzhi Wang, Yuchen Wang, Fei Tao, Ang Liu, New paradigm of data-driven smart customisation through digital twin, J. Manuf. Syst. 58 (2021) 270–280, Digital Twin towards Smart Manufacturing and Industry 4.0.

[2] Angelo Croatti, Matteo Gabellini, Sara Montagna, Alessandro Ricci, On the integration of agents and digital twins in healthcare, J. Med. Syst. 44 (2020) 161.

[3] Zinelaabidine Nadir, Tarik Taleb, Hannu Flinck, Ouns Bouachir, and Miloud Bagaa. Immersive services over 5g and beyond mobile systems, IEEE Netw. 35 (6) (2021) 299–306.

[4] Haya Elayan, Moayad Aloqaily, Mohsen Guizani, Digital twin for intelligent context-aware iot healthcare systems, IEEE Internet of Things Journal (2021) 16749–16757.

[5] Ying Liu, Lin Zhang, Yuan Yang, Longfei Zhou Lei Ren, Fei Wang, Rong Liu, Zhibo Pang, M. Jamal Deen, A novel cloud-based framework for the elderly healthcare services using digital twin, IEEE Access 7 (2019) 49088–49101.

[6] Radu Papara, Ramona Galatus, Loredana Buzura, Virtual reality as cost effective tool for distance healthcare, in: 2020 22nd International Conference on Transparent Optical Networks (ICTON), 2020, pp. 1–6.

[7] Barbara Rita Barricelli, Elena Casiraghi, Daniela Fogli, A survey on digital twin: definitions, characteristics, applications, and design implications, IEEE Access 7 (2019) 167653–167671.

[8] Adil Rasheed, Omer San, Trond Kvamsdal, Digital twin: values, challenges and enablers from a modeling perspective, IEEE Access 8 (2020) 21980–22012.

[9] Spyridon Symeonidis, Sotiris Diplaris, Nicolaus Heise, Theodora Pistola, Athina Tsanousa, Georgios Tzanetis, Elissavet Batziou, Christos Stentoumis, Ilias Kalisperakis, Sebastian Freitag, Yash Shekhawat, Rita Paradiso, Maria Pacelli, Joan Codina, Simon Mille, Montserrat Marimon, Michele Ferri, Daniele Norbiato, Martina Monego, Anastasios karakostas, Stefanos Vrochidis, xr4drama: enhancing situation awareness using immersive (xr) technologies, in: 2021 IEEE International Conference on Intelligent Reality (ICIR), 2021, pp. 1–8.

[10] Mythreye Venkatesan, Harini Mohan, Justin R. Ryan, Christian M. Schürch, Garry P. Nolan, David H. Frakes, Ahmet F. Coskun, Virtual and augmented reality for biomedical applications, Cell Reports Medicine 2 (7) (2021) 100348.

[11] Martin Eckert, Julia S. Volmerg, Christoph M. Friedrich, Augmented reality in medicine: systematic and bibliographic review, JMIR mHealth uHealth 7 (4) (2019) e10967.

[12] Carl Laverdière, Jason Corban, Jason Khoury, Susan Mengxiao Ge, Justin Schupbach an dEdward, J. Harvey, Rudy Reindl, Paul A. Martineau, Augmented reality in orthopaedics: a systematic review and a window on future possibilities, Bone Joint J. 101-B (12) (December 2019) 1479–1488.

[13] Melanie Romand, Daniel Dugas, Christophe Gaudet-Blavignac, Jessica Rochat, Christian Lovis, Mixed and augmented reality tools in the medical anatomy curriculum, Stud. Health Technol. Inform. 270 (June 2020) 322–326.

[14] Kyeng-Jin Kim, Moon-Ji Choi, Kyu-Jin Kim, Effects of nursing simulation using mixed reality: a scoping review, Healthcare 9 (8) (2021).

[15] Ryszard Wierzbicki, Maria Pawłowicz, Józefa Job, Robert Balawender, Wojciech Kostarczyk, Maciej Stanuch, Krzysztof Janc, Andrzej Skalski, 3d mixed-reality visualization of medical imaging data as a supporting tool for innovative, minimally invasive surgery for gastrointestinal tumors and systemic treatment as a new path in personalized treatment of advanced cancer diseases, J. Cancer Res. Clin. Oncol. 06 (2021).

[16] Gary White, Anna Zink, Lara Codecá, Siobhán Clarke, A digital twin smart city for citizen feedback, Cities 110 (2021) 103064.

[17] Zaheer Allam, David S. Jones, Future (post-covid) digital, smart and sustainable cities in the wake of 6g: Digital twins, immersive realities and new urban economies, Land Use Policy 101 (2021) 105201.

[18] Omar López Chávez, Luis-Felipe Rodríguez, J. Octavio Gutierrez-Garcia, A comparative case study of 2d, 3d and immersive-virtual-reality applications for healthcare education, Int. J. Med. Inform. 141 (2020) 104226.

[19] Somasekhar R. Chinnadayyala, Jinsoo park, Afraiz Tariq Satti, Daeyoung Kim, Sungbo Cho, Minimally invasive and continuous glucose monitoring sensor based on non-enzymatic porous platinum black-coated gold microneedles, Electrochim. Acta 369 (2021) 137691.

[20] Ismaeel Al Ridhawi, Ouns Bouachir, Moayad Aloqaily, Azzedine Boukerche, Design guidelines for cooperative uav-supported services and applications, ACM Comput. Surv. 54 (9) (Oct. 2021).

[21] Rogelio Gámez Díaz, Fedwa Laamarti, Abdulmotaleb El, Saddik Dtcoach, Your digital twin coach on the edge during Covid-19 and beyond, IEEE Instrum. Meas. Mag. 24 (6) (2021) 22–28.

[22] Samah Aloufi, Abdulmotaleb El Saddik, Mmsum digital twins: a multi-view multimodality summarization framework for sporting events, ACM Trans. Multimed. Comput. Commun. Appl. 18 (1) (Jan 2022).

[23] Moayad Aloqaily, Ouns Bouachir, Ismaeel Al Ridhawi, Blockchain and fl-based network resource management for interactive immersive services, in: 2021 IEEE Global Communications Conference (GLOBECOM), 2021, pp. 01–06.

[24] F.E. Fadzli, M.S. Kamson, A.W. Ismail, M.Y.F Aladin, 3d telepresence for remote collaboration in extended reality (xR) application, IOP Conference Series: Materials Science and Engineering 979 (2020) 012005.

[25] Linchao Bao, Baoyuan Wu, Wei Liu, Cnn in mrf: Video object segmentation via inference in a cnn-based higher-order spatio-temporal mrf, in: Proceedings of the IEEE Conference on Computer Vision and Pattern Recognition (CVPR), 2018, pp. 5977–5986.

[26] Hongje Seong, Junhyuk Hyun, Euntai Kim Fosnet, An end-to-end trainable deep neural network for scene recognition, IEEE Access 8 (2020) 82066–82077.

[27] Arzu Çöltekin, Ian Lochhead, Marguerite Madden, Sidonie Christophe, Alexandre Devaux, Christopher Pettit, Oliver Lock, Shashwat Shukla, Lukáš Herman, Zdeněk Stachoň, Petr Kubíček, Dajana Snopková, Sergio Bernardes, Nicholas Hedley, Extended reality in spatial sciences: a review of research challenges and future directions, ISPRS Int.l J. Geo-Inf. 9 (7) (2020).

[28] Keisuke Hattori, Tatsunori Hirai, Inside-out tracking controller for vr/ar hmd using image recognition with smartphones, in: ACM SIGGRAPH 2020 Posters, SIGGRAPH'20, Association for Computing Machinery, New York, NY, USA, 2020.

[29] Ouns Bouachir, Moayad Aloqaily, Lewis Tseng, Azzedine Boukerche, Blockchain and fog computing for cyberphysical systems: the case of smart industry, Computer 53 (9) (2020) 36–45.

[30] Vahideh Hayyolalam, Moayad Aloqaily, Öznur Özkasap, Mohsen Guizani, Edge intelligence for empowering iot-based healthcare systems, IEEE Wirel. Commun. 28 (3) (2021) 6–14.

[31] Ouns Bouachir, Moayad Aloqaily, Öznur Özkasap, Faizan Ali Federatedgrids, Federated learning and blockchain-assisted p2p energy sharing, IEEE Trans. Green Commun. Netw. 6 (1) (2022) 424–436.

[32] Haya Elayan, Moayad Aloqaily, Mohsen Guizani, Deep federated learning for iot-based decentralized healthcare systems, in: 2021 International Wireless Communications and Mobile Computing (IWCMC), 2021, pp. 105–109.

Challenges of Digital Twin in healthcare

Hazim Dahir[a], Jeff Luna[b], Ahmed Khattab[b],
Kaouther Abrougui[c], and Raj Kumar[b]

[a]Cisco Systems, Research Triangle Park, NC, United States, [b]Cisco Systems, San Jose, CA, United States, [c]Voltron Data Inc., Mountain View, CA, United States

4.1 Introduction

Digital Twin in healthcare is getting considerable amount of attention in view of new challenges faced by caregivers, patients, and providers in the healthcare field.

The term Digital Twin and concept were first coined in 2002, however NASA has been using it in some form during Apollo mission. With Digital Twin, NASA engineers were able to create accurate simulations with action plan for multiple what-if scenarios. This preparation helped NASA bring back its crew members to earth safely when Apollo 13 equipment failed.

The ability to simulate, experience and adjust the real-world environment, process or product based on learning from working on a digital replica of an equipment/product/process played a huge role in helping scientists and engineers improving the success rate, enhancing the understanding and predictability of positive outcome in various verticals.

The success experienced by stakeholders in other verticals is adding to the motivation from patient advocates, payers, clinician practicians and researchers to improve experience, monitoring and adhere to compliance and reduce cost. The potential of implementing Digital Twin technology in healthcare to improve quality of service and understanding, monitoring and reducing risks in many use-cases are tremendous. In this chapter, we

73

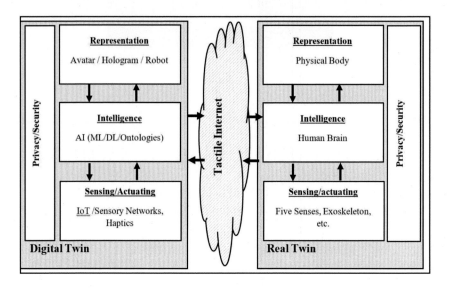

FIGURE 4.1   High level conceptual Digital Twin architecture [1].

attempt to take a deeper look at that. We start by explaining some potential reference architectures, identify the main components and technologies, while identifying and focusing on the challenges.

One Digital Twin framework that we like is shown in Fig. 4.1. Currently, there are no standards or adapted single reference architecture, however, Fig. 4.1 shows a sample of Digital Twin architecture that applies to healthcare. As we go through the chapter, we will discuss the components and the relationships among them and how they contribute the cohesive picture of digital and real twins.

The promise of technological and humanity advances of the Digital Twin is enormous and exciting. Maturity, and with it comes *trust*, is dependent on meeting current challenges. The challenges do exist in a variety of areas, such as infrastructure, communication, security and privacy. This chapter is dedicated to address all known challenges, for each area of the architecture, while covering in detail each component individually for a complete understanding of the challenges and any specific solutions for addressing them or mitigating risk factors.

4.2 Representation

In the physical world of a real twin, the physical body is the main representation form. By contrast, in a Digital Twin realm, representational forms can vary to encompass data-driven software, virtual 3D avatars and holograms, in addition to humanoid social robots as well. The decision to select the type of representation is mainly driven by the use-case or the application itself. Furthermore, with the evolution and realization of emerging enabling technologies, such as 6G, more opportunities are available for horizontal scaling by the way of supporting coverage of different technology platforms. For example, high fidelity mobile holograms can be a form of Digital Twin representation. In 2017, Gartner reports predicted billions of IoT-related Digital Twins' things would have Digital Twin representations within three to five years [2].

As with any form of technology, there are not only challenges but a wide range of opportunities that can be had with imagination and innovative solutions. The same applies to any form of Digital Twin representation. The virtual digital representation must be able to interact with its virtual world with more capabilities and methods that are not available in the physical world. The main driver behind that is that a Digital Twin must be able to interact with a wider range of disparate set of clients or devices in a virtual IoT environments.

A Digital Twin representation must capture attributes and behaviors of the real twin entity it represents. In order to deliver tangible and intangible benefits to organizations employing Digital Twins' solutions, there's a need for standards to conform to in order to ensure product reliability and effectiveness. The International Organization for Standardization (ISO) covers industrial data in TC 194 SC 4 [3]. While there are no similar ISO standards for Digital Twin representation and interoperability standards, recently IPC has launched what it calls the first international standards for Digital Twin product, manufacturing, and lifecycle frameworks [4].

4.2.1 Types of virtual digital representation

As indicated, the virtual representation in a Digital Twin world can take different forms. The choice of the representation is driven by the use-case and its requirements. Recent advancements in virtual reality technologies offer an effective suite to select from. Those technologies include Virtual Reality (VR), Extended Reality (XR), Augmented Reality (AR) and Mixed Reality (MR) technologies [5]. We refer to this suite in general as XR technologies and their hyper-connected XR devices. There are different XR devices available for each of the mentioned technologies and the list keeps on growing. Those products include Head-Mounted Displays (HMD) with AR/VR support, as well as AI-powered AR 3D glasses [6].

4.2.1.1 Avatars

A digital avatar, or a product avatar, is the digital counterpart of a physical entity. It is a combination of modeling and simulation with sensors and big data, with a specific graphical representation and visualization of the data. It can represent a human who can interact with humans, other digital avatars or Digital Twins. For example, it can represent a Digital Twin of an airplane prototype, which is a full 3D digital representation. That representation includes every single part of the airplane, with its accurate attributes and behavior so a human pilot and test engineers can virtually operate the airplane the exact way they would operate a physical real airplane. The relevance of having a product equivalent digital counterpart derives from the expected benefits. Stated briefly, it should allow defining, simulating, predicting, optimizing and verifying the product along its lifecycle, from conception and design, then industrialization and realization, and to usage and servicing [7].

A product avatar is characterized by its capabilities to process a unique identity; communicate effectively with its environment (including counterparts Digital Twins); create, retain, access, transfer and operate upon information about itself; deploy a language to display its features and requirements; and of participating in or making decisions relevant to its own destiny [8].

4.2.1.2 Holograms

Holograms are digital representations of physical entities that exist (or can be imagined) into our physical space. Holographic type communications (HTC) are expected to digitally deliver 3D images from one or multiple sources to one or multiple destinations in an interactive manner [9]. In other words, that type of presentation bridges—simultaneously—the physical and virtual space. Sources for holograms can be obtained by means of real-time capture, transmission, and 3D rendering techniques [10]. It is a technology that enables using facial expressions and gestures, which is a step up when considering 'warm-technologies' or a positive human interaction model. The market size for the hologram displays is expected to be $7.6 billion by year 2023 [11].

4.2.1.3 Robots

A Digital Twin representation is not limited to a manufacturing type of robots, but it definitely can include social robots and wearable robotics as well (wearable devices that are used to enhance a person's motion or physical abilities). Furthermore, many unarticulated systems that utilize Digital Twins might also be manipulated or modified by robotic systems that may themselves utilize Digital Twins [12]. Robots have been employed in manufacturing to deal with manufacturing complexity. Coping with present day manufacturing challenges promotes man-machine system interaction, known as a human-robot collaboration (HRC) that aims to use the best skills of humans and robots in a combination to achieve a common task [13]. A combination of intelligence, precision, dexterity and sensing skills.

4.2.2 Requirements (and challenges)

While the advances in all mentioned emerging technologies contribute to having a solid core building blocks to realize Digital Twin architecture, there are still main challenges in meeting all requirements for a successful experience. The following enumerates the current top challenges.

4.2.2.1 Hyper-fast data rate

Considering the complexity of the landscape of a Digital Twin representation in healthcare in general, there are tremendous amounts of data that needs to travel between different devices or nodes. Additionally, it needs to be in near real-time as well. That creates an unpresented constrain on the network. Consider the use-case of remote auto-surgery or life-threatening condition on the field that needs immediate response. A federated model of devices, entities and globally distributed models and infrastructures by itself spike the level of complexity in an exponential magnitude.

4.2.2.2 Extremely low-latency communications (ultra-low delay)

As mentioned above, there are various use-cases in the health care domain that makes it extremely crucial to process and communicate data in near real-time. There are so many moving parts and building blocks, that makes this a challenging stack of various technologies to work with. When we address the network infrastructure it must be clear that this doesn't include only the physical infrastructure, but also the most optimized accurate configurations, security mandates, encryption suites and transmission protocols. The promise of 6G technology is real and can fill a current gap when considering low latency.

4.2.2.3 Comprehensive end-to-end AI

While AI is already an integral part of the current software development paradigm, and it does play a crucial part of any Digital Twins architecture, there are also challenges that are not related just to the accuracy and predictability of Machine Learning algorithms used. The challenge arises when considering a federated model of different devices employing different types of AI yet must be working together in complete harmony with the most level of accuracy. AI researchers and practitioners are aware of the importance of the datasets used to train any machine learning model and improve on its accuracy. Again, the challenge is amplified when considering an end-to-end AI pipeline in a complicated IoT architecture.

4.2.2.4 Realistic and accurate trained AI (i.e., avatars)

The presentation layer of the Digital Twins architecture is where the interaction happens with humans. As humans, we have perception, judgment and experience, to name a few given skills, that must be convinced that any type of representation (not only Avatars, but Robots and Holograms) is succeeding in communicating and presenting to the human eye and mind.

4.2.2.5 Security

Healthcare domain deals with personal and sensitive information. It mandates, by definition, applying strict security measures and best practices for the entire architecture. IoT security challenges apply here as well. This is complicated by a network of homogeneous devices and a plethora of different configurations, vendors and device capabilities.

4.2.2.6 Reliability and trust

Meeting all mentioned challenges above is a minimum requirement to achieving reliability and trust. It is very important to note that all this is a requirement for all actors of any implemented healthcare use-cases. This is inclusive of patients, physicians, researchers, surgeons, technicians, and any actor interacting with the system.

4.3 Sensing/actuating

Sensing relies on one or more sensors to detect a certain parameter, transport the data, and subsequently process the data for informational purposes or for an action to be taken. We will briefly discuss both.

4.3.1 Sensing

Digital Twin depends on data as a source of fuel to be meaningful and of practical use thus Intelligent sensing with context is very critical. While challenges are specific to use-cases, some of them are discussed below and are common challenges that should be kept in mind by use-case designers and authors of use-case related to Digital Twin.

Connecting all the dots (or sensors) is a challenging and requires careful design. Designing and building systems that truly work in the real world—and deliver maximum value—is very difficult and requires careful planning. Beyond the social and psychological ramifications, there are several technical, practical challenges and blind spots. Among them: interruptions to connectivity, component failures within systems, software bugs that generate errors and noise, data sharing across systems and organizations, coping with proprietary and competing systems, and dealing with upgrades, patches, and obsolescence [14]. Some of the common sensing related challenges are explained below.

4.3.1.1 Context

Context is a key to building connected systems that work in the real world. To build a usable Digital Twin specific to Health care representing a real-world challenge or issue faced by a Patient or clinician—each with millions or billions of objects, IP addresses, and data generation points— it's necessary to take current data management techniques to an entirely different level.

When billions or trillions of devices stream data to computers—with processing taking place at various points along the way— the context of data changes drastically and should be well understood and captured by clearly separating noise and biases from data collections. Context of the data changes because data capture, collection, storage, and analysis changes on continuous basis. Within such scenarios, traditional business intelligence and analytics tools simply cannot accommodate datasets of such large and complex context-aware sensing capabilities, which are driven by the next generation of software and algorithms. This would change the way machines operate and people view and use personal devices. For example, a smartphone could recognize when it is stowed in a purse or a pocket—or if the phone holder is running to catch a flight—and adjust its settings, accordingly, including ring tone level and do-not disturb functions, automatically. It could also recognize when an individual is asleep or needs an alert to stay awake. Similarly, sensors woven into clothing, shoes, and physical objects could determine states of use-cases —by measuring heart rate, perspiration, calorie burn, and other factors, and again adjust its configurations and start executing an intelligent workflow and actuators.

It is important to note that not all bias and noise can be removed from the sensing, but care should be taken by designers and architects to have an understanding that biases, noise, and gaps will exist in the data collection, which requires proper checks and balances to be incorporated into the implementation to make sure that recommendations and outcome from the Digital Twin are well understood by end users and are identified and highlighted.

4.3.1.2 Events

Not All data collected from sensors are equal and needs to properly be categorized as such. Some data has no value after a period of time and this type of data should be archived or destroyed to prevent misuse in future. Whereas other data elements requiring storage and future use should be managed with utmost care using best practices and centralized data management systems.

4.3.1.3 Data ownership, privacy, and security

Data ownership and governance should be considered as data is collected from sensors. Questions revolving around who owns the data, how organizations verify the accuracy of the data, how many organizations charge to use the data, how long they can keep the data, and how data

Five Vs	Characteristics
Volume	Terabytes Petabytes Records Tables Files Images Distributed
Variety	Structured Unstructured Multi-Factor Probabilistic
Veracity	Trustworthiness Authenticity Origin Reputation Availability Accountability
Value	Statistical Co-relational Events Hypothetical
Velocity	Batch Real-Time Episodic Stream

FIGURE 4.2   5V of big data [23].

is formatted for use among multiple users from a variety of groups (current and future) and third parties. Furthermore, do we want the consumer of current and future data to have a say regarding data privacy? APIs and other tools that connect data raise underlying questions about ownership and interoperability. Proper care needs to be taken to address **Veracity** (Accuracy of data), **Variety** (Different Type of Data), **Velocity** (speed of data generation) and **Volume** (the amount of data generated). (See Fig. 4.2.)

As the variety and velocity of data grow—increasingly accumulated by new data sources from sensors, machines, cameras, storage, and data processing systems—several critical issues emerge, largely revolving around data personalization; data de-identification, and re-identification; and data persistence, including how it is stored and retained. "Computational capabilities now make 'finding a needle in a haystack' not only possible, but practical. Thus, care should be taken to prevent misuse of data in the future when currently ammonized data can be used to identify the source with additional algorithms or third-party data.

4.3.1.4 Reliability

For Digital Twin to deliver dependable and predictable results, it is critical to build pathways that allow data to flow in small pipelines with proper control points, like well-designed process plant used to produce a product—all while keeping critical data encrypted and secure. When one system or communications protocol isn't functioning or isn't available, the data—can detour or bypass the blockage point and continue to the destination. In some cases, this means embedding multiple communications systems within devices, caching data locally on a device until a connection is available and incorporating peer-to-peer device capabilities that allow data to flow even when an Internet connection isn't available.

4.3.1.5 Compliance and jurisdiction, legal

Digital Twin promises to heap additional layers of complexity onto an already complex environment. Attempting to understand where data originate, how data are altered or changed along an electronic pathway, creates huge challenges. In fact, as more and more Personal Health Records (PHR) are collected by devices, stored, and processed in the cloud for current and future applications, a few key questions arise: Who exactly is responsible for a problem, breakdown, or outage, particularly if it results in damage, injuries, or deaths? What happens when a country or jurisdiction won't cooperate with the rest of the international community? And what happens when deeply personal or private information goes public due to a series of unfortunate events— none of which is specifically the cause? Additionally, there are practical and regulatory issues to sort out, besides a need to examine everything from online contracts and user agreements to surveillance and the extent of privacy protections. Perhaps only one thing is certain: the years ahead present enormous technical and practical challenges. As a globalized and interconnected world takes shape, society and the legal system will be hard-pressed to manage a technology framework that is advancing rapidly and changing so many things in such profound ways. The ultimate challenge is balancing risks and protections with basic rights and freedoms.

4.3.1.6 Interoperability, propriety software and standards

The need for standards and protocols in fact includes everything from the way small devices, for example an Apple Watch, uses battery power, sense and collect data to the way devices communicate and exchange data. It touches on topics as diverse as maturity of sensing technology, data parsing principles, and adding identifiers to address future monetization opportunities. It also includes the way companies slot all the data into massive databases and which security standards they use. Without these common standards— and clear policies for managing data governance and other issues—the vast economic and practical potential of the Digital Twin will never be realized.

4.3.1.7 Usability and convenience

All the various devices and systems must deliver convenience through a combination of affordability, easy setup, functionality, effective power management, a high level of flexibility and customization, integration with legacy hardware and software systems and other connected devices, and effective security and privacy protections. This is easier said than done and is going to be biggest barrier to overcome before Digital Twin use can be expanded and generalized.

4.3.1.8 Data misuse

Data misuse is one of the biggest challenges to overcome because once data is collected with context, and it is combined with other datasets then there is a possibility that this information is misused by the organization or person having access to the data. The same health-monitoring systems that might motivate a person to exercise and eat well could be used by an insurer to increase rates or exclude so-called high-risk patients. Likewise, an employer might use a person's genome or health data—say, a genetic predisposition to heart disease, cancer, or an early indicator for a stroke—to avoid hiring or promoting an individual.

4.3.2 Actuation

Actuation is defined as action taken by people, process, or application to achieve an objective. But in healthcare this is currently not possible due to so many unknowns, noises and biases as described in the previous section (Sensing related Challenges in Implementation of Digital Twins in HC) because human life is at stake. Hence there is a need for an eco-system of adoption capabilities which in turn can be consistent to provide needed confidence to clinician or other stakeholders to take the right decision or improve the probability of the right decision.

Another point to consider, while technology often removes human judgment, decision-making, and the real-world risk of inattentiveness, it also introduces new hazards and replaces the potential for smaller scale accidents and breakdowns with larger scale problems.

Thirdly, human factors experts refer to this as the "automation paradox." As automated systems become increasingly reliable and efficient, the more likely it is that human operators will mentally "switch off" and depend on the automated system. And as the automated system becomes more complex, the odds of an accident or mishap may diminish, but the severity of a failure is often amplified.

Don Norman, professor emeritus of engineering and computer science at Northwestern University, co-founder of the Neilson Norman Group, and author of The Design of Future Things, says: "Designers often make assumptions or act on incomplete information. They simply don't anticipate how systems will be used and how unanticipated events and consequences will occur." Today there's no shortage of instances where humans encounter trouble with automated systems. For instance, motorists blindly follow incorrect directions provided by an automobile navigation system, even though a glance at the road would indicate an obvious error exists. In a few instances, motorists have even driven off a cliff or collided with oncoming traffic on a one-way street after following directions rather than using their eyes and minds. What's more, studies show that many motorists tend to use automation features, such as adaptive cruise control, incorrectly. In some cases, Norman says, these automated systems cause the vehicle to speed up as motorists exit a highway because there's suddenly no car in front of them. If a driver isn't paying attention, a collision may ensue. Motorists, airplane pilots, and train operators are all prone to becoming overly reliant on automated systems— and growing more complacent about using their skills and alertness to avoid dangerous situations. Worse, designers sometimes rely on a wrong set of assumptions or an incomplete universe of facts to build a system. They may not fully understand the way people use individual devices or tools, or cultural differences. They may also overlook the way a combination of devices alters performance or behavior. In fact, Norman, one of the world's leading design experts, argues that machine logic doesn't always jibe with the human brain. "If you look at 'human error' it almost always occurs when people are forced to think and act like machines," he warns. The Internet of Things ratchets up the stakes substantially. Dozens, hundreds, or thousands of devices create a multitude of real-world intersection points. Moreover, with devices and algorithms communicating with each other— and different standards and quality control criteria applied by different developers and companies—there's a real-world risk of building systems that do not deliver a desired level of machine-to-human communication. As Sidney W. A. Dekker, a professor in the school of humanities at Griffith University in Australia and author of the book Behind Human Error explains: "There is often a great deal of human intuition involved in a process or activity and that's not something a machine can easily duplicate.

The maturation of any technology takes time; tweaking, adjusting, and fixing. The Digital Twin capabilities are only the beginning and yet to reach a critical threshold of usability and practicality. As computing power keep increasing closer to the source, mobility keeps advancing (5G), cloud computing keeps maturing, and big data and analytics keep progressing with new algorithms, engineers, developers, and designers will begin to build connected systems that actually work. Many of these systems will reach a design sophistication that is currently available in other fields, which makes them plug-and-play. Yet the Digital Twin, particularly in the realm of the Healthcare, must reach a level of dependability that fosters trust. It's one thing for a single connected Apple Watch to malfunction. It's entirely another for a decision tree depending on data collected from multiple devices to predict future pandemic. The latter would result in massive injuries, fatalities, widespread chaos, and severe economic consequences.

All clinicians need to use the system otherwise it can result in digital divide. For example, in the healthcare arena, microscopic, connected sensors inside the body and wearable devices on the wrist or in clothing could provide an almost unimaginable level of medical diagnostics. Physicians could identify conditions and monitor diseases in real time—and dispense medicine at an optimal level. These sensors could detect the early stages of a heart attack, stroke, or cancer and increase the odds that an individual gets help before a medical emergency occurs. Obviously, those who aren't connected to these systems—and countries where the technology isn't available—won't benefit. They may have to rely on older and far fewer effective procedures.

So, the question arises, what is the best way to mitigate challenges around actuation? One possible solution is to design a constantly evolving, innovative set of digital offerings. Digital offerings deliver new value propositions.

Digital offerings— information-enriched solutions wrapped in a seamless, personalized customer experience. Digital offerings rely on software and data to create new revenue-generating value propositions. For example, Uber offered not just a ride, but also a solution to the uncertainties, inconveniences, and anxieties associated with navigating a busy city. In fact, Uber prompted some city-dwellers to abandon automobile ownership completely.

For digital offering to succeed, there must be seamless integration and interaction among the people, processes, and technology and without this clinician will not be able to use the learning, this will limit their ability to experiment with, learn from, discard, enhance, reconfigure, and scale up new offerings to provide new value propositions.

4.4 Connectivity

As we look at the functional blocks of the Digital Twin ecosystem (Fig. 4.1), it becomes clear that communication among the layers (North/South) is the most important element. If the end-goal is to simulate and make real-time decisions, then timely and secure data communication

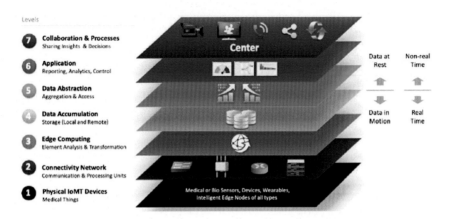

FIGURE 4.3   Cisco's IoT Framework [15].

needs to be a top priority. Looking at the above depiction, it's normal for us to zoom-in and focus on the "Digital Twin" side thinking that this is where "data" is being generated, captured, and analyzed and that is where the connectivity challenges exist. That may be true if our attention is focused on how IoT sensors generate data and how it is pushed upstream to an analytics engine that generates insights from the data. But our attention for connectivity and communication challenges is multidimensional where the interaction between the Digital Twin and the Real Twin (RT) is of significant importance as we develop that feedback loop that explains the data from the patient's point of view in real time as well.

To further break down the interaction and communication model above, we want to introduce the IoT Framework, depicted in Fig. 4.3, developed by Cisco with participation from one of the authors [14]. This model helps clarify the data elements and the data path, the analysis, and the transmission of actions generated. This model makes it easier for us to understand the connectivity requirements and the possible challenges for fulfilling them.

This point will be clearer as we briefly discuss the IoT model and the functions of each of the layers. We can then align it with the Digital Twin functional blocks:

1. **Physical Devices**: This layer shows the physical devices should be capable of:

   a. Analog to digital conversion
   b. Generating data01

2. **Connectivity**: This layer provides communication with and between IoT devices (level 1) and should be capable of:

   a. Reliable delivery across the network or possibly multiple networks to destination

   b. Supporting multiple protocols

   c. Translation between the various communication protocols

   d. Network level security

   e. Routing and switching traffic between networks (wired and wireless)

   f. Discovery, self-learning, deep packet inspection, and network level analytics

3. **Edge Computing**: This is the layer where "data" is converted into "information" and should be able to handle:

   a. Data filtering

   b. Content inspection

   c. Monitoring for, or detecting thresholds

   d. Generating events

4. **Data Accumulation**: this is the layer where data-in-motion (analyzed and inspected as it moves through the network) into data-at-rest. This layer should be cable of:

   a. Converting data from "network packets" to relational database tables

   b. Transitions from event-based to query-based computing

   c. Reduces data through filters or selective storing allowing for event comparison and aggregation

5. **Data Abstraction**: This is a very important and interesting area allowing us to scale to enterprise, regional or even global levels. This layer should be capable of:

   a. Creating schemes and views of data in the manner that applications require

   b. Combining data from multiple sources and simplifying the application

   c. Filtering, selecting, projecting, and reformatting the data to serve the client applications

   d. Reconcile differences in data shape, format, semantics, access protocols, and security

6. **The Application**: Information interpretation based on the business function or the business vertical we're concerned with (e.g., Healthcare, Manufacturing, etc.). For example, some applications will focus on monitoring device data. Some will focus on controlling devices, and this is a scenario where "action" must be sent downstream to the device. As we build a list of our challenges of Digital Twin models, the bi-directional nature of the communication model brings great oppor-

tunities for increased intelligence of the system and several challenges to also keep in mind.

7. **Collaboration and Process**: Few years ago, before the term IoT became a common or popular term, there was the idea of the Internet of Everything (IoE) and it was explained as the intersection of "things, data, people, and process." And level 7 is where we have the ability to bring the people and the process of making decision together. The IoT system, and the information it creates, is of little value unless it yields action, which often requires people and processes. Applications execute business logic to empower people. People use applications and associated data for their specific needs.

4.4.1 Sensors, sensory networks, and IOT

IoT or IoMT (Internet of Medical Things) has come a long way recently where sensors (wearables or otherwise) have experienced advancement in chip design, Operating systems, and communication protocols allowing them to be widely used. And of course, with the wider adoption comes a wider set of solutions addressing a multitude of problems. In addition to advancements in chip design also came the advancement in power systems and batteries allowing for a wider reach for the IoMT sensors taking them outside the medical facility and allowing us to monitor anything almost anywhere.

Simply speaking, When the sensors and sensory networks are operating within the boundaries of the medical facility and communicating within the capabilities of the wired or wireless networks (i.e., Wi-Fi), then the connectivity challenges are manageable or easily overcome. In addition, the Digital Twin use-case may be limited to data instantaneously captured while the patient is in the medical facility. As we start widening the circle of diagnostics or care, the data available for the Digital Twin scenario becomes scarce, intermittent, or unreliable. Usually, heart rate, sleeping patterns, or fitness activities, are recorded onto a device that the patient carries, and data is subsequently downloaded into a medical system the following day or week. If continuous monitoring is required, then another layer of communication is added (e.g., mobile phone) to receive data from the sensor and transmit it to its cloud or data center destination for further processing.

Connectivity at the IoMT layer has the following requirements (or characteristics):

The Device (Sensors):

- Small form factor
- Low Power requirements
- Battery operated (with a simple or continuous charging (Solar, Thermoelectric, etc.)

- Limited CPU or processing power
- Low or limited Memory
- Lightweight "client" for connectivity

A great deal of research and innovation has been dedicated to ideas like System on Chip (SoC), System in Package (SiP) that have scaled down the printed circuit size board, decreased power consumption, and made it possible to design wearable sensors in a variety of shapes [16]. We also see an emergence of new and advanced ways for providing continuous power for sensors and reducing reliance on what may be considered as "bulky" batteries. Recent research has focused on Thermoelectric Generation (TEG) where the body temperature is used for power generation, solar energy harvesting, or even kinetic sources (from movement), among other emerging technologies.

The Network (Sensory Network):

- Light-weight communication protocol
- Secure connectivity
- High Availability (self-healing)
- Scalable and can accommodate a large number of devices

For Digital Twin, the network must be consistent. Continuous data streams are needed and the more of the relevant data is captured, the easier, faster, and more reliable it becomes for us to build a Digital Twin model. If the data is poor and inconsistent, it runs the risk of the Digital Twin underperforming as it will be acting on poor and missing data [17].

As mentioned earlier, the connectivity challenges are greater when we decide to capture data continuously and outside the boundaries of the medical facility. And this is precisely where a plenty of research effort has been focused. For example, 5G is promising high speeds and densities and unified platform for connectivity for a wide variety of use-cases.

It's clear that if the creation of Digital Twin requires continuous data streams, then we need to make carefully engineered choices of the senor type, power, communication protocol, and the biomarker of interest for the Digital Twin calculation.

4.4.2 Connectivity for the AI/ML layer (the intelligence layer)

As for this layer, the challenges are a little bit more basic or maybe trivial since we are probably working with a single communication protocol and data-at-rest (databases). We used the term "basic" or trivial because this is what networks and IT stacks have been doing for the past 20 or 30 years. The connectivity challenge here revolves around securely delivering results, and receiving actions, from the "Representation" or avatar layer.

What we tend to see here is the need for various data elements that may not be present in a single database, data center, or even the cloud. We also cannot forget that, depending on the complexity of the Digital Twin, we may need to worry about communication among a number of high-performance computing clusters participating in building the Digital Twin.

Given the real-time nature of the Digital Twin, we find that Edge Computing enables the presence of compute resources as close as possible to the data sources. That allows us to clean the data, select relevant elements, and eventually making it available to other systems and applications. Edge computing allows us to reduce latency and improve response time. It is also noteworthy to remember that the edge and all that it hosts, in most of the cases, is used to extend the application that resides in the data center or the cloud. In other words, the edge is an extension of the cloud, it pre-processes data and prepares is for a broader mission represented by the application or "service" responsible for building the Digital Twin.

Keeping all that in mind, it is clear that the challenges are going to center around:

1. **Latency:** in the IoT and subsequently in Digital Twin world, when the phrase "real-time" is used, the word "latency" is almost always brought up and contested. We cannot ignore laws of physics that contribute to transmission of propagation delay of data. We cannot ignore computational delays related to the CPU, memory, and clocking design of our edge system. As you recall from our previous discussion, resources at the edge are mostly light weight and low cost.
2. **High availability:** The Cloud or the Data Center, by definition, include proper redundancy and resiliency measures to ensure continuous operations. No problem! However, when discussing edge computing, the lightweight and low-cost natures dictate lack of redundancy and lower availability systems. We cannot also ignore that edge computing may also be constrained by dependency on batteries for power supply.
3. **Computation resources:** This a well-known issue that have been around for a long time. How fast we can process the data dictates how fast we can push into the communication pipe. CPU and memory are closely related and are contributors to latency.
4. **Scalability:** The Cloud promises "elasticity" and infinite resources. Elasticity means that the cloud provider increases compute resources to your application when needed and reduce them by the amount not needed for your application to be able to assign them to a different application. At the edge, we're limited by size and resources.
5. **Private vs. Public Cloud:** Based on the four points above, we must make a decision about where to host our Digital Twin "Intelligence", the AI/ML components of the architecture. With a private cloud (our enterprise or health system data center), we have control and access,

but also the added operational expenses. The Public cloud may provide us with low-latency, highly available, and scalable computational resources at low-cost (pay-as-you-go). On top of all that, public cloud may be reachable by our users (or sensors) from wider geographies than those supported by our own private clouds.

4.4.3 The representation layer (the intelligence layer)

The real-time nature of the Digital Twin and their ability to represent and adjust with the Real Twin demands (or merely suggests) that "The Representation" layer be in proximity or even serviced from the same system. Meaning that it is most probably that the connectivity or latency of communication between the "The Intelligence" layer and the "The Representation" layers may be low or negligible. Of course, there are always exceptions.

In addition, the bi-directional aspect of communication here is not to be ignored. The representation layer allows the medical professionals to perform functions or actions on the Digital Twin where additional data may be required. That data needs to be requested by the representation layer from the intelligence layer or even from the sensing layer.

The challenges and requirements of inter-layer or inter-domain communications have been adequately discussed in the previous sections and apply to this layer.

4.5 Security, privacy, and ethical issues

Digital Twin in healthcare has considerable values in preventing treatments and understanding diseases, reducing costs, and even proliferating patient autonomy and freedom. However, Digital Twins have many ethical challenges that we will highlight in the following sections:

4.5.1 Security

Use of IOT Digital Twin requires a high level of privacy and security to keep patient data safe.

As cited by Dr. Abdulmotaleb El Saddik [3], – We need to balance convenience versus privacy.

The emerging trend for the consumerization of healthcare data, combined with the growing number of healthcare devices producing data has created concerns among patients, professionals, and security technology researchers.

Security and privacy guidance from federal entities like U.S. Department of Homeland Security-Cyber and Infrastructure and National Institute of Standards and Technology (NIST) have provided guidance for cybersecurity and privacy controls as applied to IOT but standards for Digital Twin security and privacy are still emerging.

The proliferation and growth of Digital Twin entities for industrial, healthcare and consumer applications may rapidly increase the number of cyber security attack vectors due to its interaction and integration to physical/real world entities to exploit system weaknesses to capture rich data.

Increased vigilance of privacy controls will be needed on Digital Twin entities to ensure that data is safeguarded, and physical world entities are aware of any changes in privacy controls. These efforts will have on-going costs to ensure security and privacy controls remain ahead of persistent threats and lower risk.

Additional efforts to help with security and privacy for IOT Digital Twins should also include efforts to stem contagion effect for cybersecurity breach events [18], which can provide better insights, actions and implications for cybersecurity preparedness and privacy controls. For example, traditional healthcare has struggled with on-premises security and privacy controls which has led to an alarming trend of patient data being compromised at a rate of nearly 2 healthcare data breaches per day of 500 or more patient records between March 2021 and July 2021 [19]. The timeliness of breach disclosure to users and organizations to provide timely information about cybersecurity and privacy risk management efforts could make a tremendous difference for IoT Digital Twin across industries and pose a call to action to ensure programmatic goals for security and privacy of Digital Twin and real-world objects are met while limiting access and exposure of data and objects for only their intended purposes.

Organizations leveraging Digital Twin will need to maintain the controls of transactions and data governance to establish and maintain compliance as industries like healthcare move to more patient/consumer models of value-based care. This may also result in multiple vendors to provide security, privacy, and observability solutions over real world and Digital Twin to maintain compliance for regulated industries and regional privacy laws.

4.5.2 Privacy and ethical issues

Privacy and Ethical issues are major challenges when dealing with data and personal information in general, and it is even more critical when dealing with healthcare data [21].

4.5.2.1 Ownership, content, and quality of data

There are many challenges with data in general and mainly around healthcare data. What data is collected, who owns it, and how accurate is the data collected are many aspects that interfere with privacy and rise ethical issues. The violation of privacy is one of the most important ethical risks.

In a study done in [20], there is a divergence between people stating that privacy is the main disadvantage with Digital Twin, however, others claim that this risk is overstated.

There is a common agreement that Digital Twins need to be built with extra care around privacy aspect. Digital Twins require huge amount of data to operate. The main subsequent concern is where does all that data go? and what are all the secondary usages of it? There are many risks that can unfold:

1- Who owns the data: if this data falls into the hands of private companies, they may not put the benefit of the citizen first. For example, insurance companies, need to decide on the quotes to provide to their customers, today they mainly rely on personal identifications and pre-existing conditions. However, with access to additional data points, the decision of these organizations may be altered in a way that does not necessarily benefit the citizen (the patient). Indeed, if they can determine that the person is not exercising regularly or not following healthy habits and hygiene, they can use that information to increase quotes. This may look like an only a privacy issue, however, this can trigger fundamental problems around freedom. People will be enforced to follow common guidelines in order not to be impacted.

2- Security breaches: security of data is a big challenge already. Data can be lost, stolen, leaked. With Digital Twin, this risk is amplified due to the amount of data that is collected. Besides, data can be stored anywhere using third party servers and services and one of the main challenges is to locate where it is stored and who is responsible and accountable for data breaches. In fact, Digital Twin data contains genetic information. If stolen, it can be used in very troublesome situations. For example, with the availability of this genetic information, and the usage of ML, we will be able to determine with high accuracy which are the healthier genes, and this will lead to social issues and discrimination between people based on their genes' classification. In other scenarios, with access to DNA information, criminals can use the DNA information and place them in criminal scenes to tamper the investigations.

Privacy is an important challenge for Digital Twin. Too little privacy could lead to some issues if data is owned by private organization that can use the data against the benefits of the citizen, or if data is used to create discrimination in society based on a superset of genes, or if cyber-criminal get access to the data and use it to mislead investigations. On the other hand, too much privacy makes it hard to break though innovation since many areas will remain under-investigated and under-researched due to the lack of data. In addition, privacy is relative to people situations. For example, people with cancer, would not care much about privacy of their data as their main need is to find a cure and recover.

Data quality is another important challenge. Whenever we deal with machine learning and artificial intelligence (ML/AI), data quality is very important. In fact, the accuracy and correctness of the generated ML/AI models depend tremendously on the quality of data that is used to train the models. In Digital Twin, this is amplified because a large amount of data is collected to create models that directly impact humans. The quality of data is also impacted by the bias that can be incurred by whoever is creating these models: How are we gathering data, who is gathering data and creating those algorithms. Big data increases inequality, we are using historic information to make predictions about the future, there is unconscious bias and people embed their own bias into technology. People who own the code deploy it using other people's data: there is no symmetry, it is an asymmetrical power situation and there is no accountability. The scary part is that it is a black box even to the programmers. There is a need to monitor any source of data. For Digital Twin in healthcare, the quality of data takes a level up since the learning is happening from biomedical information, biomedical research and sciences that are currently fragile. Hence in Digital Twin the representation is enormously impacted by the quality of data, and the quality of analysis and processing.

4.5.2.2 Disruption of structures of institutions and roles

With Digital Twin, some patients can avoid visits to hospitals and doctors. They can rely on the Digital Twin analysis to continue or change specific treatments for specific conditions. This could be beneficial since this permit to lower the load on hospitals and doctors so that they can focus on serious conditions that require more time and attention. But to which limit can we rely on the Digital Twin analysis and accuracy. Beside the data driven analysis, doctors know more context about their patient, Digital Twin can disrupt such social contact [20].

Digital Twin in healthcare can disrupt the current structures and who is responsible while deciding on patient condition. Should the Digital Twin has the ultimate decision on condition? Should doctors leverage information from Digital Twin to make more precise data driven decisions? What if the Digital Twin is compromised? Can the physician override the Digital Twin diagnosis?

4.5.2.3 Inequality and injustice

Digital Twin can increase inequality and injustice situations in many cases. First, the access to Digital Twin can be costly and restricted to a limited category of people who are rich and able to afford the expenses or have good insurance coverage. This can create a gap in society, where richer people can have access to the latest technology to run experiments on Digital Twin and have accurate analysis to find cures and treatment to their conditions.

Second, research on Digital Twin is mostly conducted in rich countries. Thus, as another aspect of inequality, richer countries will be able to benefit more from Digital Twin and this would increase the gaps between rich and poor countries.

Third, Digital Twin can suffer the bias that we see when dealing with ML and AI. In fact, most of the emerging technology experiments target "white males", which can decrease the ability of getting equal opportunities to high quality and accurate treatment to other races and gender.

Forth, Digital Twin heavily use ML and AI models that can rely on algorithmic determinism, and make decisions based on mathematics, and scoring systems. These decisions are hard to explain and justify. If citizens cases fall in the other side of the limit, and with inability to justify the Digital Twin decision, we can fall in injustice scenarios. For example, who gets prioritized for a surgery or who can benefit from some insurance coverage. In some cases, the ambiguity while analyzing results can benefit citizen.

As Harari [22] stated in his book, titled Homo Deus, "What will happen to society, politics and daily life when non-conscious but highly intelligent algorithms know us better than we know ourselves?" The concluding remark could be that stated by Gail Carr Feldman: "There has never been a time more pregnant with possibilities."

References

[1] A. El Saddik, Digital twins: the convergence of multimedia technologies, IEEE Multimedia 25 (2) (Apr.-Jun. 2018) 87–92, https://doi.org/10.1109/MMUL.2018.023121167.
[2] Gartner's top 10 strategic technology trends for 2017, https://www.gartner.com/smarterwithgartner/gartners-top-10-technology-trends-2017, 2019.
[3] ISO, ISO/TC 184/SC 4. Industrial data, https://www.iso.org/committee/54158.html.
[4] IPC, IPC releases IPC-2551, international standard for digital twins, https://www.ipc.org/news-release/ipc-releases-ipc-2551-international-standard-digital-twins.
[5] Qualcomm Technologies, Inc., The mobile future of eXtended reality (XR), https://www.qualcomm.com/media/documents/files/the-mobile-future-of-extended-reality-xr.pdf.
[6] Kuldeep Singh, Medium. The growing list of XR devices, https://medium.com/xrpractices/the-growing-list-of-xr-devices-f102262e4a58.
[7] Jose Rios, Juan Carlos Hernandez-Matias, Manuel Oliva, Fernando Mas, Product avatar as digital counterpart of a physical individual product: literature review and implications in an aircraft, https://www.researchgate.net/publication/280234034_Product_Avatar_as_Digital_Counterpart_of_a_Physical_Individual_Product_Literature_Review_and_Implications_in_an_Aircraft.
[8] C.Y. Wong, D. McFarlane, A. Ahmad Zaharudin, V. Agarwal, Cambridge Auto-ID Centre, Institute for Manufacturing, University of Cambridge, https://www.researchgate.net/publication/3996498_The_intelligent_product_driven_supply_chain.
[9] ITU-T, Telecommunication standardization sector of ITU. Focus group on technologies for network 2030 (FG NET-2030). Representative use cases and key network requirements for network 2030, https://www.itu.int/en/ITU-T/focusgroups/net2030/Documents/Technical_Report.pdf.

[10] Samsung, 6G. The next hyper-connected experience for All, https://cdn.codeground. org/nsr/downloads/researchareas/6G%20Vision.pdf.

[11] Mordor, Global holographic display market – segmented by technology, https://www. mordorintelligence.com/industry-reports/holographic-display-market.

[12] Energid Dan Dockter, The digital twin and real-time adaptive robot control, https:// www.energid.com/blog/the-digital-twin-and-real-time-adaptive-robot-control.

[13] Arne Bilberg, Ali Ahmad Malik, Digital twin driven human-robot collaborative assembly, University of Southern Denmark (SDU), https://findresearcher.sdu.dk:8443/ws/files/169020434/Bilberg_Malik_Digital_Twins_CIRP_Annals.pdf.

[14] Samuel Greengard, The Internet of Things, Essential Knowledge Series, MIT Press, 2015.

[15] The Internet of Things Reference Model, Available online http://cdn.iotwf.com/resources/71/IoT_Reference_Model_White_Paper_June_4_2014.pdf, 24 May 2016.

[16] Haijian Sun, Zekun Zhang, Rose Qingyang Hu, Yi Qian, Wearable Communications in 5G: Challenges and Enabling Technologies.

[17] Digital twin: enabling technologies, challenges and open research, https://arxiv.org/pdf/1911.01276.pdf.

[18] https://meridian.allenpress.com/cia/article/15/2/P1/451091/How-to-Reduce-the-Cybersecurity-Breach-Contagion#:~:text=Cybersecurity%20incidents%20result%20in%20significant%20negative%20financial%20and,cybersecurity%20a%20top%20concern%20of%20the%20business%20community.

[19] https://www.hipaajournal.com/july-2021-healthcare-data-breach-report/.

[20] Popa, et al., The use of digital twins in healthcare: socio-ethical benefits and socio-ethical risks life sciences, Society and Policy 17 (2021) 6, https://doi.org/10.1186/s40504-021-00113-x.

[21] Digital Twins in Health Care: Ethical Implications of an Emerging Engineering Paradigm, https://www.ncbi.nlm.nih.gov/pmc/articles/PMC5816748/.

[22] Y.N. Harari, Homo Deus: a brief history of tomorrow, Random House, UK, 2016.

[23] Yuri Demchenko, et al., Big security for big data: addressing security challenges for the big data infrastructure, Secure Data Management (2013).

Intelligent digital twin reference architecture models for medical and healthcare industry

Zhi Wang[a] and Abdulmotaleb El Saddik[a,b]

[a]University of Ottawa, Ottawa, ON, Canada, [b]Mohamed bin Zayed
University of Artificial Intelligence, Abu Dhabi, United Arab Emirates

5.1 Introduction

The digital twin (DT) concept was introduced by Grieves in 2002 as an integrated multiphysics, multiscale, probabilistic simulation of a physical asset or system to mirror the life of its corresponding physical twin by using the best available physical models, sensor updates, history information etc. [1] DT was redefined by El Saddik in 2018 as a digital replica of a living or nonliving physical entity. By connecting the physical and the virtual world, data is transmitted seamlessly to make the virtual entity lives synchronously with the physical entity [2].

This chapter mainly focuses on DT architecture models. It is organized as follows: Section 5.2 introduces the state of the art works of DT architecture models. Section 5.3 reveals the conceptions of DT architecture and various DT architecture models. A use case of medical and health industry is demonstrated in Section 5.4. Section 5.5 gives future directions. We give the conclusion in section 5.6.

5.2 Related work

The initial definition of a Digital Twin defines it very roughly as an architecture composed of a physical and virtual space and the link among these two spaces [11]. Gradually it becomes a hot topic and recently some related works made a great progress in this area [5]. This chapter mainly chooses two categories to classify them: vertical industry and technical aspect.

In term of vertical industry, a great deal of efforts has been put into manufacturing for DT architecture. This effort covers different stages in the production and also different applications cases. A Digital Twin Architecture Based on the Industrial Internet of Things (IIoT) Technologies was proposed by Vinicius et al. [17]. This simple architecture is composed of three main structures: Physical Twin (PT) in the manufacturing processes, Internal Server running the DT and simulations providing connection inside and outside workplace, and IIoT Gateway offering cross-communication between the PT and Internal Server through IoT devices and wired connection. Behrang et al. [16] proposed an architecture for a Digital Twin and an architecture of an Intelligent Digital Twin in a Cyber-Physical Production System (CPPS). The assistance system between cyber layer and physical layer based on Anchor-Points-Method is able to identify arising changes, analyzes their relations, and adjusts changes that have occurred in the DT. This framework enables use cases such as plug and produce, and predictive maintenance. A DT-based cyber physical production system architectural framework for personalized production was developed by Kyu et al. [18]: The physical manufacturing system of the CPPS architecture assembles data through IIoT device at each facility. Data are transmitted through the IIoT network of the network layer to the database abstraction interface based on the P4R information model. The service layer generates, manages, or assembles real services including the service bus. The application layer includes a context-aware application, an advanced planning application, an advanced scheduling application, and a device control application. An ISO/IEEE 11073 (X73) Standardized Digital Twin Framework for Health and Well-Being was proposed by Laamarti et al. [17]. This architecture includes the process of DT Data collection system (communication server, data storage server) collecting data from personal health devices, DT data analysis, and DT interaction (physical and virtual actuation) sending the feedback to the user in a loop cycle. It upholds both X73 compliant health devices and noncompliant health devices via an X73 wrapper module. But it focuses on data-centric applications and does not present detail contents.

From the technical perspective, the main contributor to the DT architecture is Cyber-Physical Systems (CPS). Alam et al. [12] proposed a Digital Twin Architecture Reference Model for the Cloud-Based Cyber-Physical Systems. It is based on a smart interaction controller using a Bayesian belief network. The system is divided into three operational modes: Physical level sensors-fusion mode, Cyber-level digital twin services-fusion mode and a deep integration of sensor-services fusion mode. A system context based control decision scheme applies Bayesian network and fuzzy logic based rules to choose any of these system modes for intersystem interactions. They presented a telematics-based prototype driving assistance application for the vehicular area. A reference Framework for DT within Cyber-Physical Systems was proposed by Josifovska et al. [13]. The third level of the 5-level CPS architecture called cyber level designates the requirement for digital twins as high-reliability mirroring images of CPSs entities. The framework consists of Physical Entity Platform, Virtual Entity Platform containing Virtual Entity Models, Data Management Platform, and Service Platform. The entities consisting of the Physical Entity Platform transmit specific data to the Data Management Platform. On request, this platform directs data to the Virtual Entity Platform. The Virtual Entity Platform can call concrete services from the Service Platform. Furthermore, the underlying Physical Nodes of the Physical Entity Platform can also propel a service request to the Service Platform and get an appropriate response. A Hierarchical Digital Twin Model Framework for Dynamic Cyber-Physical System Design was proposed by Duansen et al. [14]. The physical entity was expressed as a hierarchical framework to realize the layer-by-layer decomposition of "application scenario – (sub) system – component – interface" from top to bottom. Similarly, the virtual model implements the 4-layer framework consistent with the physical entity, from the bottom to the top: Virtual Interface layer including interfaces in various fields, Virtual components layer where component models of different domains match the physical components, Virtual (sub-) system layer composed of various interrelated component function models, and Virtual scenarios layer created by the system model according to various scenarios and tasks. Minerva et al. [15] summarized a general framework for the DT. This architecture exhibits a detailed representation of many layers of the classified capabilities and functions as advised by academia and industry. It consists of application layer, middleware layer, communication layer and perception layer. The middleware layer includes simulation layer, data layer, object layer, virtualization layer, and resource layer.

5.3 Challenges

Based in the discussion in the previous section, the following challenges can be identified:

1. Scalability and flexibility

The traditional architectures proposed in the above studies tend to be monolithic, rigid, and tightly coupled module stacks that are inappropriate to support DTs. They often only focused on a specific application e.g., Operation Optimization in production etc. They are too specific to be reused or extended to other application areas. They are also difficult to scale up and maintain. We believe that a DT architecture should have the capability to replace or add a component without a substantial influence to other components and their interfaces. It also should have strong flexibility to quickly implement each module to build up the architecture that is able to adjust to changing requirements of health industry [4], support evolving customer expectations and keep up with technology innovations.

2. Composability

The literatures often only have an analysis of a group of requirements and then proposed a conceptual architecture. However, the proposed architectures are often too simple. They do not define a concrete architecture of the DT, but limit themselves to mimic the pertinent physic models and regular data pipelines. Most of the literature did not give clear, accurate and complete definition of the components and interfaces involved and leave the realization out of discussion. Consequently, they often did not cover all necessary data and functional requirements of DT. They do not address issues such as security, event, stream etc. Moreover, they did not separate services implementations with interfaces or explore relationships between components. Furthermore, they do not provide any representation of where that functionality will operate. We think a good DT architecture should have the capability to choose the optimal design models and technologies for individual modules to implement themselves and easily integrate with other modules.

3. Interoperability

DT system often need to choreograph process or composite processes or services across the boundaries of systems. However, in the Medical and healthcare industry, DTs and their data are owned by various hospitals, clinics, pharmacies, and healthcare service providers. They are often separated and stored in heterogeneous and siloed systems provided by various vendors. The DT based applications often cannot be implemented by only a centralized and monospecific DT system because they would tightly couple with the IoT platforms, services and applications. Heterogeneous DT across different platforms or various DT systems often have to work together to achieve the goal. However, to our knowledge, no paper provides concrete implementation proposals and satisfactory solution of how to integrate heterogeneous DTs across different platforms or how to interoperate various DT systems meanwhile keep their autonomy.

4. Performance

DT systems have strong requirements of real-time data synchronization between the physical and the digital worlds or twins.

DT's data store should provide a good support for transaction, analysis and reporting at the same time. However, few studies address how to perform real time analysis of huge amount data meanwhile guarantee the reliable data communication and data quality.

5. Multiexperience

DT applications often involve multiple interaction channels e.g., robot arms, voice, touch, chatbots, AR/VR interface, wearables and IoT devices. Thus users want to have natural user workflows to seamlessly transit from one app to another on multiple channels so they feel that they are still working in the same application. For example, provide a continuous experience for the doctor to browse one app to view patient history, another app to make prescriptions and the third to perform remote surgery. But few studies addressed this.

6. Operationality

These studies did not take DT as a center of the whole system and analyze DTs' define, design, development, deployment and their operation from the DT lifecycle perspective.

In a nutshell, DT applications require a reference architecture that describes what functionality is needed, where that functionality will work, and the key system interfaces. So the DT designer can clearly define involved modules, analyze data and functional requirements of those modules, streamline data and control flows, and choose the most fittable communication channel between modules. However, a standard industrial architecture model that satisfies DT's specific requirements did not exist, we defined a flexible, scalable, and compostable DT reference architecture model in the following to address these challenges.

5.4 Digital twins models

This section will show the conceptions of DT architecture and various DT architecture models. The DT architecture is a hybrid architecture that combines edge computing, event-centric architecture, data-centric architecture, and application-centric architecture. The asynchronous event-driven architecture provides minimal coupling between event producers and event consumers. It allows the DT system to scale as required since we shall be able to add new modules consuming or producing existing events without impacting existing modules. It can easily add or replace event consumers to process the events without changes to the event producers that still trigger each step of the workflow as usual. What's more, it provides a good support for the real-time flow of data flow and control flow between systems. It is able to propagate the occurrence of events or state changes rapidly to the DT system that respond to or analyze them. Data-centric architecture is good at migrating data while maintaining and improving data consistency and quality especially when dealing with a big amount of data across system boundaries. Application-centric architecture can combine existing functionalities and data in easier ways to

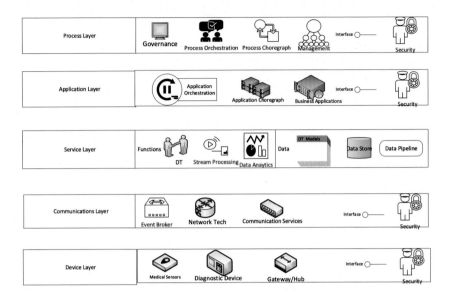

FIGURE 5.1   DT architecture layers.

automate business processes especially if the medical and health related systems have application interfaces.

Analyzing Gartner's IoT architecture [6], we concluded that it is not adaptable to DT applications as it mixes applications and services to function layer while a proper modular DT architecture requires that we separate them to application layer and service layer. In addition, it does not consider DT related requirements, components and functionalities in details for each of the layers, sublayers, and related interfaces.

From Fig. 5.1, the DT architecture is composed of the following components:

- **Tiers**: define where a module, function or process works. They are logical deployment locations.
- **Layers:** define what abilities a module, function or process must own.
- **Interfaces**: define how control and data flow get into and out through the layers and tiers.

5.4.1 Tiers' perspective

First of all, let us take a look at where we should deploy DT services in the distributed IoT System. A typical DT system can be broken down into three tiers: edge tier, platform tier, and enterprise tier:

- Edge tier is the place that data is sampled and collected from the physical environment by devices containing sensors or actuators.

TABLE 5.1 Comparison of DTs' operation locations' models.

Criteria	Model 1 IoT endpoint	Model 2 Edge gateway	Model 3 Cloud plat-form/On-premise data centre	Model 4 Enterprise business applica-tions	Model 5 Hybrid
Data aggregation	Very limited	Limited	High	High	High
Response latency	Fast	Moderate	Depend on network	Depend on network	Depend on design
Scalability	Limited	Somewhat limited	Virtually unlimited, elastic	Somewhat limited	Depend on design
Flexibility	Low close-coupled	Somewhat limited	High loose-coupled	Somewhat limited	High
Data analytics	Low	Low	High	High	High
Edge autonomy	Yes	Yes	No	NO	Depend on design

- Platform tier is the place that the DT system assembles system-wide data and events and take actions from edge tires.
- Enterprise tier is the place DT system integrates with the set of enterprise applications, processes and services needed to achieve a business goal.

There are five models to operate DT services in IOT systems according to DT's different locations:

- Model 1 – IoT Endpoint: DT is operated in an IoT endpoint e.g. a smart-watch.
- Model 2 – Edge Gateway: DT is operated in an IoT gateway to deliver services for multiple IOT endpoint at a single place e.g., edge gateway.
- Model 3 – Cloud Platform/Data Centre: DT is deployed into a cloud-based IoT platform (e.g., Azure IoT, or AWS IoT) or on-premise enterprise data center to offer DT services for many IoT endpoints across many IoT edge subsystems.
- Model 4 – Enterprise Business Applications: DT is implemented as a part of business applications on enterprise tier to offer DT services for many IoT endpoints across many IoT edge subsystems.
- Model 5 – Hybrid: DT is integrated at more than one place to meet cus-tomers' requirements by combining two or more above options. For example, you can design federated DTs that operate a DT services in each of the IoT endpoints as well as a DT services on a cloud platform.

From Table 5.1, we can see that model 1 and model 2 achieve high per-formance in processing data and performing real time streaming analysis.

However, they tightly combine data consumer (data analytics component) with data producer (data generator component). This leads to data analytics is dependent on and closely coupled with the IoT edge compute resources and infrastructure e.g. processing unit, storage space etc. Additionally, they do limit users to apply more sophisticated and scalable machine learning capabilities, such as deep learning neural network, federated machine learning etc. They are the best fit for original equipment manufacturers as they have their own product and related subject matter expertise.

Model 4 is too rigid as it closely couples DTs to each business applications. It is difficult to maintain the synchronization among various applications' DTs. In addition, data and communication are overlapping and redundant. Furthermore, it does not have dynamic scalability and high reliability as model 3. In comparison, model 3 and model 5 (mixing model 3 with other models) have high reliability and dynamic scalability by decoupling IoT platform with devices, applications, and services, and having strong computing resources. Therefore, model 3 or model 5 (mixing model 3 with other models) are mainstream architectures of DTs in medical and healthcare industry.

5.4.2 Layers' perspective

Then let us see what kind of logic capabilities the DT system can provide. We divide the logic capabilities of DT system into five layers: device, communication, service, application, and process. Security functionalities are available on all layers.

5.4.2.1 Device layer:

This layer gathers and transforms real-time analog signals from physical twins to digital information by sensors or other IoT things. It comprises sensors, devices, IoT gateways or Hubs etc. DT systems should enable automatic discover and integrate devices without any manual configuration (Plug and Play).

- Layer interfaces: Device layer interfaces typically emphasis on the interaction between the physical environment and the things at the IoT edge.
  - ✓ Physical-twin-to-sensor interface specifies telemetries, measurement recurrence, and measurement precision.
  - ✓ The sensor-to-controller interface specifies the physical link and message format.
- Security: Device layer security usually emphasis on device cataloging, patch management and embedded trust.

5.4.2.2 Communication layer

This layer specifies the event brokers, message brokers, network technologies (Wi-Fi, bluetooth, 5G etc.), and communication service vendor. DT system should be able to make communications between DTs and physical twins reliable, real time, and seamless.

- Layer interfaces: Communication layer interface emphasis on communication protocols and networking technologies. The messaging protocols (HTTP(S), MQTT, AMQP etc.) provide the channel connecting client applications with the event broker to publish events or receive events. The networking technologies dictate the local communications technology (e.g., Ethernet, X.509) within the IoT edge tier, the long-range communications technology (e.g., LoRa, Websocket, Webhook) from the edge tier to remote destinations and the physical connections to those technologies (e.g., RJ45).
- Security: communication layer security comprise authentication, authorization and encryption systems to protect the information flow between systems. It may include identity and access management (IAM) services, SSL/TLS and Simple Authentication and Security Layer (SASL) protocols.

5.4.2.3 Service layer

This layer defines the service implementations offering the back-end abilities required by the DT applications. They are internally built and operated in the tire or provided as part of a SaaS offering. These services can be divided into two categories:

Data sublayer

This sublayer defines data's flow including how DT data is collected, transformed, store and consumed. It includes data pipeline, data models, data storage (Database, data warehouse, Data Lake, Hadoop Distributed File System (HDFS) etc.), DT systems should have a single interface to access numerous data stores and strong data processing capabilities.

- Custom DT service data – Utilize custom services to store and provide access to the data that is specific to the DT application. They include IoT sensor-based time-series data, system log, telemetries from devices or their contextual data, and DT's metadata e.g., DT models. DT models are the platform neutral metadata describing DTs' physical components, behaviors and topological relationships with other TDs including state properties, telemetry events, commands, and other specifications.
- DT platform data storage – Use the data storage capabilities in your DT platforms to store state and context data that is specific to the DT app and utilized by DT platform's competencies.

- Custom DT integration cache – Utilize caching approaches in custom integration services, middleware and platforms to optimize access of data for the DT applications.
- Layer Interface: The data pipeline interface may embrace APIs from an edge gateway that execute data conversion e.g., JDBC/ODBC, OData. A data model interface may embrace a depiction of an IoT device in a DT model e.g., DTs Definition Language (DTDL). Data storage interfaces for an edge gateway may define the requirement that it supports very fast writes for big data from IoT endpoints to DB e.g., SQL, etc.
- Security: Data sublayer security mainly involve with how to protect data to avoid data polymorphism, data poisoning, data theft, data linkage and other inference attacks [6].

Function sublayer

This sublayer offers the functionalities of the DT system including stream processing, data analytics, Machine Learning and other intelligent algorithms etc.

- Custom DT services: use DT back-end-for-frontend (BFF) approach to build custom DT services to avoid introducing dependencies, vulnerability and limitations into data services. The functions include creating, registering, managing, and extracting insights from DT model and DT graph.
- DT Platform Services: These are the back-end abilities that support the DT development. It can be Argument Reality (AR)/Mixed Reality (MR)/Virtual Reality (VR) backend engines, conversational platforms, robot control platform, and multiexperience development platforms etc. including their configuration and implementation.
- DT Gateway: DT Gateway is the policy enforcement point the place that incoming requests or event is assessed against a group of policies and then decides if the access is denied or authorized. If get authorized, it will route authorized requests or event to DT service implementation. Next it will monitor traffic and report usage metrics back to the control console.
- Layer interfaces: function sublayer interface describes how to integrate DT nondata services or integrate them with data services. For example, defining REST APIs, or calling SOAP or gRPC to those services
- Security: service layer security emphasis on authentication, authorization and encryption mechanisms to guard services [6].

A DT Service instance at the platform tier is often across data sublayer and function sublayer at platform layer because it not only provides DT functions but also stores DT data such as DT models and DT graphs.

5.4.2.4 *Application layer*

This layer is comprised of business applications (e.g. ERP, CRM, EHR etc.) and DT applications. DT applications should be flexible, loose coupled, and easily integrated with business applications including application orchestration and application choreography.

- Layer interfaces: Application Layer interface defines how to integrate DT application with other business applications. For example, define the REST APIs, GraphQL or event APIs to integrate them.
- Security: Application layer security emphasis on authentication, authorization and encryption mechanisms to guard the use of applications.

5.4.2.5 *Process layer*

The process layer emphases on workflow management, operation and governance. It includes process orchestration and process choreography. This layer is the place to represent business rational of DT initiatives.

- DT control console: it is composed by administration portal and developer portal to control and configure DTs.
  - ✓ Administration portal: the administration point as part of the control plane. It is designed for DT providers to add and register DTs to the DT repository, configure DT products and manage security, traffic and monitoring policies.
  - ✓ Developer portal: the DT repository for discovery and selfservice committed of DTs as part of the control plane. It is designed for DT consumer developers to discover, apply, subscribe to and analyze usage of DTs from multiple DT suppliers.
- Layer interfaces: Process layer interface define how to integrate with the various systems that provide operations, management, and governance applications for the DT workflow and human/machine interaction with DT control console. For example, Enterprise Service Bus, Integration platforms as a service (iPaaS) etc.
- Security: security of this layer emphasis on security governance, operations and management. Security governance is to define the strategic needs of the business and guarantee the security mechanism satisfactorily meets those needs. Security operations regularly perform security-related processes. Security management set up and run the security mechanism to meet these strategic business needs [6].

The following are the details of models of DT architecture when deploy DT service on one or multiple Cloud Platforms or On-Premises Data Centers.

5.5 DT architecture models

DTs are divided into the following types: discrete DT and composite DT. Discrete DT is a DT usually used to monitor and optimize the use of granular resources, such as person, single product, device, or individual process. Composite DT is a DT normally designed to monitor and optimize the use of a related group of discrete DTs and their granular resources [3].

5.5.1 Model 1: single centralized DT management solution instance

In this model, a single DT management solution instance with a centralized DT gateway that manages, routes, and controls all accesses to DT services hosted in on-premise data centers or cloud environments. This model can scale up by adding new secured connections from the DT gateway to the DT services implementations in the on-premise data center or remote cloud.

This model offers consistent governance, security and developer experience, but it may generate high network latency and become difficult to maintain when scaling up. It also cannot isolate and secure internal interfaces from external interfaces. This model is a good choice to implement a relative simpler DT solution for the first time, most or all workloads in a single location or region.

5.5.1.1 Discrete DT on single IoT platform

Discrete DTs are deployed in the single uniform platform and they are directly physical connected to external data sources e.g. devices, sensors, partner's IT systems, mobile apps enterprise applications, and external data consumers etc. It is a good fit for building DTs for relatively simple IT-connected products [7]. As to original equipment manufacturer, the data integration efforts can be less than other options. However, as to DT owner-operator, to integrate DTs with external data sources or data consumers, they have to integrate various external heterogeneous data sources and support a variety of communication protocols. Their efforts would be much more. Additionally, original equipment manufacturer may not share their product's IoT data. This also increased DT owner-operator's difficulties. Therefore, incorporating and interoperating discrete DTs as subcomponents of a complex composite DT become more effective (3.2).

5.5.1.2 Composite DT on single platform

Discrete DTs are assembled to a composite DTs of the whole product or service and all DTs are deployed in the single uniform platform. External data sources e.g. devices, sensors, partner's IT systems, mobile apps, enterprise applications, and external data consumers etc. are directly physical

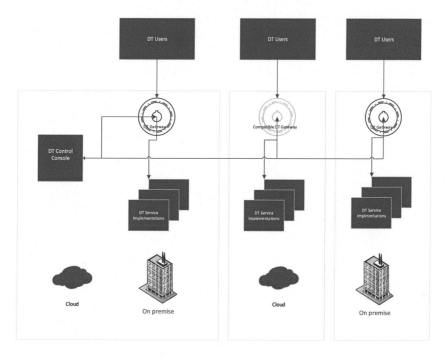

FIGURE 5.2 Distributed DT gateway.

connected to discrete DTs. It is a good fit for building DTs for a relatively complicated IT-connect products in a homogeneous environment [7]. It is easy to integrate data among discrete DTs because they can use the same service communication protocol and event architecture in the same platform. But it still needs efforts to transform the data format and communication protocols between proprietary external data sources and the common data standards of the platform. In addition, discrete DTs can only expose part of their data to the composite DT meanwhile share the computing and processing. Therefore, this model can easy DT integration and reduce composite DT's burden.

5.5.2 Model 2: distributed DT gateway

In Fig. 5.2, a decoupled DT management solution with a centralized control console, and a group of homogeneous or compatible DT gateways distributed near the service implementations across data centers and cloud environments. The centralized control console authorizes and maintains mediation definitions and governance policies and the policies and metrics collection are enforced to distributed local DT gateways at run time. This pattern can scale up by instantiating a new homogeneous safety

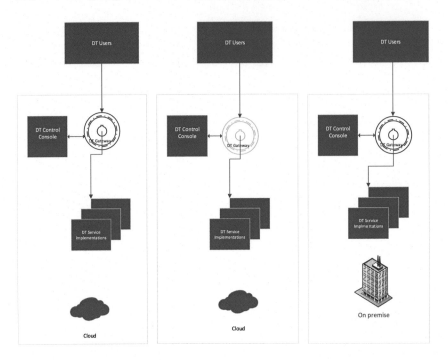

FIGURE 5.3 Multiple instance of one DT management solution.

gateway instance from the DT service implementations to the control console.

This model is able to manage all your DT services in one place but at the cost of setting up and supporting the same DT gateway at the place DT services are hosted. This pattern is a good choice when you want to offer local autonomy to the DT system, while retaining consistent governance and monitoring across all DTs. It is also a good fit for separating internal traffic from external or public access.

5.5.3 Model 3: multiple instance of one DT management solution

In Fig. 5.3, multiple instances of the same DT management solution exist and each of them are responsible for local interface and services within the same data center or cloud environment. This model no longer enforces all DT platforms traffic through a single DT gateway instance but allows each platform uses their compatible DT gateway instead. Since the DT management solutions are the same, the instances can share common base configurations and UI customization of the control consoles. DT consumers and DT providers can also appreciate an akin user experience on

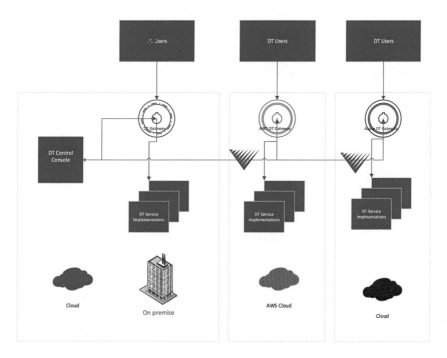

FIGURE 5.4 Federated DT gateways.

control consoles. This pattern can scale up by adding a new instance of the same DT management solution within the environment of the new DT service implementations.

One common example is to instantiate independent instances of the same DT management solution for development, test and production environments, or for internal- and external-facing interfaces, so that different environments are segregated from each other. This model is a good choice when you need to have segregated nonproduction and production environments for interfaces and services or you need consistent enforcement of DT governance policy while still keeping flexibility of policy enforcement customization where suitable while retaining a consistent experience on the DT developer portal and administration portal.

5.5.4 Model 4: federated DT gateways

In Fig. 5.4, a decoupled DT management solution with a centralized control console, and a federation of heterogeneous DT gateways located at different data centers or cloud platforms. The difference between federated DT gateways model differs and distributed DT gateways model is that this pattern makes good use of in-place aboriginal DT gateways provided by the corresponding providers, via a software adapter that can mediate the communication between the control console and heterogeneous gateways. This pattern scales up by adapting the aboriginal DT gateway of the environment where the DT services are hosted.

This is the desirable architecture to implement DT management solutions on hybrid and multiple platforms, since it provides the best combination of centralized control, local autonomy and plug-and-play flexibility and scalability. However, in reality, DT gateways are often coupled with control console from the same vendor but incompatible with another vendor's control console. Therefore, we have to develop the custom code of DT gateway adapter to integrate the incompatible DT gateway with the centralized control console.

For example, participant DTs in their own IoT platform are assembled to a composite DTs that has its product or service in its own IoT platform but their IoT platforms are different. The difference between this case and the previous case is that the composite DT integrate with other DTs operating on their own IoT platforms rather than they all run on the single IoT platform. Model 4 is a good fit for building DTs for a very complicated IT-connect products or services in a heterogeneous and complicate environment with a centralized control console.

This model is a good fit for building DTs for the most complicated IT-connect products or services in a heterogeneous and complicate environment

5.5.5 Model 5: multiple DT management solutions

In Fig. 5.5, Mutliple DT management solutions often exist in hybrid or multiple platform scenarios and each DT management solution is in charge of local interface and services within the same data center or cloud environment. Different DT management solutions operate independently of each other and thus build their own siloed governance mechanism, developer portal, and DT gateway without shared components. This model can scales up by instantiating a new seperate DT management solution exclusively for services hosted from a different vendor in a new environment.

In this model, local autonomy and optimization to meet local requirements carry more heavy than unified governance, consistency and uniformity. This model is a good choice when you need to use the aboriginal DT gateway to minimize operational complexity and cost at the cost of consistent governance and user experience. It is also a good fit when you want to provide complete local autonomy, reduce potential struggle or incompatibility between the DT gateway and the control console in heterogeneous platforms or separate internal interface from external or public access in separate locations.

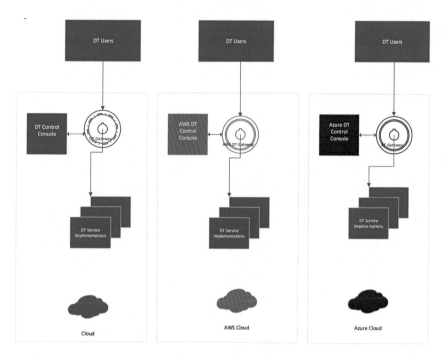

FIGURE 5.5 Multiple DT management solutions.

Architecture model	# of DT control consoles	# of DT gateways	Vendors of DT control consoles	Vendors of DT gateways
Single DTMS instance	1	1	Same	Same
Distributed DT gateways	1	N	Same	Compatible
Multiple DTMS instances	N	N	Same	Compatible
Federated DT gateways	1	N	Same	Different
Multiple DTMSs	N	N	Different	Different

5.5.6 Model summary

From above, we can see that a single DT Management solution can implement a strong centrally governance and maintain consistency in policy enforcement, but limits local autonomy for local systems. Applying separate aboriginal DT gateways simplifies operation but increase integration efforts and reduces consistency in security control and user experience.

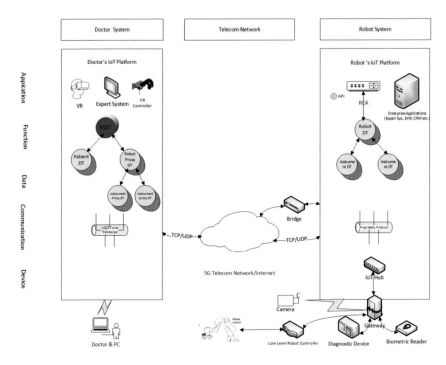

FIGURE 5.6   Case study: automatic remote surgeon using robot, DT and VR.

5.6 Case study: automatic remote surgeon using robot, DT and VR

In this section, we will dive into a detail use case in medical and health industry: Because of limited medical care resource, many patients cannot get timely treatments and have to travel to other regions or countries for surgeries. Nevertheless, it is expensive, inconvenient or impossible for many patients to travel to remote locations. We can make remote operations possible and automatic by combining VR and robotics via DTs over telecom networks.

Fig. 5.6 depicts a Remote Surgery DT (RSDT) is a composite DT. When the doctor considers performing a remote operation, a RSDT DT will be instantiated to bring the related DTs such as DT of the patient, DT of the robot, and DTs of medical instruments etc. together to accomplish the task. Pattern 4-Hybrid pattern is adopted to design the system: we implement DT of the patient using Pattern 1 Discrete DT on single IoT platform in the doctor's site first; Then apply pattern 2 Composite DT on single IoT platform to implement the DT of robot and DT of medical instruments in remote site; Next apply pattern 3.3 – federated pattern – Hierarchy to implement RSDT in the doctor's site: RSDT acts as the root of the hierarchy to coordinate and control the other DTs. It is connected to the DT of the patient and the proxy DT of the robot that is linked to several proxy DTs of medical instruments.

Initially the DT of the patient is a replica of a physical patient. RSDT can read the patient's historical medical records and the patient's current status form the DT of the patient, and develop the plan of the surgeon and medical treatments using machine learning analytics and recommendation functions. Doctor has a VR view to maneuvers or train a remote robotic arm by using DT of the robot. The DT of the robot serves productively as a broker between the doctor and the robot arm. The doctor controls the DT of the robot while the DT of robot controls the robot arm in the remote physical site to perform the surgery.

RSDT system is comprised of four parts:

1) Robot System: the DT of the robot is deployed in a cloud IoT platform near the real robot. This DT is connected with Robot Control Application (RCA) that produces transition points between the human controls and corresponding robot trajectory, and translates human controls into robot commands [8]. Thus the human control loop is decoupled from the control loop of the robot to achieve low latency and short reaction time.

2) Doctor System: A proxy DT cloning the DT of robot is deployed in a cloud IoT platform near the doctor. The synchronized proxy DT of the robot has the same dimension, state, and features as the real robot in the Virtual Reality (VR) system. The doctor can see it in a VR head-mounted display (HMD) and steer a mobile remote controller with haptic feedback function to drive it to control the remote robot [8].

3) Telecom Network: VPN on 5G mobile network connects the IoT platform near the doctor with the IoT platform near real robot that RCA resides on. The VR software layer translates the movements of the controller to the location and rotation information and transmits them to the robot's IoT platform. The two different IoT platforms exchange information via the bridge.

4) Feedback: The video of the robot arm on the remote site is captured and streamed via the gateway and the telecom network back to the doctor's IoT platform [8]. The doctor utilizes it as visual feedback to adjust his hand movement to continue controlling the robot.

The RSDT system can be useful in the following scenarios:

1) DT-DT: Plan and Training

The DT of the patient is further developed to a detail 3D digital representation of the real patient refreshed by different treatment scenarios in real time. Taking this DT of the patient as the virtual visual input, the doctor manipulates the proxy DT of the robot in VR system while the DT of the robot controls the real robot. It is a good fit for optimizing the doctor's decision making. He can try every possible option of medical treatments to see the respective outcomes and then choose the best one. He can also run simulations and evaluate different strategies for the actual surgery before performing a real operation.

2) RT-DT: Automatic Remote Surgery

Taking real video of the patient as the visual feedback, the doctor manipulates the DT of the robot in VR system while the DT of the robot controls the real robot. It is a good fit for the doctor to perform manual remote surgery [9]. By calibrating the composite AI models with the real patient in the remote operation, RSDT can be trained to gain learning experience so it can control the robot to automatically perform the surgery at the remote site without human intervention. In case of an emergency, the doctor can be noticed by RSDT about such situations and thus take over the handling of the circumstances.

Let us dive into the capabilities of the RSDT system in term of tiers and layers.

A) Device Layer

The layer deploys a wide range of devices or sensors including optical, audio, haptic etc. to generate the most cognitive experience for each doctor. It also includes a flexible and reliable industry robotic arm and medical instruments. X.509 protocol is used to connect devices and IoT gateway.

B) Communication Layer

The system is able to establish the communication between devices automatically, comprehend each other, and collaborate without human engineering. Inside local site, the communication of edge to IoT platform and intra-IoT-platform is based on MQTT over WebSocket to guarantee the required reliability and error check. However, we use User Datagram Protocol (UDP) between the sites to quick transit the video packets over the mobile network. The system also supports wireless technologies such as Wi-Fi, Bluetooth, GPS and 5G.

C) Data Layer

The system is able to collect data from various medical equipment, devices, medical instruments, computers, and external or internal sensors. In addition, it has strong data processing capabilities including ingestion, transformation, and batch loading. Moreover, it can effectively store and manage useful information.

D) Function Layer

The system can perform stream and event processing in the cloud IoT platform. It is also able to use complex AI and machine learning analytics in the cloud IoT platform to not only detect object, classify terrain, and tackle human's hand movement but also create rule based expert system and knowledge graph to optimize the doctor's decision making and perform automatic operation.

E) Application Layer

The system performs simulations by means of VR system with a HMD. It uses various sensors to create an immersive virtual operation room with virtual patient, virtual robot arm and virtual instruments. On the other hand, it is able to execute actuation commands to control physical objects. The video feed can be shown directly in front of the user's eyes as if the doctor stands on the place where the robotic arm stands [9]. Mixed Reality (MR) experience is generated by propagating a video feed from the remote operation room into the VR world. Telemetries and parameters can be visualized and embedded into DTs in flat screens, 3D scene or video stream in VR system.

F) Security

Remote surgery needs high levels of reliability and security across layers. The system can provide authorization, authentication, and encryption functions to protect the data flow between layers, tiers, the stored data and functional services. It includes access management, device catalog, and SSL/TLS and SASL protocols supporting to protect gateway software, IoT platform, applications, and service from misuse. Especially it verifies if the operator is the doctor who have enough capabilities and right to perform the remote surgery through the DT system.

By using RSDT system, we can deliver high quality of surgery disregarding the patients' location. Patients requiring operations in remote areas could get timely and the same high quality surgery as that with experienced specialists in big hospitals.

5.7 Future direction

In this section, we will discuss the lesson that we learned and future direction: When the complex composite DT wants to incorporate discrete DT from different sources, it faces huge integration and security challenges. The reason is that there do not exist widely recognized common DT standards to integrate different DT Platforms. The future direction is to establish common DT standards of DT platform cooperation and develop prebuilt DT templates [10]. The templates will offer a consistent method to develop and drive DTs with uniform DT models and data standards. These standard-based DTs generated by these templates can be easily identified, incorporated into various DT systems, and cooperate with other standardized DTs to accomplish the task. It will significantly reduce the efforts of DT platform integration.

Every DT project can use this reference architecture models to describe the required functionalities, the locations that functionalities will operate, and the key system interfaces according to the following process:

1. Analyze business requirements you are trying to meet. Mapping the business requirements to technical requirements to compose three tiers, five layers and interfaces between layers or tiers to design your DT architecture from seeing a big picture.
2. Assess locations that the functionalities will be deployed. Plan how each of the functional components appropriate in one of the three tiers. Evaluate the tier-specific interfaces and security issues.
3. Choose one of the five architecture models and determine the topology and integration methodologies of DT platforms if it needs.
4. Devise necessary functionalities in each layer at each tier. Begin with the device layer and then move to the next layer till you complete the first round of coarse-grained functionalities design for every layer. Evaluate the layer-specific interfaces and security issues [6]. For example, in process layer at edge tire, clearly describe and specify user workflows to align the multiexperience apps to provide a seamless experience as they navigate from one app to another.
5. Verify and refine the design. Iteratively and incrementally repeat the above steps to continuously review, revise and improve the design e.g. refine the coarse-grained functionalities to fine-grained functionalities from top to down or add new functionalities etc. by using Agile methodology.

The discussed DT reference architecture models break down the complexity of DT architecture and thus accelerate creation of a DT system Design. It enables continuous delivery and simplifies testing. Additionally, it eases technical and business cooperation but also simplify evaluating different vendor's DT products and services.

References

[1] M. Grieves, J. Vickers, Digital twin: mitigating unpredictable, undesirable emergent behavior in complex systems, in: Transdisciplinary Perspectives on Complex Systems, Springer, Cham, Switzerland, 2017, p. 85113.
[2] A. El Saddik, Digital twins: the convergence of multimedia technologies, IEEE Multimedia Mag. 25 (2) (Apr. 2018) 8792.
[3] R. Ferdousi, F. Laamarti, M.A. Hossain, et al., Digital twins for well-being: an overview, Digital Twin 1 (2022) 7, https://doi.org/10.12688/digitaltwin.17475.2.

[4] R.G. Díaz, F. Laamarti, A. El Saddik, DTCoach: your digital twin coach on the edge during Covid-19 and beyond, IEEE Instrum. Meas. Mag. 24 (6) (September 2021) 22–28, https://doi.org/10.1109/MIM.2021.9513635.

[5] R. Ferdousi, M.A. Hossain, A. El Saddik, IoT-enabled model for digital twin of mental stress (DTMS), in: 2021 IEEE Globecom Workshops (GC Wkshps), 2021, pp. 1–6.

[6] Paul DeBeasi, Designing an IoT reference architecture, Gartner (11 October 2019).

[7] Benoit Lheureux, Alfonso Velosa, Peter Havart-Simkin, Five Approaches for Integrating IoT Digital Twins, Gartner, Apr. 2018.

[8] Ievgenii A. Tsokalo, David Kuss, Ievgen Kharabet, Frank H.P. Fitzeky, Martin Reisslein, Remote robot control with human-in-the-loop over long distances using digital twins, in: 2019 IEEE Global Communications Conference, Dec. 2019.

[9] H. Laaki, Y. Miche, K. Tammi, Prototyping a digital twin for real time remote control over mobile networks: application of remote surgery, IEEE Access 7 (2019) 20325–20336, https://doi.org/10.1109/ACCESS.2019.2897018.

[10] A. El Saddik, F. Laamarti, M. Alja'Afreh, The potential of digital twins, IEEE Instrum. Meas. Mag. 24 (3) (May 2021) 36–41, https://doi.org/10.1109/MIM.2021.9436090.

[11] M. Grieves, J. Vickers, Digital twin: mitigating unpredictable, undesirable emergent behavior in complex systems, in: Transdisciplinary Perspectives on Complex Systems, Springer, 2017, pp. 85–113.

[12] Kazi Masudul Alam, Abdulmotaleb El Saddik, C2PS: a digital twin architecture reference model for the cloud-based cyber-physical systems, IEEE Access 5 (Jan 2017).

[13] Klementina Josifovska, Enes Yigitbas, Gregor Engels, Reference framework for digital twins within cyber-physical systems, in: 2019 IEEE/ACM 5th International Workshop on Software Engineering for Smart Cyber-Physical Systems (SEsCPS), 2019.

[14] Duansen Shangguan, Liping Chen, Jianwan Ding, A hierarchical digital twin model framework for dynamic cyber-physical system design, in: ICMRE'19: Proceedings of the 5th International Conference on Mechatronics and Robotics Engineering, February 2019.

[15] Roberto Minerva, Gyu Myoung Lee, Digital twin in the IoT context: a survey on technical features, scenarios, and architectural models, Proc. IEEE 108 (10) (Oct 2020).

[16] Behrang Ashtari Talkhestani, Tobias Jung, Benjamin Lindemann, Nada Sahlab, Nasser Jazdi, Wolfgang Schloegl, Michael Weyrich, An architecture of an intelligent digital twin in a cyber-physical production system, Automatisierungstechnik 67 (9) (2019) 762–782.

[17] V. Souza, R. Cruz, W. Silva, S. Lins, V. Lucena, A digital twin architecture based on the industrial Internet of things technologies, in: 2019 IEEE International Conference on Consumer Electronics (ICCE), 2019, pp. 1–2.

[18] Kyu Tae Park, Jehun Lee, Hyun-Jung Kim, Sang Do Noh, Digital twin-based cyber physical production system architectural framework for personalized production, Int. J. Adv. Manuf. Technol. 106 (2020) 1787–1810.

6

Artificial intelligence models in digital twins for health and well-being

Rahatara Ferdousi[a], Fedwa Laamarti[b,a], and Abdulmotaleb El Saddik[b,a]

[a]University of Ottawa, Ottawa, ON, Canada, [b]Mohamed bin Zayed University of Artificial Intelligence, Abu Dhabi, United Arab Emirates

6.1 Background and introduction

Digital Twin (DT) enables the monitoring, understanding, and optimization of health parameters of humans, and provides constant health insight to improve quality of life and well-being [1]. From 2018 to 2021, the DT has been studied vastly in contemporary research [2] as depicted in Fig. 6.1.

In the beginning, the DT models were designed to collect, manage, and visualize data [3]. For example, to collect and visualize real-time data of heart conditions. As soon as DT models started to integrate Artificial intelligence (AI) capabilities, attention started to be directed to well-being systems [4]. This is because AI empowers DT to utilize vast amounts of data for decision making, prediction, estimation, classification etc. For instance, a health monitoring DT system enables real-time heartrate monitoring of an emergency patient [4], while A can enable such systems to detect abnormalities in the heartrate [5,6].

Although there are several research works on AI-based DT models, it is still explored very little in the context of health and well-being. Therefore, in this chapter, we aim at providing an overview of AI models in the context of DT for health and well-being. The key contribution of this chapter

121

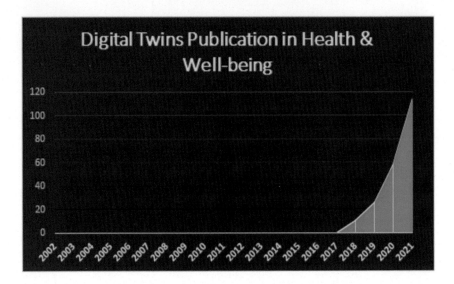

FIGURE 6.1   Growth of digital twins publication in health and well being.

is to summarize the scope, model types, learning technologies, benefits, and learning models for DT AI-based models.

The rest of the chapter incorporates the following subsections. In Section 6.2, we discuss the necessity of AI in the DT models. Then, in Section 6.3, we discuss various machine learning and deep learning techniques, and types of learnings used in DT AI-based models. In Section 6.4, we discuss our findings from the literature survey. Finally, in Section 6.5, we conclude and suggest future scope.

6.2 AI in DT models

The AI in healthcare is not new. To improve the quality of human life, AI is already in practice for individuals' well-being. Several technologies like the Internet of Things (IoT) and Cyber-Physical Systems (CPS) have adopted AI for decision-making and prediction [7]. The DT is no exception. In this section, we will review the use of AI in DT for healthcare.

At the beginning of the healthcare DT revolution, the focus of the DT model was real-time health monitoring. In [8], the authors proposed a DT healthcare model for elderly health monitoring that deals with data collection, transmission and analysis using IoT technology. Although this model supports real-time virtual monitoring of seniors' health, it cannot support automated decision-making. More specifically, to understand the early warning of crisis, the method requires some human-intervention to visualize the data relevant to a patient's health (e.g., heart rate, pulse rate, etc.). Here the lack of a predictive mechanism limited the DT model to only real-time data monitoring.

In another study [3], the authors have considered the heterogeneous data collection issues and proposed a standardized ISO/IEEE 11073 DT framework architecture for health and well-being. This model provides the process of:

• Collecting data from personal health devices.
• Processing that data.
• Providing feedback to the user in a closed feedback loop.

In general, the process is obtaining a balanced dataset for decision support [9]. Now, to support decision-making, machine learning and data mining mechanism are required to integrate into DT models [10].

A contemporary study in [10] highlighted the necessity of AI empowered DT models to broaden the scope of DT models beyond health monitoring. The authors proposed a DT model for precision medicine. They presented a behavioral AI model as a component of predictive DT models to predict accurate drugs and treatments. They conceptualized this behavioral AI model to include pattern-based machine learning prediction. For instance, to predict glucose fluctuation to provide early warnings of a patient's condition.

Gradually some research proposed DT for classifying heart-rate abnormalities [11], and disease [12]. In addition, it has been proposed for fitness management [13], and hospital management [14]. These works utilized deep learning techniques like Convolutional Neural Networks (CNN) [11,13] and machine learning techniques like Bi-directional long short-term memory (Bi-LSTMS) [15] to design the DT model with AI capabilities.

The above studies found that the AI through DT amends several scopes for health and well-being, as illustrated in Fig. 6.2.

From Fig. 6.2, we observe that a health monitoring system can provide the activity data summary from an individual's smartphone or smartwatch data. However, the monitoring using an AI-based DT model can determine if a patient has obesity, insomnia, etc. Another example shows that a monitoring system, without AI, can monitor heart rate in real-time, while DT with AI capabilities can conclude heart abnormality rate, and has the potential to predict the risk of heart disease, etc. We detail the purpose and prose of AI model in DT for health in Table 6.1.

In a nutshell, the necessity of integrating AI with monitoring, to obtain a DT model, demands predictive mechanisms to advance various scopes of healthcare.

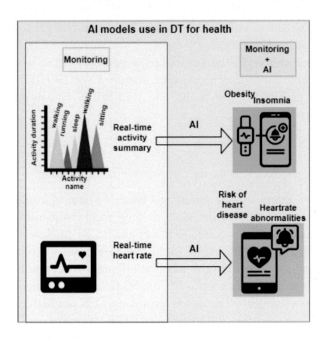

FIGURE 6.2   AI models use in DT for health.

TABLE 6.1   Scope of AI-based model for health and well-being.

Scope	Purpose	Pros of AI use in DT models
Precision health	To identify patient's trajectory for testing treatment	DT models allow testing of virtual model of health [16,17].
Preventive health	Prediction of early risk	The multimedia convergence of IoT and the AI inference engine enables forecasting risks [6].
Predictive analysis	Prediction of unseen patterns, anomaly etc.	Machine learning and Data Mining in DT model enables pattern-based prediction [18].
Explainable AI	Prediction with explanation	Explainable AI DT models, enables prediction with explanation [12]. ISO/IEEE 11073 DT framework provides meaningful health information [3].

6.3 Types of AI models in DT for health

Overall, we observe from Table 6.1, that AI models in DT support various health applications. Therefore, we need to understand the types of AI models that can support numerous healthcare applications. In this section, we discuss the types of AI models proposed in the literature of DT for health. Based on our literature survey, we found the AI models in DT for

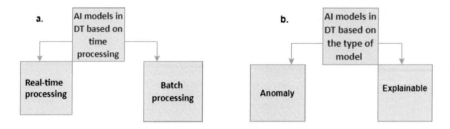

FIGURE 6.3   Types of DT healthcare model.

health can be categorized based on time of processing and model's type (Fig. 6.3). The details are given in the following sections.

6.3.1 Real-time processing

Real-time data visualization is a typical application of DT [19]. However, real-time prediction enhances the application area [20]. In general, DT real-time predictive model works as follows.

- A predictive model is built on vast amounts of data.
- A continuous stream of data is fed to it.
- Then the model makes predictions in run time.

For example, a model trained with patient's condition and waiting time can be used to predict live waiting time at emergency care. Real-time data of incoming patients will be fed to the model for forecasting waiting time for a patient.

In [5], the authors-built classifiers for real-time classification of heart issues. They applied CNN, LSTM, Support vector machine (SVM), and Logistic regression (LR) to classify ECG heart rhythm in real-time. The DT model in [21], conducted LSTM for real-time vulnerability prediction of lung cancer patients. The real-time prediction has also been studied for fitness management [19]. We categorize potential real-time techniques as illustrated in Fig. 6.4.

The model in [5] focused on classifying data to detect a health issue. In comparison, the studies in [19] and [21] focused on predicting health outcomes before those happen.

6.3.2 Batch processing

The purpose of batch processing is the opposite of real-time processing, built statically on periodic data [22].

An ML model is trained with a vast amount of data and predicts the arrival of test data. For instance, a model can be trained with activity data of obsessed people and then expect the model to predict the obesity of the in-

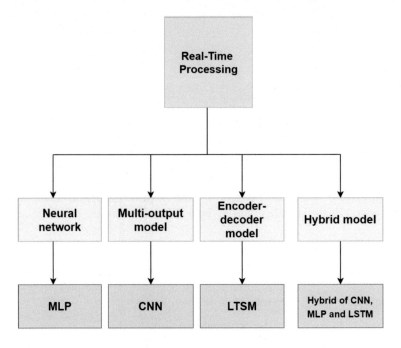

FIGURE 6.4   Categorization of real-time processing.

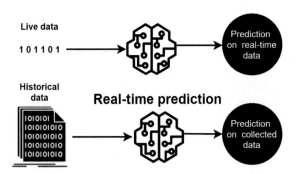

FIGURE 6.5   Processing techniques: real-time processing vs batch processing.

dividuals. The batch pr processing models are built on vast data, the batch prediction is more precise than the real-time prediction. Real-time processing is more focused on speed than accuracy [23]. It can be observed from Fig. 6.5, that real processing is conducted on real-time data, while batch processing is done on historical data. Throughout our study, we could not

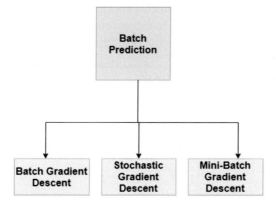

FIGURE 6.6   Batch selection techniques for batch processing.

identify any specific categorization of batch prediction. However, the difference in selecting batch size changes the learning type as demonstrated in Fig. 6.6. The learning techniques are selected based on the availability of training sample size. So far, we found the following criteria of different learning approaches for batch processing [24].

- Batch Gradient Descent Learning. If the ML algorithm considers batch size equal to the training set.
- Stochastic Gradient Descent Learning: If the ML learning algorithm considers the batch size equal to the size of the one sample.
- Mini-Batch Gradient Descent Learning. The learning algorithm is called mini-batch gradient descent if the ML learning algorithm considers the batch size more between one to the total sample size.

6.3.3 Anomaly

The anomaly detection has been applied to the DT model in various contexts. However, in the context of healthcare anomaly prediction has been proposed. There is a subtle difference between anomaly detection and anomaly prediction. As illustrated in Fig. 6.7, the detection process identifies anomalies from real-time data, while the prediction process depends on an ML model built using historical data [25].

For instance, the DT model in [6] used CNN for heart condition classification, while the DT approach in [14] detected anomalies in the hospital's data. Since the anomaly prediction depends on the ML model, we need to know the existing approach available. There were limited examples in literature as predictive DT models for healthcare are at the early stage of adoption. Therefore, to categorize existing models for anomaly, we considered both the techniques that have been already applied and the techniques that can be applied in DT predictive model.

To train a supervised DT model for anomaly prediction requires a training set where anomalous and nonanomalous data are labeled. The Neural Network (NN) [19], Support Vector Machine (SVM) [12], K-nearest neighbor (KNN) [19], and Bayesian network (BN) [26] have been proposed widely for anomaly prediction models.

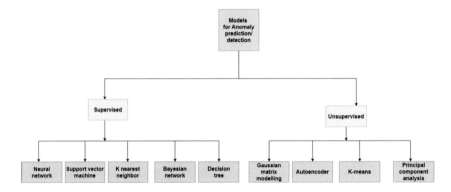

FIGURE 6.7   Difference between anomaly detection and anomaly prediction.

By contrast, the unsupervised models do not require labeled data. Instead, it assumes most data has a normal pattern, and a little of the data has an anomalous pattern. Then the model predicts the group for test data based on the pattern or characteristics similarity with the presumed groups. The K-means, Autoencoders, Gaussian Matrix modeling (GMM), Principal Component Analysis (PCA) algorithms have been proposed widely for unsupervised anomaly prediction models [18].

6.3.4 Explainable model

This type of model is a very recent addition to DT predictive models. The trust in AI prediction has been discussed as one of the key challenges in literature [27,28]. The Explainable AI (XAI) was proposed to bring interoperability and explainability to the black-box predictive models, such as the deep neural network model [29]. The XAI predictive model can bring transparency to the model by providing a contextual explanation of the prediction [30].

In Fig. 6.8, we illustrate an example of a predictive model that predicts diabetes risk with higher accuracy. Such a model cannot tell us why the person has a risk of developing diabetes. In comparison, the XAI predictive model uses XAI algorithms to extract reasons for each instance belonging to a particular class. The following illustration demonstrates a comparison of regular predictive models and XAI predictive models.

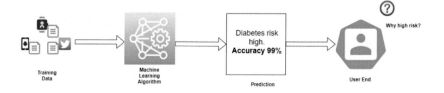

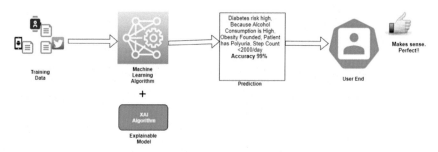

FIGURE 6.8 General predictive model vs XAI predictive model.

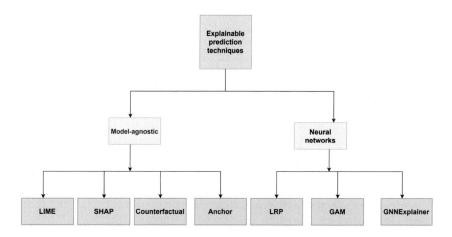

FIGURE 6.9 Categorization of explainable models.

There are a few examples of XAI DT predictive models in the literature. In Fig. 6.8, we illustrate the categorization of explainable models for health applications.

In Table 6.2, we have summarized our findings on various types of prediction that can be applied to Predictive DT models for healthcare. The authors in [12], have used Lime algorithms to extract the explanation of

TABLE 6.2   Summary of the DT models for healthcare.

Type	When to apply	Requirements	Limitations	DT for health application
Real-time [18]	For faster and instant prediction	Live data	Complex to implement. Accuracy is compromised	Predictive and Personalize healthcare (prediction for preparedness)
Batch [19]	For historical prediction	Vast amount of historical data	The model can be outdated and the model may not support new types of data	Precision health and Preventive health (Diagnostic prediction)
Anomaly [15]	For pattern-based prediction	Historical data Labeled groups of anomalous and nonanomalous data	Complex to implement	Preventive health (prediction for preparedness)
Explainable [12]	For reliable prediction When the rationale of prediction is necessary	Easily explainable model like decision tree.	Still on progress may not fit all classifiers well (e.g., Kernel-based SVM)	Personalized healthcare Preventive healthcare Participatory healthcare

liver disease prediction. They used SVM classifiers to train their model on liver disease detection data. The health parameter values of a patient to be diagnosed with liver disease were extracted by the Lime algorithms. In another research [19], the authors used a counterfactual algorithm for recommending fitness precautions based on the user's activity history. For instance, if the model finds that the activity duration is responsible for obesity of someone, it can recommend increasing activity for activity duration. Apart from the counterfactual and lime algorithms, there exist other techniques for building explainable predictive DT models. (See Fig. 6.9.)

Model-agnostic approaches are designed to provide the interoperability of trained machine learning models. In contrast, the model specific methods are designed to interpret the neural network models. The model agnostic XAI algorithms for explainable prediction models are as follows [31].

- The Local Interpretable Model-Agnostic Explanations (LIME) works as the name reflects. The explanation targets extracting characteristics of the classifiers for a specific prediction. The LIME algorithm provides interoperability of the model. For example, the explanation of the SVM algorithm for classification. The algorithm modifies single data points based on the feature values.

- SHapleyAdditive exPlanation (SHAP) designed based on LIME. This algorithm finds out the correlated features for a prediction. As the base of SHAP is LIME, it provides properties and local accuracy like LIME. Mathematically SHAP provides a theoretical guarantee of consistent accuracy. However, the SHAP is slower than LIME when used for Kernel-based classifiers. In addition, SHAP is still under development and does not support all the ML models.

- The LIME and SHAP are widely in practice due to the availability of Python libraries. There is another type of technique known as the Counterfactual algorithm. The key idea of this algorithm is to provide values of different attributes for a particular class prediction. In comparison to LIME and SHAP, the counterfactual algorithm cannot show any range or threshold to understand how various parameters contribute to a decision. In addition, the counterfactual algorithm may provide multiple explanations for each instance.

The model specific XAI algorithms for explainable prediction models are as follows.

- The Layer-wise Relevance Propagation (LRP) [32] technique explains the importance of deep neural networks. LRP propagates backward prediction in a neural network. However, the propagation rules need to be designed with manual effort.

- The generalized additive model (GAM) [33] is a statistical approach to extract an explanation of the variables of the linear predictors. The benefits of GAM models are that it suits nonparametric models. However, it's in less practice due to the computational complexity. In addition, the nonparametric approach causes overfitting. Therefore, the model may lose interoperability.

6.3.5 Learning types

In the previous section, we have seen several types of DT models. In this section, we summarize diverse styles of learning models used for those predictions in literature. This is by no means restrictive as to which types of models can be used in DT. Actually, it summarizes the models that were already put in use in existing research. Table 6.3 shows details of various learning types for building DT models in the healthcare domain.

6.4 Discussion

In this section we discuss our findings on AI models for DT in the literature.

TABLE 6.3   Learning types for building DT models in healthcare domain.

Reference	Learning type	ML algorithm/ technique used	Data source	Data type	Application domain	Model's goal	Experiment result
[18]	Deep learning	CNN	Exercise video	Timeseries	Fitness management	Pose estimation	Improvement in R̂2 score ranging between 18%–135%
[19]	Supervised learning	KNN(voting method) SVM (Kernel mode) Counterfactual algorithm	Athletes' behavior record	Historical	Fitness management	Classifying athletes' condition during training	Classification loss – 0.01
[6]	Deep learning	CNN	ECG record	Timeseries	Heart issue detection	Heart's condition classification	Accuracy: 85.77% Precision: 95.53% Recall: 86.32%
[5]	Deep Learning	LSTM	ECG record	Timeseries	Heart issue detection	Heart's condition classification	Accuracy: 97% Precision: 98% Recall: 98%
[15]	Deep learning	LSTM	Blood circulation information of patient's data	Timeseries	Blood circulation analysis	Inverse analysis of neural network	Accuracy: 90.58%
[12]	Supervised learning	SVM LIME	Liver disease dataset	Historical	Liver disease detection	Liver disease classification	Accuracy: 76%
[34]	Transfer learning	Inductive transfer learning	Various activity dataset	Historical	Fitness management	Activity recognition	Accuracy: 95.9%

1. Three types of learning, namely Deep Learning [5], Supervised Learning [19], and Transfer Learning [34], were used to build the DT models in healthcare. DT in healthcare is still in its infancy and more types of learning to build DT models are likely to be introduced in the future.

2. We found that Deep Learning algorithms are mostly used. Deep Learning algorithms can support the real-time prediction [6,33] of the digital twin predictive model. The use of Deep Learning is also well fitted to process the vast amount of healthcare data, as Deep Learning algorithms reduce the overheads of data labeling and human intervention [35]. Therefore, a study in [11], which currently uses IoT technology, has also proposed Deep Learning as the future scope of their work.

3. Apart from the application domains mentioned in the above table, some other applications of DT in healthcare were suggested, such as critical condition prediction for lung cancer patients [21], management of multiple sclerosis patients [36], trauma center management, and Hospital management [14]. However, no models were built for these applications yet.

4. Supervised learning has been used in very few models. The supervised learning was required mainly when the application domain was diagnosis or decision-making. This is logical as diagnostic models need be labeled historical data.

5. The transferred learning has been hardly considered for the DT healthcare models. Only in [34], the authors used it for activity recognition. As an application of transfer learning and DT is still growing in healthcare, this can be a reason for having fewer examples with transfer learning models.

6. The XAI algorithms, for instance lime and counterfactual algorithms, have been used with supervised learning algorithms. The study with transfer learning used inductive transfer learning that builds a model with labeled data. The reason behind selecting supervised classification algorithms or labeled data to build these predictive models could be that integrating the newest technology is easier with the supervised algorithms.

7. CNN, LSTM, MLP are some algorithms mostly used for DT Deep Learning models. By contrast, the SVM, KNN, Decision Tree have been used in the supervised learning models. For XAI, Lime and Counterfactual algorithms were used.

8. We can observe from Table 6.2 that the DT models, obtained accuracy above 85%. This higher accuracy is promising because the DT models targeting a sensitive domain such as healthcare, require an acceptable level of accuracy.

9. There is a scarcity of predictive DT models in the literature. One challenge that contributes to this is the unavailability of vast amounts of data required for building DT predictive models [19]. Another chal-

lenge is the complexity of preprocessing data from heterogeneous sources [10].

6.5 Conclusion

In the context of DT for healthcare, we found that deep learning models built with time-series data are widely used for real-time processing. In addition, supervised models with historical data are used in the case of diagnostic prediction. The newly amended technologies like XAI and transfer learning have already been considered for DT models. Several works have proposed DT with AI capabilities for future health applications. However, there is a scarcity of literature on DT predictive models to determine the requirements for such models. Multiple challenges need to be overcome to have more DT predictive models proposed in research. Such models are important to move from DT anomaly detection in healthcare, to anomaly prediction and eventually prevention.

References

[1] A. El Saddik, H. Badawi, R.A.M. Velazquez, F. Laamarti, R.G. Diaz, N. Bagaria, J.S. Arteaga-Falconi, Dtwins: a digital twins ecosystem for health and well-being, IEEE COMSOC MMTC Commun. Front. 14 (2019) 39–43.

[2] R. Fardousi, F. Laamarti, M. Hossain, C. Yang, A. El Saddik, Digital twins for well-being: an overview [version 1; peer review: 1 approved with reservations], Digital Twin 1 (7) (2022), https://doi.org/10.12688/digitaltwin.17475.1.

[3] F. Laamarti, H.F. Badawi, Y. Ding, F. Arafsha, B. Hafidh, A. El Saddik, An iso/IEEE 11073 standardized digital twin framework for health and well-being in smart cities, IEEE Access 8 (2020) 105950–105961.

[4] L.F. Rivera, M. Jiménez, P. Angara, N.M. Villegas, G. Tamura, H.A. Müller, Towards continuous monitoring in personalized healthcare through digital twins, in: Proceedings of the 29th Annual International Conference on Computer Science and Software Engineering, 2019, pp. 329–335.

[5] T. Erol, A.F. Mendi, D. Doğan, The digital twin revolution in healthcare, in: 2020 4th International Symposium on Multidisciplinary Studies and Innovative Technologies (ISMSIT), IEEE, 2020, pp. 1–7.

[6] R. Martinez-Velazquez, R. Gamez, A. El Saddik, Cardio twin: a digital twin of the human heart running on the edge, in: 2019 IEEE International Symposium on Medical Measurements and Applications (MeMeA), IEEE, 2019, pp. 1–6.

[7] A. El Saddik, F. Laamarti, M. Alja'Afreh, The potential of digital twins, IEEE Instrum. Meas. Mag. 24 (3) (2021) 36–41.

[8] Y. Shen, J. Colloc, A. Jacquet-Andrieu, Z. Guo, Y. Liu, Constructing ontology-based cancer treatment decision support system with case-based reasoning, in: International Conference on Smart Computing and Communication, Springer, 2017, pp. 278–288.

[9] S.M. Schwartz, K. Wildenhaus, A. Bucher, B. Byrd, Digital twins and the emerging science of self: implications for digital health experience design and "small" data, Front. Comput. Sci. 2 (2020) 31.

[10] S. Gochhait, A. Bende, Leveraging digital twin technology in the healthcare industry–a machine learning based approach, Eur. J. Mol. Clin. Med. 7 (6) (2020) 2547–2557.

[11] R. Martinez-Velazquez, D.P.V. Tobón, A. Sanchez, A. El Saddik, E. Petriu, et al., A machine learning approach as an aid for early Covid-19 detection, Sensors 21 (12) (2021) 4202.

[12] D.J. Rao, S. Mane, Digital twin approach to clinical dss with explainable AI, arXiv preprint, arXiv:1910.13520.

[13] R. Gámez Díaz, Q. Yu, Y. Ding, F. Laamarti, A. El Saddik, Digital twin coaching for physical activities: a survey, Sensors 20 (20) (2020) 5936.

[14] A. Karakra, F. Fontanili, E. Lamine, J. Lamothe, Hospit'win: a predictive simulation-based digital twin for patients pathways in hospital, in: 2019 IEEE EMBS International Conference on Biomedical & Health Informatics (BHI), IEEE, 2019, pp. 1–4.

[15] N.K. Chakshu, I. Sazonov, P. Nithiarasu, Towards enabling a cardiovascular digital twin for human systemic circulation using inverse analysis, Biomech. Model. Mechanobiol. 20 (2) (2021) 449–465.

[16] G. Ahmadi-Assalemi, H. Al-Khateeb, C. Maple, G. Epiphaniou, Z.A. Alhaboby, S. Alkaabi, D. Alhaboby, Digital twins for precision healthcare, in: Cyber Defence in the Age of AI, Smart Societies and Augmented Humanity, Springer Nature Switzerland AG, Cham, Switzerland, 2020, pp. 133–158.

[17] J. Corral-Acero, F. Margara, M. Marciniak, C. Rodero, F. Loncaric, Y. Feng, A. Gilbert, J.F. Fernandes, H.A. Bukhari, A. Wajdan, et al., The 'digital twin' to enable the vision of precision cardiology, Eur. Heart J. 41 (48) (2020) 4556–4564.

[18] R.G. Díaz, F. Laamarti, A. El Saddik, Dtcoach: your digital twin coach on the edge during Covid-19 and beyond, IEEE Instrum. Meas. Mag. 24 (6) (2021) 22–28.

[19] B.R. Barricelli, E. Casiraghi, J. Gliozzo, A. Petrini, S. Valtolina, Human digital twin for fitness management, IEEE Access 8 (2020) 26637–26664.

[20] H.-J. Jiang, Y.-A. Huang, Z.-H. You, Predicting drug-disease associations via using gaussian interaction profile and kernel-based autoencoder, BioMed research international, 2019.

[21] J. Zhang, L. Li, G. Lin, D. Fang, Y. Tai, J. Huang, Cyber resilience in healthcare digital twin on lung cancer, IEEE Access 8 (2020) 201900–201913.

[22] J. Čuklina, P.G. Pedrioli, R. Aebersold, Review of batch effects prevention, diagnostics, and correction approaches, in: Mass Spectrometry Data Analysis in Proteomics, Springer, 2020, pp. 373–387.

[23] S. Shahrivari, Beyond batch processing: towards real-time and streaming big data, Computers 3 (4) (2014) 117–129.

[24] N. Deepa, B. Prabadevi, P.K. Maddikunta, T.R. Gadekallu, T. Baker, M.A. Khan, U. Tariq, An AI-based intelligent system for healthcare analysis using ridge-adaline stochastic gradient descent classifier, J. Supercomput. 77 (2021) 1998–2017.

[25] M. Fahim, A. Sillitti, Anomaly detection, analysis and prediction techniques in iot environment: a systematic literature review, IEEE Access 7 (2019) 81664–81681.

[26] C. Ossai, N. Wickramasinghe, A Bayesian network model to establish a digital twin architecture for superior falls risk prediction.

[27] M. Braun, Represent me: please! Towards an ethics of digital twins in medicine, J. Med. Ethics 47 (6) (2021) 394–400.

[28] K. Bruynseels, F. Santoni de Sio, J. van den Hoven, Digital twins in health care: ethical implications of an emerging engineering paradigm, Front. Genet. 9 (2018) 31.

[29] D. Gunning, Explainable artificial intelligence (xai), Defense advanced research projects agency (DARPA), nd Web 2.2 (2017) 1.

[30] M. Nazari, A. Kluge, I. Apostolova, S. Klutmann, S. Kimiaei, M. Schroeder, R. Buchert, Explainable AI to improve acceptance of convolutional neural networks for automatic classification of dopamine transporter spect in the diagnosis of clinically uncertain parkinsonian syndromes, Eur. J. Nucl. Med. Mol. Imaging (2021) 1–11.

[31] N. Gandhi, S. Mishra, Explainable AI for healthcare: a study for interpreting diabetes prediction, in: International Conference on Machine Learning and Big Data Analytics, Springer, 2021, pp. 95–105.

[32] C. Zucco, H. Liang, G. Di Fatta, M. Cannataro, Explainable sentiment analysis with applications in medicine, in: 2018 IEEE International Conference on Bioinformatics and Biomedicine (BIBM), IEEE, 2018, pp. 1740–1747.

[33] K. Ravindra, P. Rattan, S. Mor, A.N. Aggarwal, Generalized additive models: building evidence of air pollution, climate change and human health, Environ. Int. 132 (2019) 104987.

[34] Y. Ma, A.T. Campbell, D.J. Cook, J. Lach, S.N. Patel, T. Ploetz, M. Sarrafzadeh, D. Spruijt-Metz, H. Ghasemzadeh, Transfer learning for activity recognition in mobile health, arXiv preprint, arXiv:2007.06062.

[35] A. Fuller, Z. Fan, C. Day, C. Barlow, Digital twin: enabling technologies, challenges and open research, IEEE Access 8 (2020) 108952–108971.

[36] I. Voigt, H. Inojosa, A. Dillenseger, R. Haase, K. Akgün, T. Ziemssen, Digital twins for multiple sclerosis, Front. Immunol. 12 (2021) 1556.

7

COVIDMe: a digital twin for COVID-19 self-assessment and detection

Roberto Martinez-Velazquez[a,c], Fernando Ceballos[b], Alejandro Sanchez[b], Abdulmotaleb El Saddik[a,c], and Emil Petriu[a]

[a]University of Ottawa, Ottawa, ON, Canada, [b]Department of Information Technology, University of Colima, Colima, Mexico, [c]Mohamed bin Zayed University of Artificial Intelligence, Abu Dhabi, United Arab Emirates

7.1 Introduction

A Digital Twin is a digital clone of an object or person in the real world, commonly referred to as Real Twin (RT). In this work, we present a Digital Twin (DT) of a person, more precisely, a DT of the health state of a person within the context of COVID-19 the DT is inspired in [1], [2]. We also present a proof-of-concept implementation of the DT for COVID-19, dubbed COVIDMe. We also open the case that COVIDMe is a specialization of a DT for health and well-being.

A DT for health and wellbeing has many components, some of which have the purpose of extracting "Inference Level Information" (ILI) by leveraging Machine Learning (ML) on "Base Level Information," often compiled from sensor data. In the case of COVIDMe, the platform takes self-reported symptoms from the RT and leverages ML to assess the risk of a COVID-19 infection, creating, in turn, a digital copy of the health state of the RT.

137

We have seen important advances in DT technology; however, we must try to answer critical questions when applying this technology to health and well-being. The first question is how to implement a DT for health and wellbeing? We address this question by taking the reference model presented in [2] and proposing an extra layer necessary for data preprocessing. Given the sensitive nature of a DT for Health and wellbeing, we also need to be able to answer the question of How accurate the Inference Level Information generated by the DT is? Mistakes in the ILI may lead to misleading feedback, impeding the RT from taking proper action to address any problems detected by the DT. In this work, we take COVIDMe as a use case to address this question and provide an insight into the answer. Finally, we also contribute with an architecture overview of COVIDMe.

The remainder of the document is distributed as follows: section 7.2 introduces the reader to the concept of DT, section 7.3 presents the motivation and use case that prompt the design and implementation of COVIDMe, section 7.4 gives an overview of the design and implementation of COVIDMe, section 7.5 provides an insight into the limitations of this work and preliminary results. Finally, section 7.6 presents our conclusions and future work on the platform.

7.2 Computer-aided diagnosis

Computer-Aided Diagnosis (CADx) is a mature field of research existing in the overlap between Computer Science and Medicine; studies in this interdisciplinary field are ultimately aimed at effectively emulating diagnosis processes and strategies currently observed and followed by physicians [3]. A CADx will take tests results, and symptoms collected from patients and produce a diagnosis in simple terms. Researchers have been working in CADx systems for a few decades now; in the late '50s, we can find some pioneering studies in the field, although CADx were usually described as "expert systems in medicine" [4–6].

Over the past few years, CADx systems have been attracting more attention from the scientific community thanks to advances in medicine and computer science, specifically in Machine Learning (ML). Various types of data may be used as input for the CADx system. However, we can divide it into three major groups: time series (sound and electrical signals), medical imaging, and laboratory test results [3].

A widespread use case for a CADx system is the automatic detection of tumors in medical imaging. In [7], the authors introduce a CADx system that detects lung cancer nodules on Computer Tomography scans by leveraging the predictive power of Convolutional Neural Networks, Deep Belief Networks and Stacked Denoising Autoencoder, obtaining accuracies between 0.71 and 0.82. Oversimplifying the process, the authors took a set of images to train a classifier; once the training was over, the classifier took new pictures to predict whether lung cancer nodules were present on the new images. That, in essence, is what a CADx system does. This approach is replicated through several medical imaging datasets to detect various health conditions like breast cancer or pneumonia.

The purpose of a CADx is clearly to generate diagnostic information; on the other hand, the main objective of a Digital Twin is to create an exact digital copy of an entity existing in the real world, and if the entity happens to be a person, the health state of that person would be in the scope of things a DT would copy. Therefore, the relation between a DT and a CADx system is such that a DT is not considered a CADx system. However, a DT can benefit from advances in CADx systems to improve its capacity to copy a person's health state better. In the next section, we further explain the concept of DT and the DT of a person.

7.3 Digital twin

Digital Twin (DT) Is widely known in the industry, and we can explain it as an exact digital copy of an object in the real world. A DT will replicate the characteristics and behavior of the thing in the real world to the extent that we can simulate the outcome of an object being in a specific scenario; the concept expanded over the years.

David Gelernter introduced the concept of Mirror Worlds (MW) as "software models of some chunk of reality" [8]. In Gelernter's vision, one would create a software model of the real world through computer-assisted design. The model would receive a vast amount of data from the real world to a point where the mirror world would be an exact copy in real-time. Gelernter describes the mirror world to be so thorough and detailed that one could have a mirror world of a university and know the precise location of each student, faculty member and staff member at any given time while being able to assess the structural integrity of the buildings in campus and every little detail in between.

The general idea was then picked up by Michael Grives, who took it further by introducing the Digital Twin (DT) concept in 2002 [9]. Grieves also dubs the DT counterpart (the real-world object) as Physical Twin (PT). There is a giant leap between MW and DT concepts: Grieves explained that a DT is actionable while an MW is limited to "reflecting" the physical object that is being copied. The most significant implication for Grieves DT is that if it's actionable, we can simulate the effects of external forces applied to it. According to Grieves, a DT needs to be inside a Digital Twin Environment (DTE), which is a "sandbox" of sorts containing both the DT and its environment [9], [10]. A DTE can be predictive or interrogative; the first one will allow one to understand how the DT will behave and per-

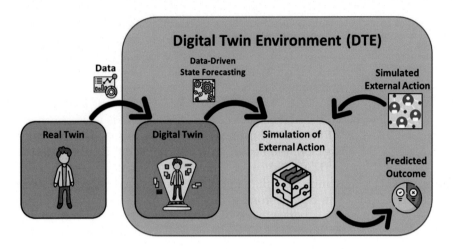

FIGURE 7.1   Overview of the Digital Twin platform.

form under certain circumstances, while the interrogative DTE can show current and past states [10].

Fig. 7.1 shows an essential representation of a DT, and we can see its base characteristics and features. The DT uses data collected from an object in real life and reproduces the object and its state in a virtual environment (namely DTE). We allowed ourselves some creative license in the naming of some components. For instance, just recording data from an object would not be a virtual representation on its own. We need a way to make sense of that data. Let us further explain this with a hypothetical use case. Let us imagine that we are collecting temperature, rpm and compression inside the car engine cylinders; we need the DT to use this data to detect an engine overheating or about to fail when the temperature, RPM and compression are off the normal operating parameters. In this diagram, we represent this functionality with Data-Driven state forecasting. Further elaborating by using this state and simulating an external action to be performed on the DT, perhaps deaccelerating in our case, the DT would be capable of predicting an outcome with some degree of accuracy.

In the previous example, as sophisticated as a modern vehicle engine can be, it falls short of the complexities of a human body.

7.3.1 Digital twin of a person

Alam et al. [11] take the concept of DT a step further by having multiple Digital Twins model the relationships between their physical counterparts. In Alam's vision, the DT exists in the cloud and communicates with other DTs instances, pushing the DTE to the cloud. Fig. 7.2 showcases a concep-

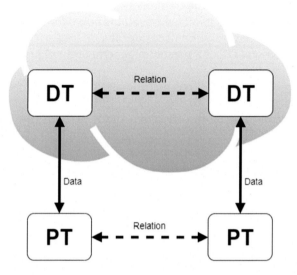

FIGURE 7.2 A conceptual overview of a Digital Twin architecture modeling relationships between their Physical counterparts.

tual overview of the DT as presented in [11] with Physical Twins sending data to the DT, which clones the PT and its relationships with others.

The next step in DT came in the form of a question: What if a DT could interact with the real world on behalf of the person, not as a delegate or assistant but as if it was the person itself interacting? For that to happen, we would need to model and clone behavior, likes, dislikes, preferences, physiology, relationships, and several other things that define a person. El Saddik introduced the DT of a person in [4]. According to El Saddik, the DT for humans presents unique technical challenges; simultaneously, special measures must be implemented to protect the person's privacy. A DT of a person must:

- have a unique identifier to prevent identity theft,
- have Sensors to clone its human counterpart and its environment,
- have Actuators interact with the real world,
- leverage Artificial Intelligence so it can make sense of the data collected with the sensors,
- have Communication capabilities such as 5G, Bluetooth, Wi-Fi to communicate with other DTs and entities,
- use a Representation platform (Multimodal Interaction) to leverage recent advances in robotics and VR,
- be Trustworthy to delegate important tasks and,
- have Privacy and Security by design to protect sensitive information.

El Saddik in [2] presented a reference framework for implementing a DT of a person. Although the framework is designed to implement a DT for health and wellbeing, it does check all the boxes for what a DT of a person should be; therefore, we can consider it a reference framework for the implementation of a DT of a person in general and not just in the context of health and wellbeing.

The DT ecosystem has three main layers of implementation, namely Data Source, AI-Inference Engine and Multimodal Interaction. The Data Source layer collects data directly from the Real Twin (the person). The data is sent to the AI-Inference Engine to extract Higher Level Information. Finally, the Multimodal Interaction layer will provide feedback to the Real Twin based on the High-Level Information supplied by the AI-Inference Engine.

We used the ecosystem presented in [2] and adapted it to design COVIDMe as a DT for health in the context of COVID-19 (Fig. 7.3). The following section will further elaborate on the use of Digital Twins for health.

7.3.2 Digital twin for health

A DT of a person may find purpose in many fields and domains of application; however, in this paper, we focus on health and wellbeing in the context of COVID-19. There is a growing trend in healthcare where therapies and treatments are tailored to the patient, considering intervariability among individuals; this relatively new trend is precision medicine, also referred to as personalized medicine. The premise of precision medicine is that everyone may respond differently to the same treatments, therapies, or medications. Therefore, treatments, therapies and drugs must be tailored to the physiology and circumstances of each patient. This level of customization implies the collection and analysis of massive amounts of data, genome sequencing being part of it. One approach to precision medicine is using common mutations in the genome of multiple individuals with the same disease or condition to assess the risk for new patients presenting the same disease or condition. We can reasonably assume that a large portion of the human population will require access to medical care at one point during their lifetimes. Apart from data management (collection, analysis, aggregation, privacy, storage, and recovery), precision medicine has a predictive aspect. The Digital Twin of a person ticks all the boxes and is almost ideal for supporting this new approach in medicine. It looks like the future of medicine has a digital component to it. In the following chapters, we present a brief overview of the topic of Digital Twin for Health [12–14]. This document will address a DT of a person intended to copy one or more aspects of a person's health state as DT for health. The Digital Twin for health is still in its infancy; regardless, we can find various examples of it in the scientific literature.

The bio-medical DT presented in [15] combines multiple sources of information, namely transcriptomic (RNA sequencing), cellular (related to metabolism, protein synthesis, replication and motility), organ information (related to organ function as part of systems) and "exposomic" (related to an individuals lifestyle, diet, exposure to toxicants, and more). The bio-medical twin is intended to be a general-purpose DT for health since it provides an overview of several aspects of the human body and its ailments. The bio-medical DT possesses two main features; it detects diseases by leveraging Graph Neural Networks. It can also forecast results in clinical scenarios by leveraging GAN Networks. In this instance, we observe that the proposed model is a "general-purpose" DT for health, meaning that its design is intended to create virtual copies of various organs and aspects of a person's health. We can also have DTs for health that replicates particular body functions, organs, or even particular clinical conditions, such as diseases or any other aspect of an individual's health. We can also have aggregates of DTs for health, then we can have a public health DT, observing the same things as a personal DT for health, but at a population scale [13].

Another use case for DT for health is presented in [16] with the DT for trauma. In case of an accident, a DT would collect all data from the patient and relay it to the trauma physicians in the ER at the hospital; the premise is that by the moment the patient arrives at the hospital, the treating physicians have an action plan already to help the patient. Another DT for health meant to help in hospital settings is HospiT'Win [17]. It supports clinical decisions by forecasting outcomes by applying Machine Learning to data collected from hospitalized patients.

The DT for health and wellbeing is still in its infancy. However, we can see the first "iterations" reported in the literature. With this work, we are adding our "two cents" by writing our implementation of it and identifying challenges to be addressed in this topic.

7.4 COVIDMe and the spread of COVID-19

Testing is of the utmost importance in the fight against the COVID-19; since the start of the crisis back in 2020, health authorities around the globe continue to implement aggressive testing campaigns to get a more accurate estimation of the number of infected people as part of their strategies to stop the spread of the virus [18–23]. After all, how can any government help people without knowing who is infected? There are two types of tests to detect current infection with SARS-CoV-2 (the virus causing the COVID-19 disease), the antigen test and the Nucleic Acid Amplification Tests (NAATs), the latter are considered by the Centers for Disease Control and Prevention (CDC) as the most sensitive tests [24–26]. The reverse transcription-polymerase chain reaction (RT-PCR) test is likely the most popular test among the NAATs family. Unfortunately, testing rates in developing countries such as Mexico are significantly lower than in developed countries; underlying reasons may be multifactorial. The costs associated with the RT-PCR tests are very likely to factor into the mix. The fact remains that as of November 2021, Mexico reported having performed under 90 thousand tests per 1 Million people, while countries such as Germany, Canada, the UK, USA and others reported close to 1 Million tests per 1 Million people and above; the country with the highest testing rate among the countries most affected is the UK with over 5 Million tests per 1 Million population [27]. As of November 2021, if every person in the UK got tested for COVID-19, each person would have been tested for COVID-19 close to 5 times on average, much less than once a month. The UK is among the best-performing countries in terms of testing. Yet, health authorities in that country would have a problem figuring out the exact number of infected people at any given time because not everyone is being tested regularly, so asymptomatic and people experiencing mild symptoms of COVID-19 would be very hard to detect. Instead, to assess community transmission of the disease, health authorities make their estimations in terms of Percent Positive (PP), the percentage of positive results from the total number of tests realized inside the health jurisdiction. The interpretation of PP may vary depending on the circumstances of the jurisdiction being monitored. A high PP in a jurisdiction with low testing rates may indicate that more testing is needed. In contrast, a high PP in a jurisdiction with a high testing rate may indicate high levels of community transmission [28]. Sadly enough, some developing countries present high PP rates, in some cases 20% and above, in combination with significantly lower testing rates when compared to developed countries [29].

As explained, a higher testing rate is preferred. However, figures show that we may not be able to test 100% of the population periodically. We make the argument that COVIDMe would be helpful to address the issue of continuous COVID-19 testing by capturing the health state of the Real Twins and screening individuals for COVID-19 indirectly in the process. In turn, the collective data coming from each COVIDMe instance may indirectly provide an insight into the progress of COVID-19 infections by leveraging the relationships between the Real Twins. The two main questions of this work remain. We discuss the implementation of COVIDMe to validate the DT for the health ecosystem presented in [2]. Secondly, given that COVIDMe is potentially a tool with the potential to detect COVID-19 infections automatically, we provide an insight into the effectiveness and accuracy of the implementation itself. The effectivity of COVIDMe hinges on its capacity to accurately assess the current state of health of the RT.

7.4.1 Automatic detection of COVID-19

The effectiveness of COVIDMe to assess the current health status of the RT relies mainly on the accuracy of the COVID-19 classification component, which has a machine learning classifier at its core designed explicitly for this purpose (detect COVID-19). Researchers around the globe have adopted several strategies to develop and train classifiers in the task of detecting COVID-19; next, we present some examples and group them by type of data used during training, namely, images, sound or signal-based, text and combinations of the previous.

A convolutional neural network (CNN) is a particular type of neural network initially introduced by LeCun et al. in [30]. Convolutional neural networks have proven to be exceptionally effective when used to tackle tasks involving image recognition; it's not a surprise that teams all over the globe turn to these networks to train classifiers to detect the presence of CVOVID in Computer-Aided Tomography (CAT) Scans, conventional X-ray Scans, Ultrasound Scans or any other form of medical imaging technique. Many works focus on using medical imaging, primarily on scans from the chest area from both individuals suspected of being infected with the virus and healthy people. The argument to justify the focus on chest scans is that the SARS-CoV-2 virus creates lesions in the lungs of affected persons (primarily in severe cases). According to the task for which the algorithms were trained, we can group the works into segmentation and direct classification (the classifier identifies COVID-19 lesions without properly delimiting these within the images) [31–36]. With the image approach, we observe the accuracies of 0.8 and above. In some cases, we identified a potential risk of overfitting a dataset, most likely due to small datasets being used for training and testing, which is not ideal for implementing deep learning or CNNs.

Although less popular, another approach to automatically detecting COVID-19 by applying machine learning is the use of sounds like coughing or abnormal breathing (e.g., whizzing sounds) captured with digital stethoscopes. In [37], Dash et al. present a classifier to identify possible viral infections by extracting cepstral features from coughing and breathing sounds. The classifier takes the cepstral features to train a Support Vector Machine (SVM) classifier and achieves an accuracy of 0.85. However, the classifier reported in this work was not trained to classify COVID-19 specifically. Instead, it is trained to detect flu-like viral infections.

In other instances, classifiers are trained with features extracted from cough sounds o detect the presence of COVID-19 [38,39].

Yet another approach is the use of data extracted directly from medical laboratory tests (i.e., blood tests). The results with this kind of data are promising, and it seems that the features obtained from the lab tests are well suited to be used with more conventional machine learning algorithms (not deep learning) [40–43].

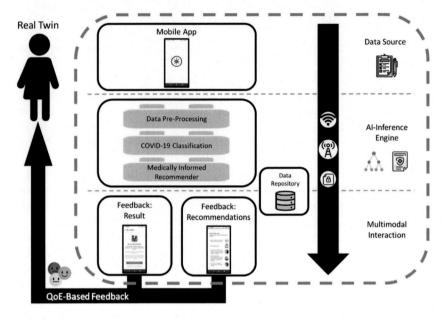

FIGURE 7.3   COVIDMe, a DT for health in the context of COVID-19.

We took a different path for the COVIDMe classification module and used self-reported symptoms as features to assess the health of the Real Twin. While we acknowledge that other types of data may lead to more accurate predictions, we took into consideration that COVIDMe requires frequent data collection from the Real Twin, which makes it inviable to use other types of data such as CAT Scans, X-ray scans or medical laboratory test results.

7.5 An overview of the COVIDMe software architecture

COVIDMe draws directly from the DT architecture introduced in [2]. However, we decided to streamline the overall ecosystem because COVIDMe is only concerned with copying a single aspect of the human body: the health status concerning COVID-19. Our platform also has three primary modules: "Data Source Module," "AI-Inference Engine Module," and "Multimodal Interaction Module."

Fig. 7.3 shows the final approach to COVIDMe. Much like any digital twin platform, we have an object from the real world, an individual that we will refer to as a Real Twin (RT) throughout the paper. A real twin is an individual of whom we are creating a digital clone (Digital Twin); our digital twin will create a digital copy of a small portion of the real twin, more precisely, a set of symptoms required for the classifier to predict a chance for the RT to be infected with SARS-CoV-2 (COVID-19). We used the classifier described in [44] to implement the "COVID-19 Classification" component in the AI-Inference Engine (Fig. 7.3). The classifier required an array of true/false values representing the presence of symptoms associated with COVID-19 and other symptoms which were not directly associated; the complete list is fever, cough, sudden onset of symptoms, known covid contact, odynophagia, rhinorrhea, irritability, chest pain, cephalea, chills, diarrhea, fatigue loss appetite, dyspnea, myalgia, arthralgia, conjunctivitis, abdominal pain, polypnea, vomit, anosmia, dysgeusia, cyanosis.

The mobile application collects the data by prompting the user to report the presence or absence of symptoms via an in-app survey. The data collection module transmits the data to the AI-Inference Engine through the network. The data is preprocessed to be fed into the classifier, producing a prediction. The prediction is required to provide feedback to the real twin. The feedback is presented in text and images in seconds after capturing the complete list of symptoms. Alternatively, the Multimodal Interaction Module can show the result of previous assessments thanks to COVIDMe using a data repository. It is essential to mention that COVIDMe it's not meant to replace a PCR test which is the gold standard to diagnose COVID-19; instead, among the objectives of this platform is to support the decision process for individuals and health jurisdictions when dealing with the spread of the COVID-19 by creating a DT of a person within the context of the COVID-19. The feedback produced for the individual is one of three types, namely low probability, inconclusive and high probability; COVIDMe will then prompt the user with the corresponding recommendations. In the next section, we present a kite level use case diagram supporting our implementation of the DT for health and elaborating on the COVIDMe architecture.

7.5.1 Use-case diagram

We identify two actors for COVIDMe, the Real Twin and the Digital Twin. Next, we explain the main flow for the five use cases shown in Fig. 7.4.

Start assessment

The use case starts when COVIDMe prompts the RT to answer the survey of symptoms associated with COVID-19. The RT will immediately proceed to answer the survey and tap on the button "Send" when finished. Once the RT has responded to the survey, COVIDMe (DT) will format the answers into a tuple of symptoms. At this point, the use case for "Pre-Process Data" will receive the tuple of symptoms to generate a feature array used in the "Screen for COVID-19" use case to create an assessment

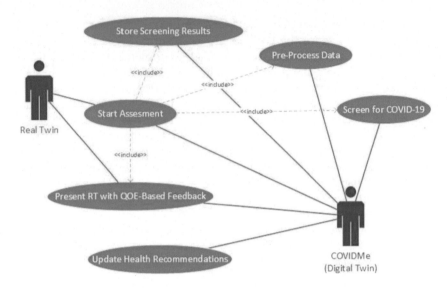

FIGURE 7.4   Kite level use case diagram for COVIDMe.

of the RT health status. The assessment is stored immediately in the "Store Screening" use case and used to generate appropriate feedback to the RT in the "Present RT with QOE-Based Feedback" in parallel.

Preprocess data

The use case starts after COVIDMe receives the tuple of symptoms collected from the survey and encodes it into an array of features. The use case ends with the creation of an array of features.

Screen for COVID-19

The use case starts when COVIDMe feeds the feature array into the COVID-19 classifier to calculate the RT's probability of being infected with SARS-CoV-2 (COVID-19). The result is equivalent to screening for COVID-19, and it can be expressed in one of three states: the RT is likely sick of COVID-19, the RT is not likely sick of COVID-19 and Undetermined. The use case ends with the screening being created.

Store screening results

The use case starts when COVIDMe obtains a screening for COVID-19, which is sent to a repository in the cloud for further analysis.

Present RT with QOE-based feedback

The use case starts with COVIDMe presenting the screening process results in the form of Quality of Experience-Based feedback. COVIDMe

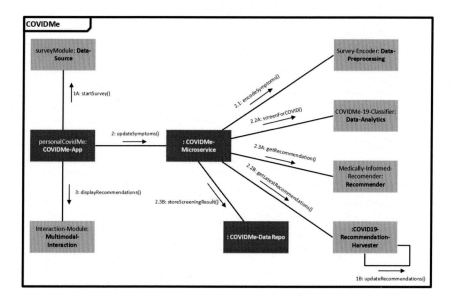

FIGURE 7.5   Low-granularity communication diagram for COVIDMe.

will select proper health guidelines (recommendations) to advise the RT to follow in accordance with the screening results.

Update health recommendations

The use case starts with COVIDMe harvesting guidelines/recommendations for the general population to follow regarding COVID-19. The use case ends with COVIDMe updating its internal repository with the most recent guidelines.

7.5.2 Communication diagram

A communication diagram is helpful to describe the dynamics between the different parts and components in a system. Fig. 7.5 shows nine lifelines (boxes), each representing an element of COVIDMe in a communication diagram. The whole platform is deployed in three different places. The RT smartphone runs the mobile app that holds the "Data Source" and "Multimodal Interaction"; represented in the diagram by the "COVIDMe-App" lifeline. The "AI-Inference Engine" module is deployed in a cloud computing service. We are not opinionated as to what service exactly. It can be any if it supports the deployment of Docker©containers; The lifeline corresponding to this bit of COVIDMe is "COVIDMe-Microservice" The data repository is also deployed in the cloud.

TABLE 7.1 Detailed message descriptions for the communication diagram.

Numeral	Message	Description
1A	startSurvey()	The COVIDMe-App will trigger the Data-Source module to start the survey to collect the symptoms from the RT.
1B	updateRecommendations()	The Recommendation-Harvester will look for the most up-to-date recommendations for self-managing COVID-19 in different situations; it will look at official sources.
2	updateSymptoms()	The COVIDMe-App will send the COVIDMe-Microservice a new tuple with the most recent symptoms as reported by the RT.
2.1	encodeSymptoms()	The COVIDMe-Microservice will preprocess the tuple of symptoms to transform it into an array of features.
2.2A	screenForCOVID()	The COVIDMe-Microservice will send the array of features to the COVIDMe-19-Classifier to classify the features and calculate the RT's probability of having COVID-19.
2.2B	getLatestRecommendations()	The COVIDMe-Microservice will recover the latest set of recommendations and guidelines to manage COVID-19 from the Recommendation-Harvester.
2.3A	getRecommendation()	Based on the prediction from 2.2A, the COVIDMe-Microservice will get the corresponding recommendation to display to the RT.
2.3B	storeScreeningResult()	The COVIDMe-Microservice will send the prediction from 2.2A to the COVIDMe-DataRepo for storage.
3	displayRecommendations()	The COVIDMe-App will send the corresponding recommendations from 2.3A to be displayed by the Multimodal-Interaction lifeline as a form of human-readable feedback for the RT.

Lifelines have interactions between them, and in most cases, these interactions happen in a specific order. The diagram shows the interactions in the form of messages. One way to read it is by following the sequence of each message. Table 7.1 describes the interactions triggered by each message.

Over the implementation and pilot-testing of COVIDMe, we identified some challenges that remain open. In the next section, we'll discuss the contributions and open challenges for a DT for health in the context of COVID-19.

7.6 Discussion and future work

A Digital Twin is a digital copy of an entity of the real world, be it a machine, an inanimate object or, in this case, a person. However, creating a digital copy of a person is no trivial task; a person is far too complex and has psychological, behavioral, and physiological components that are, at the very least, challenging to clone in its entirety. In this work, we focused on a small portion of a person by capturing an aspect of the Real Twin health, self-reported symptoms, which for the sake of the discussion in this section, we are dubbing "base-level information." From base information, the platform can infer "high-level information" about the person's health, in this case, an infection of COVID-19.

In this work, we outlined the software architecture for a DT for health in the context of COVID-19. However, the same approach may be used in different contexts. In our initial iterations, we realized that the accuracy of the platform to detect COVID-19 was highly dependent on the accuracy of the module in charge of predicting the health state of the RT (COVID-19 infection). Our approach to this was to make sure that the Data Preprocessing component of the platform produced an array of features as opposed to simply an array of symptoms or answers. The difference is subtle but necessary. At some point, the most accurate way to detect COVID-19 is to record coughing sounds. Then instead of sending a time series to the classifier component, the data-preprocessing component would extract the features, reducing the coupling between the classifier and the preprocessing data components. The Data-Analytics component (Classifier) may be retrained and improved without updating code related to the data-preprocessing. This is also beneficial for the DT for health in general as this approach may be used to implement a DT for health in the context of other diseases and conditions.

The range of health conditions that may be detected and "cloned" by the DT for health would depend mainly on the data that the DT can capture. The newest smartphones are equipped with a wide range of sensors, namely, accelerometer, digital compass, GPS and camera; if necessary, modern devices are also equipped with Bluetooth capability to connect to external sensors and self-reported data and medical data tests results. When virtually almost any kind of user data can be collected and recorded with a smartphone, it is safe to assume that there is a vast range of high-level information to be extracted from base-level information. Consequently, we may obtain a more accurate and complete "picture" of the health state of the Real Twin by extending the data source module in our DT for health (Fig. 7.3). To the best of our knowledge, this work would be among the first implementations of the DT for the health ecosystem presented in [2] for COVID-19.

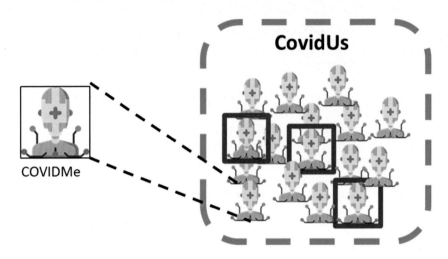

FIGURE 7.6   COVIDUs concept.

People connect and relate to others, and those connections influence a person's health state in many ways. In the case of COVID-19, physical contact with infected people can potentially lead to the person becoming infected with SARS-CoV-2 (COVID-19). The same applies to any number of transmissible diseases. We believe that aggregation of DTs for health is viable and necessary to understand the dynamics of an outbreak or a community scale and not just on an individual level. This kind of knowledge could lead to better control of the spread of diseases such as the one cause by SARS-CoV-2 (COVID-19). We have taken the first steps in that direction by introducing a data repository into our architecture; in this way, we can have several instances of COVIDMe push their data into a single repository, which would allow for an implementation of a "COVDUs" (Fig. 7.6) platform. However, several challenges need to be addressed before such a platform can be implemented. There are still privacy and authentication concerns we need to solve to prevent any data leak from the platform, given the sensitive nature of the data. We followed an agile approach to this implementation; in the next iteration, we plan on adding support for different types of data and classifiers; towards this objective, we know the design needs the use of interfaces that ensure an optimal coupling and cohesion. The question remains What data do we need to support? And, what would those specific interfaces look like to better support the use of multiple data sources?

One of the limitations of this study is that COVIDMe is at a proof-of-concept stage and has been pilot-tested only with a handful of users. We are now focused on using self-reported symptoms to detect COVID-19 mainly because of how easy it is to have participants send data without the need to leave their homes, thus keeping participants safe. However, we need to explore more data types in the future and find which approach for a classifier is best.

COVIDMe needs the use of standards; the self-reported symptoms, for instance, are not organized in any particular way. That means that the developer needs to be sure in what order the classifier is expecting that input. By standardizing the data in the platform, we may ensure compatibility between different components even if these are updated or swapped for other implementations, for instance, a better classifier. Also, the results from the classifier have no particular format, so before elevating that data to a COVIDUs platform, we might need to implement the use of standardized data.

If COVIDUs is implemented, we need to allocate a great deal of time and resources to design and develop an Ai-Inference Engine for it, a population-level AI-Inference Engine.

The Mobile App format the answers for the survey before sending them to the Microservice. That bit may be different depending on the type of data to be used, which implies using an extra layer in the platform to format and possibly encode the data collected before sending it to a microservice.

To address the security and privacy concerns, we are thinking that we need to address this by authenticating the users. In [45], the topic of authenticating an individual with biometrics, specifically with ECG signals, is broadly studied, and biometric authentication can be one way to authenticate the RT; this approach needs to be studied for COVIDMe.

7.7 Conclusions

We successfully implemented a DT for health in the context of COVID-19 using the reference model presented in [2]. The platform (COVIDMe) is designed so that we can replace or improve different components with relative ease and minimum effort. We made an argument regarding the accuracy of the AI-Inference Engine by explaining that accuracy depends mainly on the classifier's performance and created a design that supports different classifiers. We also outlined an architecture for a DT for health in the context of COVID-19 on which we can build and improve.

References

[1] A. El Saddik, Digital twins: the convergence of multimedia technologies, IEEE Multimed. 25 (2) (Apr. 2018) 87–92, https://doi.org/10.1109/MMUL.2018.023121167.

[2] A. El Saddik, et al., Dtwins: a digital twins ecosystem for health and well-being, IEEE COMSOC MMTC Commun. Front. 14 (2019) 39–46.

[3] J. Yanase, E. Triantaphyllou, A systematic survey of computer-aided diagnosis in medicine: past and present developments, Expert Syst. Appl. 138 (Dec. 2019) 112821, https://doi.org/10.1016/J.ESWA.2019.112821.

[4] S.G. Vandenberg, Medical diagnosis by computer: recent attempts and outlook for the future, Behav. Sci. 5 (2) (Apr. 1960) 170, [Online]. Available https://www.proquest.com/scholarly-journals/medical-diagnosis-computer-recent-attempts/docview/1301267195/se-2.

[5] R.S. Ledley, Digital electronic computers in biomedical science, Science 130 (3384) (Dec. 1959) 1225–1234, [Online]. Available http://www.jstor.org/stable/1757193.

[6] H. Weinrauch, A.W. Hetherington, Computers in medicine and biology, J. Am. Med. Assoc. 169 (3) (Jan. 1959) 240–245, https://doi.org/10.1001/jama.1959.03000200038008.

[7] W. Sun, B. Zheng, W. Qian, Computer aided lung cancer diagnosis with deep learning algorithms, Proc. SPIE 9785 (Mar. 2016), https://doi.org/10.1117/12.2216307.

[8] D.H. Gelernter, Mirror worlds, or, The day software puts the universe in a shoebox... : how it will happen and what it will mean, Oxford University Press, New York, 1991.

[9] M.W. Grieves, Virtually intelligent product systems: digital and physical twins, in: Complex Systems Engineering: Theory and Practice, American Institute of Aeronautics and Astronautics, Inc., Reston, VA, 2019, pp. 175–200.

[10] M. Grieves, J. Vickers, Digital twin: mitigating unpredictable, undesirable emergent behavior in complex systems, in: Transdiscipl. Perspect. Complex Syst. New Find. Approaches, Jan. 2017, pp. 85–113.

[11] K.M. Alam, A. El Saddik, C2PS: a digital twin architecture reference model for the cloud-based cyber-physical systems, IEEE Access 5 (2017) 2050–2062, https://doi.org/10.1109/ACCESS.2017.2657006.

[12] G. Coorey, G.A. Figtree, D.F. Fletcher, J. Redfern, The health digital twin: advancing precision cardiovascular medicine, Nat. Rev. Cardiol. 18 (12) (Oct. 2021) 803–804, https://doi.org/10.1038/S41569-021-00630-4.

[13] M.N. Kamel Boulos, P. Zhang, Digital twins: from personalised medicine to precision public health, J. Pers. Med. 11 (8) (Jul. 2021) 745, https://doi.org/10.3390/JPM11080745.

[14] J. Corral-Acero, et al., The 'Digital Twin' to enable the vision of precision cardiology, Eur. Heart J. 41 (48) (Dec. 2020) 4556–4564, https://doi.org/10.1093/EURHEARTJ/EHAA159.

[15] P. Barbiero, R. Viñas Torné, P. Lió, Graph representation forecasting of patient's medical conditions: toward a digital twin, Front. Genet. 12 (Sep. 2021) 1289, https://doi.org/10.3389/FGENE.2021.652907/BIBTEX.

[16] A. Croatti, M. Gabellini, S. Montagna, A. Ricci, On the integration of agents and digital twins in healthcare, J. Med. Syst. 44 (9) (Sep. 2020) 1–8, https://doi.org/10.1007/S10916-020-01623-5/FIGURES/2.

[17] A. Karakra, F. Fontanili, E. Lamine, J. Lamothe, HospiT'Win: a predictive simulation-based digital twin for patients pathways in hospital, in: 2019 IEEE EMBS Int. Conf. Biomed. Heal. Informatics, BHI 2019 – Proc., May 2019.

[18] R. Ait Addi, A. Benksim, M. Amine, M. Cherkaoui, Asymptomatic COVID-19 infection management: the key to stop COVID-19, J. Clin. Exp. Investig. 11 (3) (Mar. 2020) em00737, https://doi.org/10.5799/jcei/7866.

[19] M. Salath, et al., COVID-19 epidemic in Switzerland: on the importance of testing, contact tracing and isolation, Swiss Med. Wkly. (Mar. 2020), https://doi.org/10.4414/smw.2020.20225.

[20] C.R. Wells, et al., Optimal COVID-19 quarantine and testing strategies, Nat. Commun. 12 (1) (Dec. 2021) 356, https://doi.org/10.1038/s41467-020-20742-8.

[21] B. Peng, et al., Reducing COVID-19 quarantine with Sars-CoV-2 testing: a simulation study, BMJ Open 11 (7) (Jul. 2021) e050473, https://doi.org/10.1136/bmjopen-2021-050473.

[22] A. Robert, Lessons from New Zealand's COVID-19 outbreak response, Lancet Public Heal. 5 (11) (Nov. 2020) e569–e570, https://doi.org/10.1016/S2468-2667(20)30237-1.

[23] J. Summers, et al., Potential lessons from the Taiwan and New Zealand health responses to the COVID-19 pandemic, Lancet Reg. Heal. – West. Pacific 4 (Nov. 2020) 100044, https://doi.org/10.1016/j.lanwpc.2020.100044.

[24] Nucleic Acid Amplification Tests (NAATs), CDC, https://www.cdc.gov/coronavirus/2019-ncov/lab/naats.html. (Accessed 5 December 2021).

[25] Interim Guidance for Antigen Testing for Sars-CoV-2, CDC, https://www.cdc.gov/coronavirus/2019-ncov/lab/resources/antigen-tests-guidelines.html. (Accessed 5 December 2021).

[26] Test for Current Infection, CDC, https://www.cdc.gov/coronavirus/2019-ncov/testing/diagnostic-testing.html. (Accessed 5 December 2021).

[27] COVID-19 testing rate by country, Statista, https://www-statista-com.proxy.bib.uottawa.ca/statistics/1104645/covid19-testing-rate-select-countries-worldwide/. (Accessed 5 December 2021).

[28] D. Dowdy, G. D'souza, COVID-19 testing: understanding the 'percent positive' – COVID-19 – Johns Hopkins bloomberg school of public health, https://www.jhsph.edu/covid-19/articles/covid-19-testing-understanding-the-percent-positive.html, 2020. (Accessed 30 April 2021).

[29] How does testing in the U.S. compare to other countries? – Johns Hopkins coronavirus resource center, testing hub – data visualizations, https://coronavirus.jhu.edu/testing/international-comparison. (Accessed 30 November 2021).

[30] Y. LeCun, et al., Backpropagation applied to handwritten zip code recognition, Neural Comput. 1 (4) (Dec. 1989) 541–551, https://doi.org/10.1162/neco.1989.1.4.541.

[31] D.P. Fan, et al., Inf-net: automatic COVID-19 lung infection segmentation from CT images, IEEE Trans. Med. Imaging 39 (8) (Aug. 2020) 2626–2637, https://doi.org/10.1109/TMI.2020.2996645.

[32] X. Wang, et al., A weakly-supervised framework for COVID-19 classification and lesion localization from chest CT, IEEE Trans. Med. Imaging 39 (8) (Aug. 2020) 2615–2625, https://doi.org/10.1109/TMI.2020.2995965.

[33] E.F. Ohata, et al., Automatic detection of COVID-19 infection using chest x-ray images through transfer learning, IEEE/CAA J. Autom. Sin. 8 (1) (Jan. 2021) 239–248, https://doi.org/10.1109/JAS.2020.1003393.

[34] T. Ozturk, M. Talo, E.A. Yildirim, U.B. Baloglu, O. Yildirim, U. Rajendra Acharya, Automated detection of COVID-19 cases using deep neural networks with x-ray images, Comput. Biol. Med. 121 (Jun. 2020) 103792, https://doi.org/10.1016/J.COMPBIOMED.2020.103792.

[35] M.J. Horry, et al., COVID-19 detection through transfer learning using multimodal imaging data, IEEE Access 8 (2020) 149808–149824, https://doi.org/10.1109/ACCESS.2020.3016780.

[36] S. Hu, et al., Weakly supervised deep learning for COVID-19 infection detection and classification from CT images, IEEE Access 8 (2020) 118869–118883, https://doi.org/10.1109/ACCESS.2020.3005510.

[37] T.K. Dash, S. Mishra, G. Panda, S.C. Satapathy, Detection of COVID-19 from speech signal using bio-inspired based cepstral features, Pattern Recognit. 117 (Sep. 2021) 107999, https://doi.org/10.1016/j.patcog.2021.107999.

[38] A. Tena, F. Clarià, F. Solsona, Automated detection of COVID-19 cough, Biomed. Signal Process. Control 71 (Jan. 2022) 103175, https://doi.org/10.1016/j.bspc.2021.103175.

[39] E.A. Mohammed, M. Keyhani, A. Sanati-Nezhad, S.H. Hejazi, B.H. Far, An ensemble learning approach to digital corona virus preliminary screening from cough sounds, Sci. Rep. 11 (1) (Jul. 2021) 1–11, https://doi.org/10.1038/s41598-021-95042-2.

[40] T. Xia, Y.Q. Fu, N. Jin, P. Chazot, P. Angelov, R. Jiang, AI-enabled microscopic blood analysis for microfluidic COVID-19 hematology, in: Proc. – 2020 5th Int. Conf. Comput. Intell. Appl. ICCIA 2020, Jun. 2020, pp. 98–102.

[41] V.A. de Freitas Barbosa, et al., Heg.IA: an intelligent system to support diagnosis of COVID-19 based on blood tests, Res. Biomed. Eng. (Jan. 2021) 1–18, https://doi.org/10.1007/S42600-020-00112-5.

[42] D. Brinati, A. Campagner, D. Ferrari, M. Locatelli, G. Banfi, F. Cabitza, Detection of COVID-19 infection from routine blood exams with machine learning: a feasibility study, J. Med. Syst. 44 (8) (Jul. 2020) 1–12, https://doi.org/10.1007/S10916-020-01597-4.

[43] M. AlJame, I. Ahmad, A. Imtiaz, A. Mohammed, Ensemble learning model for diagnosing COVID-19 from routine blood tests, Inform. Med. Unlocked 21 (Jan. 2020) 100449, https://doi.org/10.1016/J.IMU.2020.100449.

[44] R. Martinez-Velazquez, Diana P. Tobón V., A. Sanchez, A. El Saddik, E. Petriu, A machine learning approach as an aid for early COVID-19 detection, Sensors 21 (12) (2021), https://doi.org/10.3390/s21124202.

[45] J.S. Arteaga Falconi, A. El Saddik, Security with ECG Biometrics BT, in: N. Derbel, O. Kanoun (Eds.), Advanced Methods for Human Biometrics Cham, Springer International Publishing, 2021, pp. 65–79.

CHAPTER

8

Improving human living environment and human health through environmental digital twins technology

Zhihan Lv[a] and Dongliang Chen[b]

[a]Faculty of Arts, Uppsala University, Visby, Sweden, [b]College of Computer Science and Technology, Qingdao University, Qingdao, China

8.1 Introduction

In this era of heightened environmental awareness and tight municipal budgets, water supply and wastewater collection and treatment have become a critical part of environmental governance [1]. Digital Twins (DTs) infrastructure plays a crucial role in helping public utility corporations optimize operations and capital expenditures, reduce infrastructure inefficiencies, enhance resilience, and plan a better future [2]. DTs infrastructure has the potential to update wastewater treatment facilities by reducing overall costs, which facilitates maintaining and improving their wet infrastructure, to improve human living environment [3]. Central airconditioning is widely used in shopping malls, office buildings, hospitals, factories, and other environments, which consumes significant energy, and even emits harmful substances polluting the environment. Reducing the energy consumption of central air-conditioning is beneficial to energy conservation and emission reduction to build a low-carbon green life, thus ameliorating the subhealth state of the human body [4].

The adoption of DTs enables residents to have more control over their work area and environmental conditions, thus refining the human living experience [5]. The most important element of the living environment is water, which has a direct impact on human health. As a core content of smart city construction, smart water management systems are one of the important symbols reflecting the intelligent level of urban management [6]. Moreover, sewage treatment is an important link in the water circulation system. Thus, the high efficiency of underground pipeline monitoring and maintenance is the booster of smart city construction [7]. As a ubiquitous energy consumption system in the human living environment, the central air-conditioning is at a low load rate most of the time [8]. Therefore, a fine operation and maintenance management of the central air-conditioning is essential to a comfortable human living environment. Chen, et al. (2020) [9] conducted an evacuation design simulation based on the deep neural network model. Various faults inevitably occur in the operation of central air-conditioning, such as performance attenuation of chillers, stuck valves or frequent oscillations, and sensor drift [10–12]. The DTs space of the central air-conditioning can update the model in time through real-time receiving data from the physical objects, to achieve consistency with the running state of the physical objects [13–15]. The DTs space can realize transparency of the air-conditioning in the process of running, and assist system optimization design, system optimization, and fault diagnosis [16–18].

Here, the application of DTs to human living environment improvement is investigated through research literature and establishing models. The innovations of the present work lie in the parameter identification method of equipment models of the central air-conditioning water system based on a genetic algorithm. Besides, the interval estimation of the air-conditioning model is implemented through the K-means clustering algorithm, and the error compensation is completed by using an artificial neural network (ANN), which achieves an excellent effect. The method reported here can effectively strengthen the uncertainty parameter estimation of the central air-conditioning water system, reduce air-conditioning energy consumption, and improve the human living environment.

8.2 Parameter identification and uncertainty estimation of the DTs model for central air-conditioning

8.2.1 Construction of the DTs sewage treatment platform

In smart city management, the water supply network and drainage network occupy a core position, and an unimpeded sewage transport system is crucial to the overall operation of a city [19]. DTs technology can be applied to related links of sewage treatment, such as sewage discharge, collection, transportation, and disposal. Meanwhile, combined with other advanced technologies, such as cloud computing, Internet of Things, big data, intelligent, and mobile Internet technology, it can comprehensively collect data of citywide industrial pollution sources, sewage pipe networks, sewage pumping stations, sewage plants, water supply, self-provided water, river water level, water points, and precipitation monitoring [20]. Moreover, based on the idea of regional water balance, the whole city is divided into grids, the primary subdivision comprising a sewage pumping station and its sewage collection area and the secondary subdivision covering each lot are established. The intelligent sewage management platform can solve major problems in the operation and maintenance management of sewage pipe networks, such as inconsistent standards, multiple management, scattered data, insufficient decision support ability, and slow response and disposal of emergency events. This platform dramatically improves operation efficiency, reduces operation cost, and realizes integrated operation and maintenance [21].

Here, the dynamic DTs of the underground sewage network are established to visualize the sewage network. Moreover, Building Information Modeling (BIM) technology is employed to build the DTs base of the smart city by integrating the model data of underground, surface, and above ground [22]. The platform is connected with the dynamic data and business data of sewage intelligent sensing equipment to establish a DTs city for sewage pipelines. Pipeline maintenance personnel with a handheld mobile device can see three-dimensional images of the pump station and pipelines through the inspection application, and the virtual scene can be consistent with the scale and orientation of real objects. In addition, the Mixed Reality (MR) [23] underground pipeline management platform is built to realize the virtuality and reality combination of the underground pipeline BIM model and the inspection and maintenance site. Operators can view the spatial distribution status and information of underground pipes through the mobile terminal without excavating the road surface, realizing intelligent and intensive operation management of sewage facilities and equipment. Fig. 8.1 reveals the application of DTs to sewage treatment infrastructure.

8.2.2 Parameter identification of the equipment model of central air-conditioning water system based on genetic algorithm (GA)

The central air-conditioning water system is the most important component of central air-conditioning. Besides, effectively and accurately identifying the equipment model parameters, such as chillers, pumps, and cooling towers of the central air-conditioning water system, is the key to ensuring the accuracy of DTs of central air-conditioning [24]. Therefore, the central air-conditioning of a factory is taken as the research object of the present work to optimize each module. Fig. 8.2 displays the structure of the central air-conditioning water system of the factory.

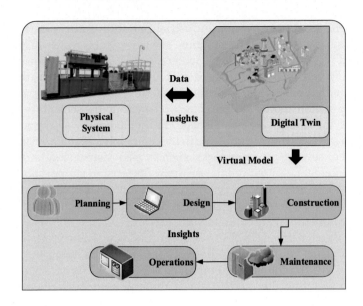

FIGURE 8.1   Application of DTs to sewage treatment infrastructure.

This factory has nine air-conditioning plant rooms, with a total of 51 combined processing units. The control module of the air-conditioning is used to monitor the process parameters of temperature and humidity in each workshop as well as the air quality and the operation parameters of the air-conditioning equipment. The Department of Energy (DOE-2) model [25] is used as the chiller model here, which is a typical model used to simulate air-conditioning and is applicable to all types of units. The DOE-2 model adopts temperature-cooling capacity curve, Temperature-Effective Interest Rate (EIR) curve, and partial load rate-EIR curve to simulate the operation characteristics of chiller under variable working conditions.

The temperature-cooling capacity relationship curve is used to calculate the chiller's cooling capacity correction coefficient CapFunT [26]. It also determines the functional relationship between the chiller's maximum available cooling capacity, chilled water outlet temperature, and cooling water inlet temperature, as shown in Eq. (8.1).

$$CapFunT = g_{11} + g_{12} * T_{chw,out} + g_{13} * T^2_{chw,out} + g_{14} * T_{chw,in}$$
$$+ g_{15} * T^2_{chw,in} + g_{16} * T_{chw,in} * T_{chw,out} \tag{8.1}$$

In Eq. (8.1), $T_{chw,out}$ represents the chilled water outlet temperature, $T_{chw,in}$ signifies the cooling water inlet temperature, and g_{11}, g_{12}, g_{13}, g_{14}, g_{15}, and g_{16} denote fitting coefficients. Then, the maximum available re-

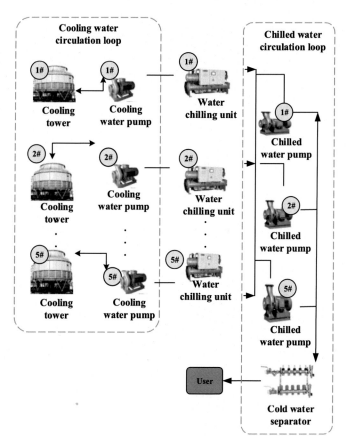

FIGURE 8.2 Structure and composition of the central air-conditioning water system in a factory.

frigerating capacity CAP_{ava} of the chiller can be expressed as Eq. (8.2).

$$CAP_{ava} = CAP_{rated} * CapFunT \qquad (8.2)$$

In Eq. (8.2), CAP_{rated} refers to the rated cooling capacity of the chiller. The temperature-EIR relation curve is used to calculate the chiller EIR correction coefficient EIRFunT and determine the functional relationship between the chiller EIR and chilled water outlet temperature and cooling water inlet temperature, as presented in Eq. (8.3).

$$EIRFunT = g_{21} + g_{22} * T_{chw,out} + g_{23} * T^2_{chw,out} + g_{24} * T_{chw,in}$$
$$+ g_{25} * T^2_{chw,in} + g_{26} * T_{chw,in} * T_{chw,out} \qquad (8.3)$$

In Eq. (8.3), g_{21}, g_{22}, g_{23}, g_{24}, g_{25}, and g_{26} also represent the fitting coefficient. The partial load rate-EIR relationship curve is used to calculate the EIR correction coefficient EIRFunPLR of the chiller and determine the functional relationship between the chiller EIR and the change of the partial load rate, as shown in Eq. (8.4).

$$EIRFunPLR = g_{31} + g_{32} * PLR + g_{33} * PLR^2 \qquad (8.4)$$

In Eq. (8.4), PLR stands for the partial load rate, that is, the ratio between the actual cooling capacity of the chiller and the maximum available cooling capacity. Besides, g_{31}, g_{32}, and g_{33} also represent the fitting coefficient. Then, the power consumption P of the chiller can be written as:

$$P = \frac{CAP_{ava} * CapFunT * EIRFunPLR}{COP_{rated}} \qquad (8.5)$$

where COP_{rated} signifies the rated Coefficient of Performance (COP) of the chiller. The actual current load of the chiller is calculated according to Eq. (8.6).

$$Q_{load} = C_{p,chw} * m_{chw} * (T_{chw,in} - T_{chw,set}) \qquad (8.6)$$

In Eq. (8.6), Q_{load} represents the refrigerating capacity of the chiller at a set chilled water outlet temperature, $C_{p,chw}$ refers to the specific heat capacity of chilled water fluid, and m_{chw} indicates the flow rate of chilled water.

When the chiller's cooling capacity is larger than the maximum available cooling capacity that the chiller can provide at the chilled water outlet temperature setting value, the maximum cooling capacity Q_{met} that the chiller can output is the actual maximum available cooling capacity.

$$Q_{met} = \min(CAP_{ava}, Q_{load}) \qquad (8.7)$$

However, when the output of the cooler cannot meet the system load demand, the outlet temperature of the cooler cannot reach the set value of outlet temperature. At this time, the COP and outlet temperature of the chiller need to be corrected. COP after correction can be presented as:

$$COP = \frac{Q_{met}}{P} \qquad (8.8)$$

All the heat Q_{reject} absorbed by cooling water is:

$$Q_{reject} = Q_{met} + P \qquad (8.9)$$

The actual outlet temperature $T_{chw,out}$ of chilled water in the chiller is:

$$T_{chw,out} = T_{chw,in} - \frac{Q_{met}}{m_{chw} \times C_{p.chw}} \qquad (8.10)$$

The actual outlet temperature $T_{cw,out}$ of cooling water of chiller is:

$$T_{cw,out} = T_{cw,in} + \frac{Q_{reject}}{m_{cw} \times C_{p.chw}} \tag{8.11}$$

where m_{cw} represents the flow rate of cooling water, and $C_{p.chw}$ denotes the specific heat capacity of cooling water fluid.

Water pump is the power equipment of the central air-conditioning water system, so it is as important as water chillers. Water pumps are divided into chilled water pump and cooling water pump according to the location of the loop. Reasonable frequency conversion adjustment of water pumps can effectively save energy. The total pump efficiency is composed of three parts, namely frequency converter efficiency η_{VFD}, motor efficiency η_m, and pump efficiency η_{pu}. After frequency conversion, the ratio of pump motor speed to rated motor speed is a speed ratio f, as shown in Eq. (8.12).

$$f = \frac{n}{n_0} \tag{8.12}$$

Eq. (8.13) indicates the relationship between pump head and pump flow rate as well as speed ratio; Eq. (8.14) describes the relationship between pump efficiency and pump flow rate as well as speed ratio.

$$H_{pu} = h_{01} * m_w^2 + h_{02} * f * m_w + h_{03} * f^2 \tag{8.13}$$

$$\eta_{pu} = h_{11} * \frac{1}{f^2} * m_w^2 + h_{12} * \frac{1}{f} * m_w + h_{13} \tag{8.14}$$

In Eq. (8.13), H_{pu} represents pump head, m_w refers to pump flow, and $h_{01}, h_{02}, h_{03}, h_{11}, h_{12}$, and h_{13} are fitting coefficients. The motor efficiency η_m and frequency converter efficiency η_{VFD} are calculated according to:

$$\eta_m = h_{21} * (1 - e^{h_{22}*f}) \tag{8.15}$$

$$\eta_{VFD} = h_{31} + h_{32} * f + h_{33} * f^2 + h_{34} * f^3 \tag{8.16}$$

where $h_{21}, h_{22}, h_{23}, h_{32}, h_{33}$, and h_{34} are fitting coefficients.

Shaft power of pump can be expressed as Eq. (8.17).

$$P_{pump,shaft} = \frac{m_w * H_{pu} * g * SG}{1000 * \eta_{pu}} \tag{8.17}$$

In Eq. (8.17), $P_{pump,shaft}$ signifies the shaft power of pump, $SG = 1$ represents the specific gravity of the conveying fluid, and $g = 9.8 \text{ m /s}^2$ indicates the gravitational acceleration. Then, the total consumption power of the frequency conversion pump is:

$$P_{pump,in} = \frac{m_w * H_{pu} * SG}{102 * \eta_{VFD} * \eta_m * \eta_{pu}} \tag{8.18}$$

Because both water chilling unit and pump models have uncertain parameters, it is essential to identify them according to the actual situation. GA is widely used in parameter identification due to its parallel computation, strong global search ability, and strong expansibility. Because the equipment parameters of central air-conditioning water system have no physical significance, it is difficult to determine the solution space, so the simple GA is easy to fall into local optimum.

Model parameter identification aims to find a group of optimal solutions in a given parameter solution space. Therefore, the initial solution space selection determines the final computational efficiency and accuracy of the algorithm. Equipment parameter identification of the central air-conditioning water system can be divided into three cases: with equipment performance coefficient provided by the manufacturer, without equipment performance coefficient but with other similar coefficients, and without coefficients. Therefore, a device model parameter identification framework is constructed by combining multistrategy initial solution space optimization (MISSO) method and GA, that is, the MISSO-GA model parameter identification framework, as shown in Fig. 8.3.

According to the above three parameter identification cases, the initial solution space optimization steps are given respectively. For the case with equipment performance coefficient provided by the manufacturer, the upper bound of the initial solution space of a parameter is the product of manufacturer's given value and $(1 + a)$, and the lower bound is the product of the manufacturer's given value and $(1 - a)$, where a represents the scaling coefficient. In the second case without equipment performance coefficient but with other similar coefficients, similarity between the same type of equipment and the equipment waiting to be identified is firstly compared, and some similar equipment is screened for standby. Then, the maximum value of performance curve coefficients of all reference equipment is taken as the upper bound, and the minimum value is the lower bound. In the third case without coefficients, upper and lower bounds of parameter solution space are determined according to literature or expert experience.

8.2.3 Prediction interval estimation of the central air-conditioning model based on the K-means clustering algorithm

The accuracy of model prediction directly affects the prediction risk of engineering personnel's quantitative model in practical application. Forecasting the interval prediction relative point can quantitatively give the prediction interval under different confidence levels and obtain more comprehensive prediction information. Here, the prediction interval estimation method based on K-means clustering [27] is proposed for the equipment model of central air-conditioning water system. Fig. 8.4 shows the

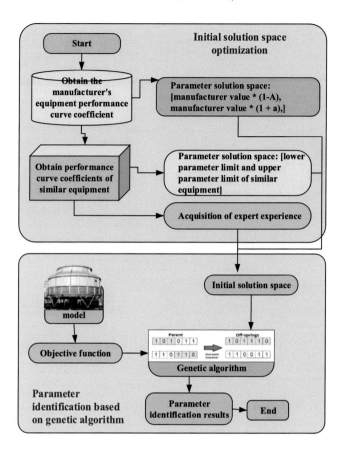

FIGURE 8.3   MISSO-GA model parameter identification framework.

flow of the prediction interval estimation method, which predicts model parameters through residual clustering and prediction interval estimation.

Model residual is the difference between the measured value and the model predicted value. Because the model predicted value depends on the model input value, model residual and model input are strongly correlated. Different model residual distribution can be obtained by changing the model input combination.

$$e = y - \overrightarrow{y} \qquad (8.19)$$

In Eq. (8.19), e signifies model residual, y represents actual value, and $\overrightarrow{y}$ refers to model predicted value. In addition, different input values have different impacts on model residuals, so the impact of input values on model residuals should be determined before clustering. Random Forest (RF) algorithm [28] can use multiple decision trees to achieve classification

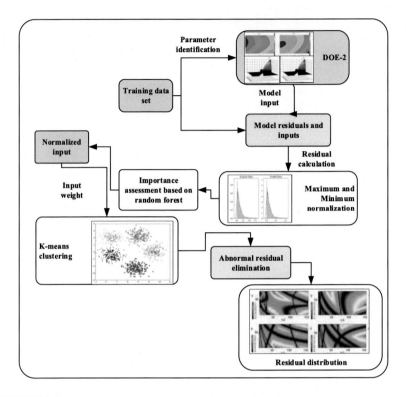

FIGURE 8.4   Procedures of model residual clustering based on K-means clustering.

or regression prediction, so it is widely used to analyze the correlation degree between variables with nonlinear relationships. The influence of an input value on the output can be quantified by calculating the mean of the change in impurity of the input value in the corresponding nodes of the entire decision tree. Then, the quantified value is normalized to get the importance of each input value. The higher the value, the greater the impact.

$$I_j = \frac{I_j}{\sum I_i} \tag{8.20}$$

In Eq. (8.20), I_j refers to the importance of the j-th input value, and $\sum I_i$ represents the sum of the importance of all input values. Here, the mean square error is taken as the index of impurity, the device model input is taken as the input of the RF algorithm, and the device model residual is taken as the output of RF. The normalized model input data is weighted, and then the K-means clustering algorithm is used for clustering. The algorithm process is shown in Fig. 8.5.

1	**Start**
2	Select the initialized K samples as the initial clustering center: $a = a_1, a_2, ..., a_k$
3	Obtain n m-dimensional data.
4	Randomly generate K m-dimensional points.
5	**While (t)**
6	**for (int i=0; i < n; i++)**
7	**for (int j=0; j < k; j++)**
8	Calculate the distance from point i to class j.
9	And it is divided into the class corresponding to the cluster center with the smallest distance
10	**for (int i=0; i < k; i++)**
11	Find all data points belonging to their own category.
12	Recalculate cluster centers:
13	$a_j = \frac{1}{|c_i|} \sum_{x \in c_i} x$
14	Modify your own coordinates to the center coordinates of these data points
15	**End**

FIGURE 8.5   Flow of the K-means clustering algorithm.

The threshold of kernel density estimation and abnormal residuals are calculated according to:

$$f(x) = \frac{1}{nh} \sum_{i=1}^{n} K \left(\frac{x - x_i}{h} \right) \tag{8.21}$$

$$\delta = \frac{f_{max}}{s} \tag{8.22}$$

where $x_1, x_2 ..., x_n$ represent the sample point of the same variable, K refers to the kernel function, and h represents the bandwidth. Besides, δ represents the outlier density threshold, f_{max} signifies the maximum kernel density function, and s stands for the scaling factor.

Prediction interval estimation is to estimate an interval, so that the probability of the actual value falling within this interval is approximately equal to a given Prediction Interval Nominal Confidence (PINC). Firstly, maximum minimum normalization and weighting are performed on the model input in an unknown sample point. Then, the Euclidean distance between the sample point and the cluster center is calculated. Finally, the upper and lower bounds of the prediction interval on the sample point are calculated by the device model.

$$U = \begin{cases} \vec{y} + q_{100(\beta+0.5(1-\alpha))\%} & \beta \geq 0.5(1-\alpha) \text{ and } 1-\beta \geq 0.5(1-\alpha) \\ \vec{y} + q_{100\%} & 1-\beta < 0.5(1-\alpha) \\ \vec{y} + q_{100(1-\alpha)\%} & \beta < 0.5(1-\alpha) \end{cases}$$

(8.23)

$$L = \begin{cases} \vec{y} + q_{100(\beta-0.5(1-\alpha))\%} & \beta \geq 0.5(1-\alpha) \text{ and } 1-\beta \geq 0.5(1-\alpha) \\ \vec{y} + q_{100\alpha\%} & 1-\beta < 0.5(1-\alpha) \\ \vec{y} + q_{0\%} & \beta < 0.5(1-\alpha) \end{cases}$$

(8.24)

In Eqs. (8.23) and (8.24), U represents the upper bound of the prediction interval, $\vec{y}$ denotes the lower bound of the prediction interval, and L signifies the predicted value of the device model under a certain sample point. Moreover, q refers to the residual percentile of the model residual cluster closest to the sample point, $100(1-\alpha)\%$ means the confidence level, and β stands for the proportion that model residual errors are less than 0 in the residual cluster.

8.2.4 Error compensation for the equipment model of central air-conditioning water system based on ANN

It is known above that if the prediction interval of the equipment model is excessively large, the reliability of the equipment model will decrease. Therefore, a optimization method is proposed for the uncertainty of model prediction based on ANN, which is primarily used to compensate the predicted value of freezing outlet temperature. The nonlinear fitting function of ANN is utilized to model the relationship between the model residual and the inputs to modify the model prediction results and reduce the uncertainty of the model prediction results. Fig. 8.6 illustrates the model residual compensation method based on ANN.

This method consists of two steps, namely residual compensation neural network training and residual compensation of the equipment model. The number of neuron nodes in the hidden layer is given by the following Kolmogorov theorem equation:

$$n = 2t + 1$$

(8.25)

where n and t represent the number of neuron nodes in the hidden layer and the input layer, respectively. Eq. (8.26) is adopted to compensate the predicted value of the model.

$$\bar{\bar{y}} = \vec{y} + \vec{e}$$

(8.26)

In Eq. (8.26), $\bar{\bar{y}}$ refers to the predicted value of the equipment model after compensation, $\vec{y}$ denotes the predicted value of the equipment model, and $\vec{e}$ signifies the predicted value of the residual of the equipment model.

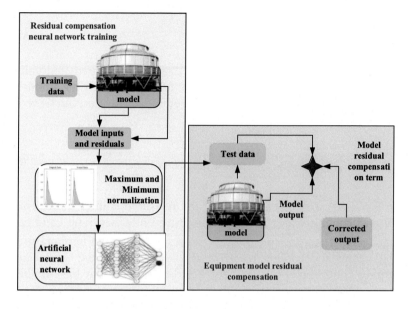

FIGURE 8.6　Flow of the model residual compensation method based on ANN.

8.2.5 Case analysis of algorithm performance

The model evaluation indexes adopted here are Mean Absolute Error (MAE), Mean Absolute Percentage Error (MAPE), Root Mean Square Error (RMSE), Coefficient of Determination R^2, and Coefficient of Variation of the Root Mean Squared Error (cV-RMSE). MAE, MAPE, RMSE, and CV-RMSE can reflect the deviation degree between the predicted value and the actual value. The smaller the value is, the higher the prediction accuracy of the model is. R^2 reflects the fitting degree of the model. The closer the value of R^2 is to 1, the better the fitting degree of the model to the observed value is. These indexes are calculated according to:

$$MAE = \frac{1}{m}\sum_{i=1}^{m}\left|y_i - \overrightarrow{y}_i\right| \tag{8.27}$$

$$RMSE = \sqrt{\frac{1}{m}\sum_{i=1}^{m}(y_i - \overrightarrow{y}_i)^2} \tag{8.28}$$

$$MAPE = \frac{1}{m}\sum_{i=1}^{m}\frac{\left|y_i - \overrightarrow{y}_i\right|}{y_i} \tag{8.29}$$

$$R^2 = 1 - \frac{\sum_{i=1}^{m}(y_i - \overrightarrow{y}_i)^2}{\sum_{i=1}^{m}(y_i - \overline{y})^2} \tag{8.30}$$

$$CV - RMSE = \frac{\sqrt{\sum_{i=1}^{m}(y_i - \overrightarrow{y}_i)^2}}{\overline{y}} \qquad (8.31)$$

where y_i represents the actual value, $\overrightarrow{y}_i$ refers to the predicted value, $\overline{y}$ denotes the average of the actual value, and m signifies the number of samples. The objective function of parameter identification can be written as Eq. (8.32).

$$f = \frac{1}{2m} \sum_{i=1}^{m} \left[(T_{chw,outi} - T_{cw,outi})^2 + (\overrightarrow{T}_{chw,outi} - \overrightarrow{T}_{cw,outi})^2 \right] \qquad (8.32)$$

In Eq. (8.32), $T_{chw,outi}$ and $T_{cw,outi}$ are measured chilled water temperature and cooling water outlet temperature under working condition i, respectively; $\overrightarrow{T}_{chw,outi}$ and $\overrightarrow{T}_{cw,outi}$ represent the outlet temperatures of chilled water and cooling water predicted by the DOE-2 model under operating condition i, respectively. Besides, m represent the number of samples.

When GA is adopted to identify model parameters of chilled water pump, the objective function is:

$$f = \frac{1}{m} \sum_{i=1}^{m} \left[(E_{chwi} - \overrightarrow{E}_{chwi})^2 \right] \qquad (8.33)$$

where E_{chwi} represents the actual measured power consumption per hour of the chilled water pump under working condition i, $\overrightarrow{E}_{chwi}$ signifies the power consumption per hour predicted by the chilled water pump model, and m signifies the number of samples. Moreover, the scaling factor a = 0.2.

The objective function of the cooling water pump can be expressed as Eq. (8.34).

$$f = \frac{1}{m} \sum_{i=1}^{m} \left[(E_{cwi} - \overrightarrow{E}_{cwi})^2 \right] \qquad (8.34)$$

In Eq. (8.34), E_{cwi} refers to the actual measured power consumption per hour of the cooling pump under working condition i, $\overrightarrow{E}_{cwi}$ denotes the power consumption per hour predicted by the cooling pump model, and m represents the number of samples. In addition, the scaling factor a = 0.2.

Here, the coping strategy of the second case is adopted to determine the initial solution space of parameter identification. A control group is introduced, and the initial solution space of each parameter is set as $[-1, 1]$.

Besides, the parameters of the chiller DOE-2 model are identified on the training set of the sample data by using MATLAB® genetic toolbox.

Furthermore, Prediction Interval Coverage Probability (PICP), Average Coverage Error (ACE), and Mean Prediction Interval width (MPIW) are adopted as the evaluation indicators of the prediction interval estimation method. PICP represents the proportion of measured data samples within the prediction interval. ACE represents the difference between PICP and PINC. The smaller the absolute value of ACE is, the more reliable the model can estimate the prediction interval. MPIW represents the average width of the forecast range. The larger the PICP, the better the effect; the smaller the MPIW, the better the effect.

$$PICP = \frac{1}{m} \sum_{i=1}^{m} c_i \qquad (8.35)$$

$$ACE = PICP - PINC \qquad (8.36)$$

$$MPIW = \frac{1}{m} \sum_{i=1}^{m} (U_i - L_i) \qquad (8.37)$$

Among Eqs. (8.35)–(8.37), L_i and U_i represent the lower and upper bounds of the prediction interval under the working condition i. If $L_i \leq y_i \leq U_i$, then $c_i = 1$. Besides, y_i represents the measured value under the working condition i, and m denotes the total number of samples used for performance evaluation. Then, the K-means clustering algorithm is used to cluster the normalized weighted model inputs. The cluster number k is set to 2, 3..., 20.

8.3 Results and discussion

8.3.1 Results of parameter identification based on GA and MISSO

Fig. 8.7 displays the prediction accuracy of chilled water outlet temperature and cooling water outlet temperature of the DOE-2 model of chiller of the central air-conditioning water system.

From Fig. 8.7, the DOE-2 model has a high overall prediction accuracy for cooling water outlet temperature of the chiller, with MAE, RMSE, MAPE, and CV-RMSE values reaching 0.275°C, 0.389°C, 0.865%, and 1.179%, respectively. Moreover, the value of R^2 is 0.986, close to 1, indicating that the model fits well. However, the overall prediction accuracy of chilled water outlet temperature is far less than that of cooling water outlet temperature, which cannot meet the actual requirements. The model is easy to fall into local optimum, and the sensor measurement error is large.

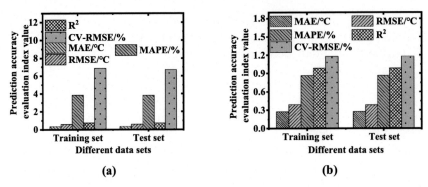

FIGURE 8.7   Prediction accuracy of chilled water outlet temperature and cooling water outlet temperature of chiller DOE-2 model ((a) prediction accuracy of chilled water outlet temperature; (b) prediction accuracy of cooling water outlet temperature).

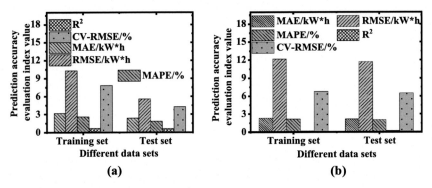

FIGURE 8.8   Parameter identification accuracy of the chilled water pump model and cooling water pump model ((a) parameter identification accuracy of the chilled water pump model; (b) parameter identification accuracy of the cooling water pump model).

Fig. 8.8 illustrates parameter identification accuracy of the chilled water pump model and cooling water pump model.

According to Fig. 8.8, the parameter identification accuracy of chilled water pump and cooling water pump models based on GA and MISSO is high. Besides, the absolute relative error between predicted value and measured value of unit power consumption of most pump models is controlled below 5%.

8.3.2   Results of prediction interval estimation of central air-conditioning model based on K-means clustering algorithm

When k = 17, the average value of the absolute value of ACE in the prediction interval of chilled water outlet temperature of the chiller model

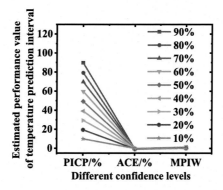

FIGURE 8.9   PICP, ACE and MPIW values of the test set at different confidence levels.

is at least 0.45%. Therefore, the cluster number k in this experiment is determined as 17. Fig. 8.9 reveals the results of PICP, ACE, and MPIW of the test set at different confidence levels when k = 17.

Through Fig. 8.9, all values of PICP are extremely close to the given PINC value, and the absolute value of ACE is all <0.80%. It shows that the prediction interval of chilled water outlet temperature of chiller established by this method is highly reliable. In addition, there is a positive correlation between PINC and MPIW. When PINC = 90%, MPIW reaches a maximum of 1.28°C; when PINC = 10%, MPIW attains a minimum of 0.13°C.

The simulation results of this the K-means clustering algorithm are compared with those of the clustering algorithm in reference [29,30], to further demonstrate the estimated effect of the chiller model on the prediction interval of chiller outlet temperature. The prediction interval of chiller model outlet temperature on a certain day with five typical PINC values of 50%, 60%, 70%, 80%, and 90% is selected for comparison, and the results are shown in Fig. 8.10.

From Fig. 8.10, the estimated MPIW increases with the growth of PINC. When the confidence level is 50%, the width of the prediction interval is small, and when the prediction error is relatively large, the prediction interval cannot contain the actual value. When the confidence level is 90%, the prediction interval is too wide, which will have adverse effects on application scenarios such as control strategy formulation based on prediction interval estimation.

Similarly, it can be concluded that when k = 12, the absolute value of ACE of the cooling water outlet temperature prediction interval of the chiller model is the smallest. Fig. 8.11 indicates the results of PICP, ACE and MPIW of the test set at different confidence levels when k = 12.

Fig. 8.11 shows that all the results of PICP and PINC are very close, and the absolute values of ACE are all less than 0.80%. It demonstrates that the

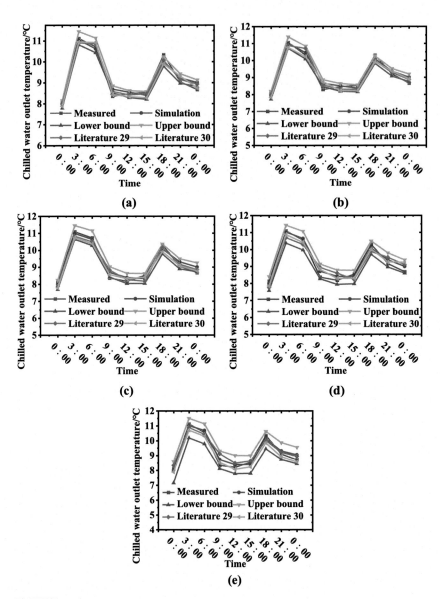

FIGURE 8.10   Comparison of prediction interval of outlet water temperature of chiller model in a certain day under different typical PINC values ((a) prediction interval results when PINC value is 50%; (b) prediction interval results when PINC value is 60%; (c) prediction interval results when PINC value is 70%; (d) prediction interval results when PINC value is 80%; (e) prediction interval results when PINC value is 90%).

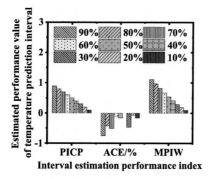

FIGURE 8.11   Results of PICP, ACE, and MPIW of the test set at different confidence levels.

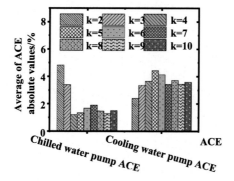

FIGURE 8.12   Comparison of the mean absolute value of ACE in the predicted interval of hourly power consumption of the chilled and cooling water pump models at different k values.

prediction interval of cooling water outlet temperature of chiller by this method is reliable. Moreover, there is a positive correlation between PINC and MPIW. When PINC = 90%, MPIW attains a maximum of 1.11°C; when PINC = 10%, MPIW reaches a minimum of 0.10°C.

Fig. 8.12 illustrates the mean average value of ACE in the predicted range of power consumption per unit hour of the cooling and cooling water pump models under different k values.

According to Fig. 8.12, when k = 4, the mean absolute value of ACE in the predicted range of power consumption per unit hour of the cooling pump model is the smallest, which is 1.21%, so k is set to 4. When k = 2, the mean absolute value of ACE in the predicted range of power consumption per unit hour of the cooling pump model is the smallest, which is 1.74%, so k = 2.

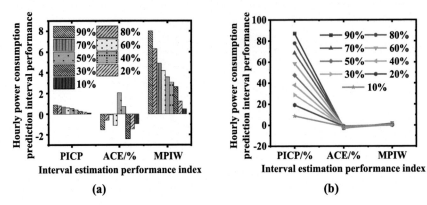

FIGURE 8.13 Results of PICP, ACE, and MPIW of the test set at different confidence levels ((a) results of prediction interval performance indexes of chilled water pump model unit power consumption when k = 4; (b) prediction interval performance of cooling water pump model per unit power consumption when k = 2).

Fig. 8.13 shows the performance comparison of unit power consumption prediction interval between chilled water pump and cooling water pump models at different confidence levels after k value is determined.

From Fig. 8.13(a), MPIW decreases with the decrease in PINC. When PINC = 90%, MPIW of the average predicted interval width is the largest, reaching 8.02 kW*h; when PINC = 10%, MPIW of the average predicted interval width is the smallest, attaining 0.48 kW*h. The absolute values of ACE are all less than 2.50%, and all PICP values are close to PINC values, so the prediction range of hourly power consumption of chilled water pump is relatively reliable. The prediction interval results of cooling water pump model obtained in Fig. 8.13(b) are similar.

8.3.3 Residual error compensation results of the model based on ANN

Fig. 8.14 indicates the comparison of chilled water outlet temperature prediction accuracy and prediction interval width under different model structures.

Through Fig. 8.14(a), the error between the predicted value and the actual value of chilled water outlet temperature by the compensated chiller model decreases significantly. The performance indexes MAE, RMSE, MAPE and CV-RMSE after compensation are 0.221°C, 0.317°C, 2.546°C, and 3.625%, respectively, which are 36.49%, 46.00%, 33.16%, and 45.73% lower than those before compensation. Besides, the value of R^2 is 0.923, almost close to 1, showing an increase by 25.75%.

According to Fig. 8.14(b), the prediction interval width of the compensated DOE-2 model is narrower. Besides, the prediction interval width of

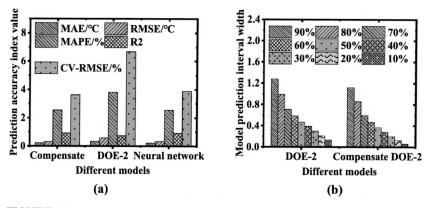

FIGURE 8.14 Comparison of model prediction performance before and after error compensation ((a) comparison of prediction accuracy before and after compensation; (b) comparison of predicted widths of chilled water outlet temperature interval at different confidence levels before and after compensation).

chilled water outlet temperature at different confidence levels decreases, with the confidence level decreasing by 53.85% to 10%. Therefore, the model residual compensation method based on ANN can effectively reduce the uncertainty of model prediction.

8.4 Conclusion

With the development of DTs technology, it is increasingly applied to improve human environment and human health. Here, a DTs pipeline detection method of groundwater is put forward. Besides, a central air-conditioning water system model is established by combining DTs and artificial intelligence, to improve fault prediction, energy conservation, and emissions reduction. Moreover, the real-time optimization of central air-conditioning operation state parameters is implemented through adjusting the prediction interval of different parameters. In addition, the MISSP-GA model parameter identification framework is constructed; the prediction interval estimation method via K-means clustering and model residual error compensation method based on ANN are proposed to reduce the uncertainty of model prediction, and result has been proved to be good. Although the present work has achieved the above findings, it did not collect the local outdoor air temperature and humidity parameters near the cooling tower. Consequently, the application effect of the method reported here on the cooling tower needs to be verified. In addition, the application of DTs only focuses on the direction of uncertainty estimation of model prediction. Future research will investigate fault prediction and maintenance of central air-conditioning water system.

References

[1] L. Lu, H. Zheng, J. Jie, et al., Reinforcement learning-based particle swarm optimization for sewage treatment control, Complex Intell. Syst. (2021) 1–12, https://doi.org/10.1007/s40747-021-00395-w.

[2] O. El Marai, T. Taleb, J.S. Song, Roads infrastructure digital twin: a step toward smarter cities realization, IEEE Netw. 35 (2) (2020) 136–143, https://doi.org/10.1109/MNET.011.2000398.

[3] K. Alexopoulos, N. Nikolakis, G. Chryssolouris, Digital twin-driven supervised machine learning for the development of artificial intelligence applications in manufacturing, Int. J. Comput. Integr. Manuf. 33 (5) (2020) 429–439, https://doi.org/10.1080/0951192X.2020.1747642.

[4] L. Xing, B. Jiao, Y. Du, X. Tan, R. Wang, Intelligent energy-saving supervision system of urban buildings based on the Internet of things: a case study, IEEE Syst. J. 14 (3) (2020) 4252–4261, https://doi.org/10.1109/JSYST.2020.2995199.

[5] A. Fuller, Z. Fan, C. Day, C. Barlow, Digital twin: enabling technologies, challenges and open research, IEEE Access 8 (2020) 108952–108971, https://doi.org/10.1109/ACCESS.2020.2998358.

[6] Y. Zhang, W. Luo, F. Yu, Construction of Chinese smart water conservancy platform based on the blockchain: technology integration and innovation application, Sustainability 12 (20) (2020) 8306, https://doi.org/10.3390/su12208306.

[7] H.M. Yasin, S.R.M.A. Zeebaree Sadeeq, S.Y. Ameen, I.M. Ibrahim, R.R. Zebari, A.B. Sallow, IoT and ICT based smart water management, monitoring and controlling system: a review, Asian J. Res. Comput. Sci. (2021) 42–56, https://doi.org/10.9734/ajrcos/2021/v8i230198.

[8] D. Li, Analysis on the monitoring system of energy conservation and comfort in office buildings based on Internet of things, Int. J. Low Carbon Technol. 15 (3) (2020) 351–355, https://doi.org/10.1093/ijlct/ctz083.

[9] Y. Chen, S. Hu, H. Mao, W. Deng, X. Gao, Application of the best evacuation model of deep learning in the design of public structures, Image Vis. Comput. 102 (2020) 103975, https://doi.org/10.1016/j.imavis.2020.103975.

[10] T. Jiang, P. Ju, C. Wang, H. Li, J. Liu, Coordinated control of air-conditioning loads for system frequency regulation, IEEE Trans. Smart Grid 12 (1) (2020) 548–560, https://doi.org/10.1109/TSG.2020.3022010.

[11] X. Qi, Z. Ji, H. Wu, J. Zhang, H. Yang, Analysis of emergency demand response levels of central air-conditioning, J. Electr. Eng. Technol. 15 (6) (2020) 2479–2488, https://doi.org/10.1007/s42835-020-00525-3.

[12] L. Kang, G. Wang, Y. Wang, Q. An, The power simulation of water-cooled central air-conditioning based on demand response, IEEE Access 8 (2020) 67396–67407, https://doi.org/10.1109/ACCESS.2020.2986309.

[13] K.J. Wang, Y.H. Lee, S. Angelica, Digital twin design for real-time monitoring–a case study of die cutting machine, Int. J. Prod. Res. (2020) 1–15, https://doi.org/10.1080/00207543.2020.1817999.

[14] J. Xu, J. Wang, Y. Tian, J. Yan, X. Li, X. Gao, SE-stacking: improving user purchase behavior prediction by information fusion and ensemble learning, PLoS ONE 15 (11) (2020) e0242629, https://doi.org/10.1371/journal.pone.0242629.

[15] G. White, A. Zink, L. Codecá, S. Clarke, A digital twin smart city for citizen feedback, Cities 110 (2021) 103064, https://doi.org/10.1016/j.cities.2020.103064.

[16] D. Yang, H.R. Karimi, O. Kaynak, S. Yin, Developments of digital twin technologies in industrial, smart city and healthcare sectors: a survey, Complex Eng. Syst. 1 (1) (2021), https://doi.org/10.20517/ces.2021.06.

[17] M. Garrido-Baserba, L. Corominas, U. Cortés, D. Rosso, M. Poch, The fourth-revolution in the water sector encounters the digital revolution, Environ. Sci. Technol. 54 (8) (2020) 4698–4705, https://doi.org/10.1021/acs.est.9b04251.

[18] Y. Liu, K.A. Hassan, M. Karlsson, O. Weister, S. Gong, Active plant wall for green indoor climate based on cloud and Internet of things, IEEE Access 6 (2018) 33631–33644, https://doi.org/10.1109/ACCESS.2018.2847440.

[19] X. Zhang, J. Shen, P.K. Saini, M. Lovati, M. Han, P. Huang, Z. Huang, Digital twin for accelerating sustainability in positive energy district: a review of simulation tools and applications, Front. Sustain. Cities 3 (2021) 35, https://doi.org/10.3389/frsc.2021.663269.

[20] R. Molinaro, J.S. Singh, S. Catsoulis, C. Narayanan, D. Lakehal, Embedding data analytics and CFD into the digital twin concept, Comput. Fluids 214 (2021) 104759, https://doi.org/10.1016/j.compfluid.2020.104759.

[21] H.F. Badawi, F. Laamarti, A. El Saddik, Devising digital twins DNA paradigm for modeling ISO-based city services, Sensors 21 (4) (2021) 1047, https://doi.org/10.3390/s21041047.

[22] H. Darvishi, D. Ciuonzo, E.R. Eide, P.S. Rossi, Sensor-fault detection, isolation and accommodation for digital twins via modular data-driven architecture, IEEE Sens. J. 21 (4) (2020) 4827–4838, https://doi.org/10.1109/JSEN.2020.3029459.

[23] S. Rokhsaritalemi, A. Sadeghi-Niaraki, S.M. Choi, A review on mixed reality: current trends, challenges and prospects, Appl. Sci. 10 (2) (2020) 636, https://doi.org/10.3390/app10020636.

[24] P.A. Lindahl, M.T. Ali, P. Armstrong, A. Aboulian, J. Donnal, L. Norford, S.B. Leeb, Nonintrusive load monitoring of variable speed drive cooling systems, IEEE Access 8 (2020) 211451–211463, https://doi.org/10.1109/ACCESS.2020.3039408.

[25] S. Yan, D. Cai, Y. Chen, Y. Xue, Y. Liu, L. Wu, Z. Song, Reliability modelling and prediction method for phase change memory using optimal pulse conditions, J. Shanghai Jiaotong Univ. (Science) 25 (1) (2020) 1–9, https://doi.org/10.1007/s12204-020-2153-8.

[26] J. Li, Z. Li, Model-based optimization of free cooling switchover temperature and cooling tower approach temperature for data center cooling system with water-side economizer, Energy Build. 227 (2020) 110407, https://doi.org/10.1016/j.enbuild.2020.110407.

[27] M. Ahmed, R. Seraj, S.M.S. Islam, The k-means algorithm: a comprehensive survey and performance evaluation, Electronics 9 (8) (2020) 1295, https://doi.org/10.3390/electronics9081295.

[28] T. Wang, X. Wang, R. Ma, X. Li, X. Hu, F.T. Chan, J. Ruan, Random forest-Bayesian optimization for product quality prediction with large-scale dimensions in process industrial cyber–physical systems, IEEE Int. Things J. 7 (9) (2020) 8641–8653, https://doi.org/10.1109/JIOT.2020.2992811.

[29] J. Miao, X. Zhou, T.Z. Huang, Local segmentation of images using an improved fuzzy C-means clustering algorithm based on self-adaptive dictionary learning, Appl. Soft Comput. 91 (2020) 106200, https://doi.org/10.1016/j.asoc.2020.106200.

[30] D. Lin, Q. Wang, An energy-efficient clustering algorithm combined game theory and dual-cluster-head mechanism for WSNs, IEEE Access 7 (2019) 49894–49905, https://doi.org/10.1109/ACCESS.2019.2911190.

9

Role of smart technologies in detecting cognitive impairment and enhancing assisted living

Devvrat Bhardwaj[a], Jeffrey Jutai[b], and Pascal Fallavollita[a,b]

[a]School of Electrical Engineering and Computer Science, Faculty of Engineering, University of Ottawa, Ottawa, ON, Canada, [b]Interdisciplinary School of Health Sciences, Faculty of Health Sciences, University of Ottawa, Ottawa, ON, Canada

9.1 Introduction

Dementia is defined as a disorder of the brain in which a person suffers from an impaired ability to perform cognitive tasks in daily life [1]. The onset of dementia is typically triggered by aging, illnesses, or injuries [1]; however, its severity increases gradually, transitioning from age-related memory impairment to a mild cognitive deterioration stage and lastly to dementia [2], which is often irreversible [3]. Mild cognitive impairment (MCI) has been described as the earliest detectable cognitive syndrome [4] and a plethora of research works in the last decade have focused on using low cost and easily available smart technologies for detecting MCI [5], as early detection can lead to effective treatment [3,6,7]. About 50 million people suffer from dementia worldwide and with the increase in life expectancy due to improving healthcare facilities, this population is expected to increase three-folds by 2050 [8]. Other than the failure to perform the activities of daily life (ADL), the patients also suffer from low memory retention [9], mood variability [10] and agitation [11], all of which lead to reduction in quality of life [1,11]. However, majority of the affected individuals receive care at home, despite deterioration in their

Digital Twin for Healthcare
https://doi.org/10.1016/B978-0-32-399163-6.00014-7

181

FIGURE 9.1   Digital health and medical technologies [15].

condition [12]; and the friends and family members, although playing a critical role in providing care, find the process of assisting the patients stressful, very costly and time demanding, and themselves experience deterioration in quality of life, both socially and financially, sometimes even trading off their jobs to provide care [13]. Therefore, digital cognitive assessment methods based on analyzing physiological signals (the natural indicators [14]) are being developed using scalable artificial intelligence techniques to track longitudinal cognitive and behavioral changes in a patient; to reduce the costs involved; to reduce the dependability on labor intensive care giving; and to assist the health provider's diagnosis [3]. This chapter explores the recent advances in detection of mild cognitive impairment based on various physiological features such as speech, vision, emotional state, hearing, brain activity, gait, handwriting and sleep patterns, and discusses the developments in providing assistive care to the affected individuals using smart technologies. (See Fig. 9.1.)

9.2 Mild cognitive impairment (MCI) detection

9.2.1 Using gait patterns and postural dynamics

It has been observed that individuals with MCI are at an increased risk of falling [16]. Determination of the decline in walking patterns (gait) can be helpful for preventing falls, detecting MCI, and measuring the rate of cognitive decline. As seen in [4], M. Gwak et al., proposed a novel algorithm for selecting optimal gait features from accelerometer and gyroscope sensors, and used it to differentiate cognitively healthy individuals from those with MCI. The researchers [4] conducted a scenario-based assessment study with both cognitively healthy and MCI participants, where their gait features were measured using Samsung smartwatch on a 60 meter "walk-turn-walk activity", at both the start and the end of a session. The authors [4] preprocessed and ranked the retrieved features using forward feature selection, and obtained an accuracy of 88% by utilizing Extra Trees, Random Forests, and logistic regression. Similarly, C. Ricciardi et al., [5] analyzed the spatial-temporal gait parameters of individuals with Parkinson's disease (PD) using machine learning, to investigate the relationship between cognition and gait. The researchers acquired gait patterns of 45 PD (23 with MCI) patients, on three different tasks (normal gait, cognitive dual-task, and motor dual-task) using SMART DX optical system. The collected data was augmented and balanced using synthetic minority oversampling technique (SMOTE), which after feature extraction (into 18 features) was analyzed using Decision Trees, Random Forest, and K-Nearest Neighbors (KNN), with the leave one out cross validation approach; and the best classification accuracy of 86.2% was achieved for motor dual task using Decision Trees, the best classification accuracy of 82.4% was achieved with Random Forests for cognitive dual task and the best classification accuracy on normal gait was achieved using KNN to be 83.8%.

In another study, S. Ikeda et al., [6], investigated age-related changes in gait patterns with the help an eyewear device equipped with triaxial accelerometer. The authors recruited 118 healthy individuals from various age groups (25–69 years) and conducted an experiment where each participant performed. The authors collected acceleration signals in x (mediolateral), y (anteroposterior), and z (vertical) directions from 118 healthy individuals during two alternating sets of resting and walking for 20 m. The researchers performed robust linear regression analysis to find association between participants' age and the triaxial variances, and observed that anteroposterior (y) and vertical (z) accelerations were significantly correlated (positively) with the age. This led them to the conclusion that with decline in gait stability due to age, the gait speed decreased, and frequency of vertical body sway increased. Similarly, M. Pau et al., in [17], observed that the accelerations in the antero-posterior and the supero-inferior directions differed significantly between the cognitively healthy and the impaired. The researchers studied alterations in gait patterns and parameters, and reduction in gait smoothness in 90 cognitively impaired adults, by using an inertial sensor attached to the participants' lower back, which collected their antero-posterior (AP), supero-inferior (V) and medio-lateral (ML) accelerations at a sampling rate of 100 Hz, during a 30 m walking task. Also, the researchers [17] found out that the participants with cognitive decline showed a significant reduction in stride length and gait speed.

A. Pantall et al., in [18] investigated the relationship between postural deterioration and cognitive impairment. Assessments for both postural and cognitive decline (MoCA and GDS) were conducted at 18th, 36th and 54th month of the study with 35 individuals suffering from Parkinson's disease. During each postural assessment, the participants were equipped with a triaxial accelerometer attached to their back, and stood straight for 120 seconds in a given area. The researchers extracted four features from the recorded acceleration data namely jerk, root mean square, frequency, and ellipsis, for the phases of initial 30 s, initial 60 s and total of 120 s. On analyzing postural dynamics using linear mixed-effect models and various statistical tests over axis, phase, and time, it was observed that time had a significant decreasing effect on ellipsis, whereas jerk and RMS were significantly affected in the antero-posterior axis of acceleration. Moreover, the researchers noted that with progression in the cognitive decline, the postural dynamics (i.e., ellipsis, jerk, and RMS) followed a decreasing trend (negative correlation).

9.2.2 Using physiological changes in ECG and EEG

S. Khatun et al., in [19] proposed a classification model to identify individuals with mild cognitive impairment by analyzing single channel electroencephalography (EEG) evoked by auditory signals. The researchers recruited 15 healthy and 8 MCI participants and collected every individual's EEG signals (twice), on correct alphabet identification task (based on contrastive vowel sounds of 'a' and 'u'). The collected signals after band pass filtration were processed in MATLAB® for event-related potential (ERP) points which led to extraction of 590 features, out of which 25 most prominent features (ranked via Random Forests) were used for classification. The researchers trained logistic regression (LR) and support vector machine (SVM) classification models using leave-one-out cross validation and different cost parameters, and observed that cognitively impaired and healthy individuals were distinguishable on the basis of their perception of clearer sounds with an accuracy of 87.9% achieved on SVM. This result is similar to the result obtained by J. Li et al., [20] where it was observed that left frontal cognitive ERP patterns differentiated healthy participants from those with amnestic MCI with an accuracy of 85%.

Z. Fangmeng et al., in [14] investigated emotional fluctuations in healthy participants by analyzing their physiological signal spectrograms and proposed continuation of this research to detect dementia patients. The researchers used short video clips to induce six different emotions in 11 individuals, and recorded their EEG and ECG signals. The researchers only focused on alpha (8–14 Hz) and beta (14–30 Hz) EEG waves, and the R-R interval (4 Hz) from the ECG waves. These signals were analyzed using SciPy and BioSPPy libraries in Python and the authors observed that beta waves in EEG and LF amplitudes in RRI were prominent in the high arousal state. Separate convolutional neural networks (CNNs) were constructed for EEG and RRI signals, with ReLU as activating function and stochastic gradient descent as the optimizer. The final binary classification into low and high arousal categories was done by combining the two signals as inputs to a merged CNN, which resulted in an accuracy of 93.33% after 300 epochs.

9.2.3 By tracking eye movement

K.C. Fraser et al., in [7] examined eye-tracking patterns to predict cognitive impairment. 30 cognitively healthy and 27 individuals with MCI were recruited in this research, and were made to read two paragraphs in Swedish, one silently and the other loudly, while their eye movements were being tracked at a sampling rate of 1000 Hz for "fixations" and "saccades" using the monocular Eyelink 1000 Desktop Mount. On processing, 13 features were extracted as baseline and augmented with word metadata (type and frequency) and 22 statistically calculated features. The data was then analyzed by implementing 3 classifiers, namely Naïve Bayes (NB), logistic regression (LR) and support vector machine (SVM). After evaluating each classifier for each task using the leave-one-out cross validation, it was observed that reading silently led to better eye tracking and better accuracy as compared to reading aloud. On analyzing the features, it was noted that healthy individuals tend to read the text in a start to finish manner whereas individuals with MCI were observed to be skipping words and returning to them later. Moreover, the authors analyzed the combined tasks and found out that better accuracies were achieved after concatenating the features between tasks, rather than individual classifications. Similarly, J. Jiang et al., [21] recorded eye movement and EEG from 152 MCI and 184 healthy individuals, on a visual stimulus experiment to differentiate between the two populations. After extracting and analyzing 40 features, the researchers achieved the best classification accuracy of 81.51% on a logistic regression model trained on combination of EEG signals, eye tracking data as well as clinical variables.

9.2.4 Sleep monitoring

B. Chen et al., in [3] conducted a real-life home monitoring experiment with 49 participants to differentiate between cognitively healthy seniors and those affected with mild cognitive impairment by studying their sleep patterns. The participants were divided into blocks of 6–7 individuals and observed for 2 months with privacy preserving sensors installed throughout their apartment to measure contact, motion, sleep, and activity. Also, wearable sensors were deployed that measured health statistics and proximity of personal items. Moreover, the Pittsburgh Sleep Quality Index (PSQI) and Geriatric Depression Scale (GDS) tests were conducted as a baseline for comparison with sensor data. From the bed sensor (i.e., the sleep mat), the authors extracted features of sleep quality such as sleep latency, duration, and variability, and analyzed them using logistic regression, keeping PSQI as the predictor variable. It was observed that only sleep state variability was significantly correlated to the PSQI score, and was able to classify correctly with a recall value of 0.71 and precision of 0.91. Also, I. Jaussent et al., [22], after performing 8-year longitudinal study, concluded excessive daytime sleeping (EDS) pattern to be an independent predictor of cognitive decline in the elderly.

9.2.5 Using handwriting

N.D. Cilia et al., in [9] predicted the cognitive status of individuals based on the features extracted from their handwriting. The researchers recruited 180 participants (90 CH and 90 with MCI), and conducted an experiment where each individual performed 6 unique graphical tasks which measured their cognitive deterioration, elementary motor function, long term motor planning, coordination of movements, attention span, and spatial dysfunction. The data was collected on Wacom bamboo folio tablet and processed using MovAlyzer software. After extracting the features, the authors trained 3 classifiers, namely Decision Trees (DT), Support Vector Machines (SVM) and Neural Networks (NN), using 5-fold cross validation for each of the six tasks. The final class was predicted by combining the performance of each classifier using the majority vote rule. It was observed that NN's were outperformed by the other two algorithms and that the approach of combining different classifiers provided better accuracy in distinguishing the classes as compared to single classification tests. Also, the option to reject was provided in the classifiers to further enhance the accuracy.

9.2.6 Using multiple signals (smart homes)

A. Lauraitis et al., in [23] constructed an Android based digital health evaluation tool called "Neural Impairment Test Suite", to detect neurological degeneration (indicative of cognitive decline), and developed a hybridized decision support system using multiple classifiers. The researchers conducted a 16-tasked digital test with 15 participants (8 healthy, 7 neurologically impaired) which measured their cognition (from identification, calculative and problem-solving tasks), hand tremor (from touch, constructive and drawing tasks), speech (from voice recordings), and total energy expenditure. This study was performed for a total of 5 rounds over time and resulted in a dataset of 150 records, with each record having approximately 238 features. The researchers first analyzed the features for

individual tasks using 14 different classifiers, with a 10-fold cross validation approach, and reported the best attribute selection and classification
methods with respect to accuracy. After that, the authors integrated the
feature set, extracted the best features using "binary grey wolf optimization particle swarm optimization" and constructed a hybridized classifier
by combining the performance of 13 different classifiers using the majority vote and average probability rules, and evaluated it using train/test
dataset split. It was observed that the hybridized classifier achieved an accuracy of 96.12% in differentiating the cognitive status, which was greater
than any individual classifier. Moreover, the authors analyzed the audio
data separately, using bi-directional Long Short-Term Memory (LSTM),
and support vector machines, and were able to achieve 94.29% and 100%
accuracy respectively, in classifying individuals according to their cognition level. R. Narasimhan et al., [24] also proposed to predict Alzheimer's
disease progression by capturing temporal patterns in longitudinal data
obtained from MCI affected patients using nonwearable sensors and by
utilizing LSTM Recurrent Neural Network (RNN) with two hidden layers to analyze them. However, this study [24] lacked analysis on a real-life
deployment scenario.

R. Chen et al., in [25] utilized an online study platform to remotely monitor symptoms of cognitive impairment using multiple smart devices, with
the goal to differentiate between healthy individuals and those with di-
minishing cognitive health. The authors enrolled 119 elderly participants
(84 healthy and 35 symptomatic) and centrally collected encrypted multimodal data over a period of 12 weeks from the sensors in iPhone 7+,
Beddit sleep monitoring device and Apple Watch Series 2, which were
provided to each individual. Also, the participants were given an iPad
pro and a smart keyboard, to report their mood and energy daily, along
with performing tasks on an assessment app biweekly. The sensor data
streams after decryption, normalization, timestamping, and time aligning
were diagnosed using "behaviorgrams" with a 1-minute resolution, to detect patterns. The time-aligned data channels obtained in "behaviorgrams"
and assessment from the psychomotor tasks through the digital app were
used to extract a total of 996 features, and the dataset was artificially augmented as the number of features exceeded the sample size. The authors
split the bi-weekly grouped data in 70/30 holdout for iterations of every
100 samples and trained extreme gradient boosted regression models using 3-fold cross validation to tune the hyperparameters. The model with
the best area under the ROC curve (AUROC) was trained on the full 'local'
training data, and tested to obtain performance metrics. The performance
metrics for 100 such iterations were averaged, and it was observed that
cognitively impaired personnel were distinguishable from their slow typ-
ing speed, reliance on Siri for suggestions and less routinely behavior as
compared to healthy controls.

E. Demir et al., [26] acknowledged the mental effects of preventive household restrictions placed on cognitively impaired patients and proposed an ambient assisted living (AAL) system to overcome such challenges. The researchers utilized 7 types of nonwearable sensors (Reed switch, temperature sensor, flame sensor, rain sensor, LDR, force resistive sensor, and motion sensor) to detect context specific activity, and send the relative information to care providers and doctors via a mobile application in real time.

9.3 Providing assisted living

9.3.1 By using augmented reality (AR)

M. Rossi et al., in [2], developed an augmented reality application, CogAR, for Android to assist cognitively impaired individuals in performing activities of daily life, by providing memory cues. The researchers associated augmented object information with visual markers (printed pictures) which were placed at various places in the home setting. This information was displayed to the user whenever the sensors on head mounted device (HMD) automatically scanned the markers. Moreover, the authors provided emergency contacts such as that of the caregivers and the ambulance on top of the HMD display screen. Also, the feature to share the user's field of view in real time with a caregiver was included via wireless screen mirroring.

Similarly, H. Ro et al., [27] proposed a projection based augmented reality (AR) system to provide assistance and therapy to dementia patients. The system, equipped with 360-degree rotatable RGB-D camera, was deployed in three versions, namely movable, portable, and robotic, and was capable of adjusting to any environment by utilizing 3D map reconstruction. The applications were supported with voice and touch interfaces, such as user monitoring, user identification, medication reminder, media access, and spatial art.

However, D. Wolf et al., [28] discovered that augmented reality (AR) based solutions in dementia care lacked flexibility and incentive mechanisms. To overcome this challenge, the authors proposed cARe, a generic user-centric framework to cover a broad range of use-case scenarios. The researchers iteratively performed pilot studies with caregivers and dementia patients and integrated activity specific configuration support (for care providers; fast and efficient) and incentive mechanisms (for patients; slow and arrow-based) in the AR application. Moreover, the developers added time-out feature (via detecting absence of head movement), to cope up with the dementia patient's forgetfulness. Also, the authors presented preliminary results from an ongoing study on patient's ability to prepare meals independently and reported that cARe led to better and extended performance than a paper-based recipe approach.

9.3.2 By managing wandering

To provide assistive care on wandering patterns, O. Yilmaz [29] proposed a cost-effective method to monitor the patients. It was reported in the paper [29] that 67% of dementia patients may exhibit wandering behavior, and if not found within a day, suffer with serious injuries or even death. However, continuously monitoring the patients is emotionally, financially, and physically taxing on the care providers. Therefore, to tackle these challenges, the author developed iCarus, a flexible, rule based AAL system to monitor wandering behavior remotely and provide context-based solutions. The system classified patient's spatial information in 3 fixed zones (defined by the caregivers), namely green as the safe zone, red as the danger zone, and an orange zone, that was only to be treated as red if a rule was violated, otherwise green. Moreover, support to include additional zones (blue) and rules was provided. Temporal rules were also included, that upon violation modified the zone status and triggered necessary protocols. If a wandering episode was detected (due to spatial, temporal, or self-defined rule violation), the iCarus app on the patient's end warned the person with text and audio prompts, and provided navigation assistance to return to their safe zone. If such an attempt was futile, then the respective caregiver was notified on their app with contextual information (geographical location, distance, speed etc.) regarding the wandering episode, and provided the options to track the patient and/or contact emergency services.

9.3.3 By analyzing emotional fluctuations

The inability to release the emotional, physiological, or environmental stress often makes the patients agitated and interpreting such changes in the mood/behavior is tedious for the caregivers/support staff [11]. Therefore, to assist the health care staff in their assessments, C. Melander et al., in [11] investigated the correlation between electrodermal activity (EDA) detected by a sensor and a patient's agitation level (as observed by nursing staff). The researchers selected 9 dementia affected individuals, and collected both rapid (short-term stimulus) and stable (long-term emotional change) EDA data using the Discrete Tension Indicator (DTI-2) wristband sensor over a period of 2 weeks. During this time period, the assistant nurses monitored the patients using an observation scheme where each behavior was rated with a specific color in a 24h time window. Stress levels extracted from the EDA data and the dichotomized classes from the observational data (agitated or calm) were tested for correlation using Spearman rho. Moreover, the researchers evaluated the goodness-of-fit on a binary logistic regression model with the observed class as the predicted variable, and the sensor values at the time of observation, 1h prior to it and 2h prior to it, as predictors. It was found out that the sensor measurements of EDA were able to correctly predict 73.5% of the observed agitations, and the researchers concluded that there existed a significant correlation between observed agitation and the increased EDA sensor values, and that the sensor values prior to the nurses' observation were predictive of agitation. However, this study was limited by the fact that there was no differentiation for increased EDA value due to positive or negative emotions.

A similar study conducted by C.L. Kwan et al., in [10] proposed a novel algorithm to intelligently assist dementia affected patients and their caregivers, based on a patient's significant mental and emotional moments derived from changes in their physiological signals. The researchers used a fingertip worn device (Triple Point Sensor) to measure electrodermal activity (EDA) signals, heart rate and skin temperature of 3 dyads (dementia patient + caregiver) for 8 weeks, along with synchronously recording audio-visual data with a ceiling camera for each 45-minute session. This data was analyzed using a custom software called Events Finder, to detect emotional arousal (from EDA), changes in accelerations (from heart rate), and changes in vasodilation and vasoconstriction (from skin temperature). A combined score was calculated for all events at different time intervals (i.e., as a function of time) and 20 second video clips during the highest scoring events were extracted. The dyads were interviewed on whether the respective events qualified as significant moments and with each interview, the Event Finder parameters were calibrated iteratively for every individual. The Event Finder after being trained and optimized for each individual's physiological reactions, was able to detect 70% of the significant moments experienced by the patients.

9.4 Conclusion

In this chapter, recent developments in detecting mild cognitive impairment (MCI) using machine learning algorithms trained on multimodal sensor data were discussed. It is evident that, for predicting MCI, the trend of relying on paper-based neuropsychological evaluation tests alone, has shifted to detecting the dynamic variations in a patient's natural physiological signals and behavioral patterns, using powerful AI algorithms. Moreover, computerized cognitive tests employing artificial intelligence, such as the language independent Integrated Cognitive Assessment (ICA) [30], have demonstrated high convergent validity, and a potential to replace conventional cognitive tests. Apart from offering high accuracy and scalability in diagnostic applications, artificial intelligence techniques along with the internet of things (IoT), are valuable in providing personalized treatment recommendations to the patients, as well as assistance to their healthcare providers. The advancements in research-grade, wireless, wearable sensor devices (such as the Hexoskin shirt and the Empatica E4 watch), have expanded the scope of practically integrating AI into the activities of daily life (ADL) of a patient, while preserving their independence, mobility, and privacy. Along with artificial intelligence, healthcare assistance (especially rehabilitation) using Augmented Reality (AR) has been on the rise.

We have learned that the state-of-the-art research on diagnosing cognitive impairment and providing assisted living to dementia patients, is highly focused on utilizing technological solutions. In the near future, smart homes equipped with unobtrusive wireless sensors and devices (both wearable and nonwearables), powered with AI and AR algorithms, will help cognitive challenged patients in leading an independent lifestyle; while getting accurate diagnosis, personalized healthcare, and timely assistance.

Acknowledgments

We gratefully acknowledge the AGE-WELL NCE (AW-PP2019-PP3), for their financial support on the PATH (Program to Accelerate Technologies for Homecare) project.

References

[1] D.S. Geldmacher, P.J. Whitehouse, Evaluation of dementia, N. Engl. J. Med. 335 (5) (1996) 330–336.

[2] M. Rossi, G. D'Avenio, S. Morelli, M. Grigioni, Cogar: an augmented reality app to improve quality of life of the people with cognitive impairment, in: 2020 IEEE 20th Mediterranean Electrotechnical Conference (MELECON), IEEE, 2020, pp. 339–343.

[3] C. Brian, T. Hwee-Pink, I. Rawtaer, T. Hwee-Xian, Objective sleep quality as a predictor of mild cognitive impairment in seniors living alone, in: 2019 IEEE International Conference on Big Data (Big Data), IEEE, 2019, pp. 1619–1624.

[4] M. Gwak, E. Woo, M. Sarrafzadeh, The role of accelerometer and gyroscope sensors in identification of mild cognitive impairment, in: 2018 IEEE Global Conference on Signal and Information Processing (GlobalSIP), IEEE, 2018, pp. 434–438.

[5] C. Ricciardi, M. Amboni, C. De Santis, G. Ricciardelli, G. Improta, G. D'Addio, S. Cuoco, M. Picillo, P. Barone, M. Cesarelli, Machine learning can detect the presence of mild cognitive impairment in patients affected by Parkinson's disease, in: 2020 IEEE International Symposium on Medical Measurements and Applications (MeMeA), IEEE, 2020, pp. 1–6.

[6] S. Ikeda, S. Ichinohe, R. Kawashima, Eyewear equipped with a triaxial accelerometer detects age-related changes in ambulatory activity, DigitCult-Scientific J. Digit. Cultures 2 (2) (2017) 1–8.

[7] K.C. Fraser, K.L. Fors, D. Kokkinakis, A. Nordlund, An analysis of eye-movements during reading for the detection of mild cognitive impairment, in: Proceedings of the 2017 Conference on Empirical Methods in Natural Language Processing, 2017, pp. 1016–1026.

[8] M. Smalls, Fall prevention: the first line of defense. Integration of innovated strategies to decrease falls for the hospitalized patient, 2021.

[9] N.D. Cilia, C. De Stefano, F. Fontanella, A. Scotto di Freca, Handwriting-based classifier combination for cognitive impairment prediction, in: International Conference on Pattern Recognition, Springer, 2021, pp. 587–599.

[10] C. Lai Kwan, Y. Mahdid, R. Motta Ochoa, K. Lee, M. Park, S. Blain-Moraes, Wearable technology for detecting significant moments in individuals with dementia, BioMed Res. Int. 2019 (2019).

[11] C. Melander, J. Martinsson, S. Gustafsson, Measuring electrodermal activity to improve the identification of agitation in individuals with dementia, Dement. Geriatr. Cogn. Disord. Extra 7 (3) (2017) 430–439.

[12] K.L. Harrison, C.S. Ritchie, K. Patel, L.J. Hunt, K.E. Covinsky, K. Yaffe, A.K. Smith, Care settings and clinical characteristics of older adults with moderately severe dementia, J. Am. Geriatr. Soc. 67 (9) (2019) 1907–1912.

[13] H. Brodaty, M. Donkin, Family caregivers of people with dementia, Dialogues Clin. Neurosci. 11 (2) (2009) 217.

[14] Z. Fangmeng, L. Peijia, M. Iwamoto, N. Kuwahara, Emotional changes detection for dementia people with spectrograms from physiological signals, Int. J. Adv. Comput. Sci. Appl. (IJACSA) 9 (10) (2018) 49–54.

[15] Designed by macrovector_official / Freepik, Free Vector / Digital health technologies flat composition, [Online]. Available https://www.freepik.com/free-vector/digital-health-technologies-flat-composition_7272631.htm.

[16] T.Y. Liu-Ambrose, M.C. Ashe, P. Graf, B.L. Beattie, K.M. Khan, Increased risk of falling in older community-dwelling women with mild cognitive impairment, Phys. Ther. 88 (12) (2008) 1482–1491.

[17] M. Pau, I. Mulas, V. Putzu, G. Asoni, D. Viale, I. Mameli, B. Leban, G. Allali, Smoothness of gait in healthy and cognitively impaired individuals: a study on Italian elderly using wearable inertial sensor, Sensors 20 (12) (2020) 3577.

[18] A. Pantall, P. Suresparan, L. Kapa, R. Morris, A. Yarnall, S. Del Din, L. Rochester, Postural dynamics are associated with cognitive decline in Parkinson's disease, Front. Neurol. 9 (2018) 1044.

[19] S. Khatun, B.I. Morshed, G.M. Bidelman, A single-channel eeg-based approach to detect mild cognitive impairment via speech-evoked brain responses, IEEE Trans. Neural Syst. Rehabil. Eng. 27 (5) (2019) 1063–1070.

[20] J. Li, L.S. Broster, G.A. Jicha, N.B. Munro, F.A. Schmitt, E. Abner, R. Kryscio, C.D. Smith, Y. Jiang, A cognitive electrophysiological signature differentiates amnestic mild cognitive impairment from normal aging, Alzheimer's Res. Ther. 9 (1) (2017) 1–10.

[21] J. Jiang, Z. Yan, C. Sheng, M. Wang, Q. Guan, Z. Yu, Y. Han, J. Jiang, A novel detection tool for mild cognitive impairment patients based on eye movement and electroencephalogram, J. Alzheimer's Dis. 72 (2) (2019) 389–399.

[22] I. Jaussent, J. Bouyer, M.-L. Ancelin, C. Berr, A. Foubert-Samier, K. Ritchie, M.M. Ohayon, A. Besset, Y. Dauvilliers, Excessive sleepiness is predictive of cognitive decline in the elderly, Sleep 35 (9) (2012) 1201–1207.

[23] A. Lauraitis, R. Maskeliūnas, R. Damaševičius, T. Krilavičius, A mobile application for smart computer-aided self-administered testing of cognition, speech, and motor impairment, Sensors 20 (11) (2020) 3236.

[24] R. Narasimhan, G. Muthukumaran, C. McGlade, A. Ramakrishnan, Early detection of mild cognitive impairment progression using non-wearable sensor data–a deep learning approach, in: 2020 IEEE Bangalore Humanitarian Technology Conference (B-HTC), IEEE, 2020, pp. 1–6.

[25] R. Chen, F. Jankovic, N. Marinsek, L. Foschini, L. Kourtis, A. Signorini, M. Pugh, J. Shen, R. Yaari, V. Maljkovic, et al., Developing measures of cognitive impairment in the real world from consumer-grade multimodal sensor streams, in: Proceedings of the 25th ACM SIGKDD International Conference on Knowledge Discovery & Data Mining, 2019, pp. 2145–2155.

[26] E. Demir, E. Köseoğlu, R. Sokullu, B. Şeker, Smart home assistant for ambient assisted living of elderly people with dementia, Proc. Comput. Sci. 113 (2017) 609–614.

[27] H. Ro, Y.J. Park, T.-D. Han, A projection-based augmented reality for elderly people with dementia, arXiv preprint, arXiv:1908.06046, 2019.

[28] D. Wolf, D. Besserer, K. Sejunaite, M. Riepe, E. Rukzio, care: an augmented reality support system for dementia patients, in: The 31st Annual ACM Symposium on User Interface Software and Technology Adjunct Proceedings, 2018, pp. 42–44.

[29] Ö. Yilmaz, An ambient assisted living system for dementia patients, Turk. J. Electr. Eng. Comput. Sci. 27 (3) (2019) 2361–2378.

[30] C. Kalafatis, M.H. Modarres, P. Apostolou, H. Marefat, M. Khanbagi, H. Karimi, Z. Vahabi, D. Aarsland, S.-M. Khaligh-Razavi, Validity and cultural generalisability of a 5-minute ai-based, computerized cognitive assessment in mild cognitive impairment and Alzheimer's dementia, Front. Psychiatr. 12 (2021) 1155.

10

Digital twins and cybersecurity in healthcare systems

Issam Al-Dalati

Cyber Security Architect, Ottawa, ON, Canada

Our future is a race between the growing power of technology and the wisdom with which we use it. **Stephen Hawking**

10.1 Introduction

One of the unique characteristics of digital twins in the health care sector not only it can represent or mirrors a physical product or medical device, but it can also digitally mirror a human body itself which is more complex and important to secure. Due to advancement in health technologies, it is possible to construct and map different parts of the human body such as liver, kidney, or event protein structures. So, it can leverage modern technologies such as smart sensor technology, big data analytics, cloud computing, and artificial intelligence (AI) to detect and prevent specific organ or system failures, improve overall health, and explore new security opportunities in the health sector [1].

Digital twins in health care are leveraging many current remote home health care systems and technologies to allow health providers access to patient's data in real time. This way it will enable quicker emergency response, reduce cost of personal health care, and improve patient's treatment plans [2]. As a result, home and health care applications will eventually use the same network to achieve digital twins' goals of providing future health care. One way to implement health care system in home-based networks is to use different types of wearable monitoring sensor devices to create a Wireless Personal Area Network (WPAN) or Wireless Body Area Network (WBAN). A WBAN is a collection of low-power and lightweight wireless sensor devices with limited computation power and

195

resource capacity [1]. Therefore, the main problem that home based networks will become a main backdoor vulnerability not only to digital twins and to the health care system and providers connected to it.

Exploitable vulnerabilities to create ransomware in the related medicate software applications can pose critical security threats to the digital twin systems, and impact on millions of users. Some tailored attacks are based on in depth knowledge of the medical monitoring device along with its control features. For example, an attack against glucose monitoring and insulin delivery system used for diabetes critical treatment can result in insulin dose error that can endanger a person with diabetes life. Attacks on digital twins may require an attack path where multiple correlated vulnerabilities are exploited before either the physical patient or the asset can be targeted. The attack paths where data is collected can pass through technologies like IoT sensors and then sent over different wireless mediums and sent back to the cloud for example where artificial intelligence quantum computing can be used on this sensitive data that can affect someone life. Moreover, due to the spread of these applications, they can offer hackers identity theft and government unethical surveillance. So, these security attacks have extreme consequences and must be analyzed and prevented. This chapter will analyze these attacks and challenges.

The chapter will also list the latest literature research around the proposed solutions that have the potential to mitigate these security attacks and risks associated with digital twins. These solutions will be gathered and grouped into cyber security and privacy frameworks. Cyber Security usually refers to protective measures put in place to ensure confidentiality, integrity, and availability of digital data. Privacy on the other hand refers to the user control rights over their personal information and how that information is used within the digital twin. Even though you can achieve security without privacy, but you can't achieve privacy without security. Therefore, a balance between cyber security and privacy is required when designing and implementing digital twins in the world. Because of this, there may be a gap around a complete research study defining the current challenges and suggested solutions for both cyber security and privacy for digital twins in health care. In this chapter, our contribution is to provide this clarity.

The proposed cyber security framework will go over enhancing current IDS in medical sensors using machine learning anomaly behavior detection and deep learning algorithms. The framework will compare different authentication methods ranging between channel characteristics variation, radio frequency finger printing, and biometric authentication. It will also cover stronger Intrusion Prevention Systems (IPS) and communication encryptions for digital twins. The proposed privacy framework will cover privacy by design principles and how blockchain can be integrated to enhance building trust within the digital twin environment within health care. Digital Twins in health care integrate different new technologies and components with different purposes. The chapter will also list the most related standards that can be used as guidance to get digital twins as securely compliant as possible.

10.2 Digital twin opportunities in cyber security

Digital twin opportunities in the cyber security field can be best defined as "A digital twin, which is used for the purpose of enhancing the security of a cyber-physical system, is a virtual replica of a system that accompanies its physical counterpart during phases of its lifecycle, consumes real-time and historical data if required, and has sufficient fidelity to allow the implementation of the desired security measure" [2].

The following subsections will cover these main opportunities that digital twin can enable in the healthcare systems.

10.2.1 Improving security design and testing

During various phases of the design of a new device or a system, digital twins can be used to reduce design risk anticipation and more accurate evaluation of its security strength against potential future attacks.

Digital twin can be used by the testing engineer to simulate targeted cyber-attacks to identify any weakness and potential security holes in the proposed design. This can be done via blue team and/or red team exercises. The blue team is the defensive security professionals who work from the inside to mimic the patient or the physical health security system to protect it and tests its performance against any kind of threats. The red team is the offensive security professionals who work against the patient or the physical health security system to test for vulnerabilities by launching simulated attacks against the digital twins without effecting the performance of the original physical entity.

The results of these exercises will be returned to the design team as a feedback loop to improve the current design of the system from the security aspects [3]. For example, analysis results can reveal weak areas in the architecture of the system and unprotected services or features that would allow attackers to create any possible backdoors and damages in the system.

Moreover, it can help designers think ahead and create a more robust incident handling techniques during the operation phase of the system when medical devices are compromised or hacked. Security architecture engineers could also test new defenses before putting them into production or train cyber analysis on how to respond to cyber incidents in an ideal Cyber Security Operation Centers (CSOC) [3].

From the cost side of things, digital twins can save time and cost to build a separate testing and training environment. It can also eliminate the risk of testing new changes in the production environment since its mirrors the latest version and state of the production version in a separate sandbox [4].

10.2.2 Support better intrusion detection

Since digital twins maintain a live copy of the physical system or the patients that it is mirror then it can be used as a baseline for anomaly behavior-based IDS to monitor the system or the patient behavior and report any deviations from the digital twin baseline. Digital twins in IDS can helped identify unknown sensors, unspecified ports, and changes in the hardware/software settings. Moreover, the IDS can take the twins baseline state as primary input for host-based IDS and audit the network traffic for network based which can yield a low false positive rate without a training phase for the machine learning algorithms [4]. As a result, an intrusion from an attacker can simply be detected by comparing the inputs and outputs of physical devices attached to the patient or the physical health device and those of the digital twins [5]. The main assumption for this case is that digital twins must be running in an isolated environment that is protected against insider malicious attacks in specific. Otherwise, attacks can tamper with the digital twins' configuration to ensure that any of the physical sensor's settings go unnoticed or undetected.

10.2.3 Enhance privacy controls

A balance between security and privacy is required to protect sensor medical devices against external attacks. Since digital twins uses machine learning techniques anyways to classify personal information, it can help complement privacy controls or standards like General Data Protection Regulation (GDPR). Enhancing the digital twin's classification technique can help implement two main GDPR requirements. One is called anonymization which is a process to remove personal identifiers that are direct and indirect that may lead to a patient being identified. The other requirement is defined as pseudonymization which is the process of changing data in a way that it can no longer be attributed to a specific patient or data subject without the use of data extension [6,7]. Finally, integrating security and legal compliance support into digital twins for healthcare might become more feasible and easier to monitor.

10.3 Digital twin cyber security framework

A typical scenario for a digital twin for health services would start with sensors attached to the body of the patient measuring and capturing vitals like heartbeat, blood pressure, and blood sugars for example. The data is streamed in real time from these sensors and sent back into a central processing unit such as gateway or the cloud for analysis using algorithms. These health-related algorithms will help predict potential health risk based on well-defined parameters and thresholds. The central processing unit used will then send alerting and events to the clinic monitoring this patient. The clinic or the doctors can then evaluate the events received and send back control commands back to the sensors for the proper medicine. This way, digital twins enable two-way communications between the digital and physical environments. Through this one scenario, the data will travel through different software and hardware platforms via different communication mediums. This presents a unique challenge for cyber security where traditional security measures won't be enough to keep the system, the data, and the patient safe. Attacks on digital twins may require an attack path where multiple correlated vulnerabilities are exploited before either the physical patient or the asset can be targeted.

Therefore, a cyber security framework is presented here to integrate threat modeling through the digital twin network and build cyber resilience that will be integrated throughout the whole digital twin process. Threat modeling will be covered first to explain these security challenges then building cyber resilience with proposed solutions is detailed to tackle each of the challenges of digital twins mentioned.

10.3.1 Digital twins threat modeling in health care

Security threat modeling is needed to analyze and defend against advanced cyber-attacks that exploit different technologies. The attack paths where data is collected can pass through technologies like IoT sensors and then sent over different wireless mediums and sent back to the cloud for example where artificial intelligence quantum computing can be used on this sensitive data that can affect someone life.

The threat modeling technique that can be used to analyze the possible attacks on digital twins in the health field is The Kill Chain. This intelligence-driven computer network defense strategy was developed by the Lockheed Martin Corporation in 2011 [8]. The Lockheed Martin researchers created this strategy to anticipate and mitigate future intrusions based on knowledge of the threat and protect against end-to-end cyber-attacks by Advanced Persistent Threats (APTs). The Kill Chain analyzes attacks from the perspective of the attacker, to drive the selection of defensive courses of action. The model, in military context, describes a chain of events that is required to perform a successful attack. Breaking a chain in this sequence of events would result in the failure of the attack [8,9]. These events can be categorized into seven main phases:

1. Reconnaissance: the attacker does a careful investigation about a selected target by researching any background information about the target through the internet and social media. In this phase, the attacker is usually looking for any potential vulnerabilities in the target network.
2. Weaponization: based on the researched findings, the attackers create a remote access trojan customized to one more vulnerability found in the target network.
3. Delivery: the created trojan or malware is sent to the target through either email attachments, website links or removable media drives.
4. Exploitation: this is the phase where the attacker acts on the target network to exploit a discovered vulnerability.
5. Installation: the attacker installs a remote trojan for example to maintain a continuous backdoor to the network target.
6. Command and Control: this phase creates a channel that is will be used for remote manipulation of the target.
7. Actions on Objective: at this phase, the attacker completes their intended goals on the target. For example, encryption for ransomware, fraud, or data exfiltration.

Exploitable vulnerabilities in the related medicate software applications can pose critical security threats to the digital twin systems, and impact on millions of users. An example of such issue is the WannaCry Ransomware, which effected the National Health Service in the United Kingdom, where many of the systems were still running unprotected Windows XP version. Some attacks are based on in depth knowledge of the physical device along with its control features. So digital twins may represent virtual replicas of physical devices that can represent valuable knowledge that might be misused for launching covert attacks. For example, tailored malware attacks called Stuxnet against the nuclear facility at Natanz aimed to manipulate the speed of the centrifuge rotors which had a deep understanding of the nuclear plant design [10].

Glucose monitoring and insulin delivery systems are used for the treatment and management of diabetes which can be used by digital twins. A lab study was done to demonstrate successful security attacks on a commercially deployed glucose monitoring and insulin delivery system used for diabetes critical treatment. The scientist was able to fully discover the password of the remote control and glucose meter. Then they were able to send a fake data packet which was accepted by the insulin pump. These fake packets lead to misleading meter measurement information which can result in insulin dose error that can endanger a person with diabetes life [11].

10.3.2 Common attacks on digital twins medical devices

Different attacks can cause major disruptions against implanted medical sensor devices within a short distance network like WPANs or WBANs. One of the common examples of attacking devices are Unmanned Aerial Vehicles (UAV) like drones where they represent an ideal device to attack due to their small size, ease of access and reach to small areas [12]

These attacks can be classified into three main categories:

1. User based attacks: a typical attack is cloud account hijacking occurs when a hacker obtains your personal information and uses it to take over your digital twin. There are various techniques for achieving this like spyware, cookie poisoning, phishing attack, ransomware, cloud account corporate credentials are exposed somewhere or using social engineering tactics.
2. Host based attacks:
   a. Device tampering: this requires physical access to the sensor where the attacker can tamper with the circuit components, replace it with a trojan device or change the programming of the sensor to compromise the device integrity.
   b. Resource exhaustion: a form of a Denial of Service (DoS) attack caused by high retransmissions of packets due to corrupted or lost packets. The aim of such an attack is to consume energy of the sensors to bring it down and stop the health application that is running on that device.
   c. Hello flood attack: the malicious sensor device used in this attack has high power and communication range. The sensor over floods all its neighboring sensors with short Hello messages to waste the energy and the battery of the sensor device [6]. A stronger authentication method is required to mitigate this attack. One example to use is called uTESLA part of the Security Protocols for Sensors Networks (SPINS) which provides a stronger broadcasting authenticated communication to the IoT sensors network scenario [13,14].
   d. False data injection attack: the attacker can temper with measured data by a small margin at each sensor by modifying their payloads for example that will mislead the decision-making system to generate inaccurate results. These types of attacks effects the integrity of data and causes lots of worldwide incidents such as the Ukrainian power grid attack in 2015 that effected more than 230K people who lost power to their homes [15]. Therefore, incorrect heart rate or blood pressure reading can jeopardize a patient treatment and health in a digital twin.
3. Network based attacks:

a. Jamming DoS: malicious nodes in the network can occupy a live communication channel to cause interference in the service and block sensors from communication between each other.
b. Link layer collision: this occurs when two sensors transmit on the same frequency at the same time. Frequent retransmission can cause delays and can affect the performance of time sensitive health applications running. Unfairness is the intermittent version of the collision attack which can cause degradation of the sensor communication quality. One common correction to use against these attacks is the use of error correcting codes that adds additional processing overhead on the other hand [16].
c. Acknowledgment spoofing: a man in the middle kind of attack where the malicious sensor node overhears packet transmissions from its neighboring sensor and sends back a false acknowledgment to deceive the sender sensor that the transmission was successful. Another type of spoofing would be identity spoofing where a malicious node impersonates the destination node to receive information for a man in the middle attack purposes. A real example in action would be called Drone in the Middle (DitM) attack where a drone intercepts communication between two monitoring sensors to redirect any transmission between them [12]. One way to help against spoofing is to append a Message Authentication Code (MAC) after the packet which can be used to verify if the message was received correctly or altered [13]. Using stronger authentication techniques like central certificate authority can help prevent such an attack [18].
d. Sinkhole attack: a malicious sensor tries to attack routing traffic through its device by advertising fake routing updates pretending that it has the shortest path to the destination node. The sinkhole attack can be used also to launch other different attacks such as selective forwarding and acknowledge spoofing attacks. This attack might be only applicable if there are number of IoT medical sensors forming a network to forward traffic within each other. One proposed approach to detect sinkhole attack nodes using a geostatistical method by detecting the energy hole that forms around each sinkhole since the routes via these nodes generate more traffic than the rest of the nodes. The second method proposed uses a distributed monitoring system to detect nodes with lower average residual energy level [17].
e. Wormhole attack: two malicious sensors are used in this type of attack to establish a low latency link between both of those sensors. This link is advertised through the network as the best path to use for communication so other sensors will use it for their routing. This attack can cause a stoppage in critical health services scenarios if a network of IoT devise is used to propagate signals to each

other. One method used to mitigate against such attacks is using the round-trip signal time and speed to estimate the distance between sensors, so they all be aware of the shortest and best path to use. Another method is a cluster-based avoidance mechanism where the network is divided into three level hierarchical clusters. This way each sensor can calculate the shortest path without the need for advertising or updates [19].

f. Eavesdropping and traffic analysis: this attack targets mostly the privacy of the user by intercepting data during the communication and can use the read data for future attacks. Strong access control and encryption methods can help mitigate this type of attack [13, 18]. Secure Network Encryption Protocol (SNEP) part of the SPINS protocols can help in providing confidentiality by using symmetric encryption algorithms which requires lower overhead [14].

g. Desynchronization: an attack that disrupt communication by providing a different timestamp to disrupt timing between two sensors. This attack will try to consume the battery of both medical sensors to bring the service eventually down. One way to prevent such an attack is to enforce strong authentication of all transmissions between sensors [13].

h. Selective forwarding: a malicious sensor only forwards data to a few selected sensors instead of all sensors and drops the rest. This attack is applicable only if a network of sensor devices is used to route data. A possible counter to this attack is using multiple routes to send the data between sensors [13,20].

i. Sybil attack: this occurs when the malicious sensor has more than one identity that is used to disrupt the data aggregation and fair resource allocations between sensors within a network of sensor devices used in a health application service [19]. One way to help against Sybil attacks is to use something called random key predistribution techniques where a random set of key related information is assigned to each sensor. This way during the initialization phase, the sensors can discover the common key related information it shares with other neighboring sensors, so the common keys are used as shared server key [13].

j. Steppingstone attack: more than one malicious device is used in this scenario from one attacker. For example, more than one drone can be used to forward commands to each other until the command reaches the closest drone to the target monitoring sensor in the digital twins [12].

k. Evil twin attack: malicious drone or device acts as the fake Wi-Fi access point to the digital twin network where starts first by deauthenticating all connected sensors and then launching a fake Wi-Fi access point so the sensors can reconnect with it [12].

10.3.3 Digital twin authentication and identification challenge

One of the biggest Distributed Denial of Service (DDOS) cyber-attack in history happened on October 21, 2016. A malware called Mirai turned on an army of zombie botnets (400K devices) on one of the world's largest DNS providers. The attack took down internet access in many of America's largest cities and impacted famous sites including PayPal, Twitter, and Netflix. Mirai gave the entire world just a glimpse of how vulnerabilities in IoT devices can be exploited. Mirai infected IoT devices by successfully attempting to log in using a table of 61 common hard-coded default usernames and passwords [21].

At exploitation phase in the kill chain process, the intruder begins executing when they enter the victim's environment. This where access control mechanism acts like the keys to the main doors. This mechanism starts by identification then authentication then authorization to access certain features within digital twin. Digital identity within digital twin's key challenge is that it requires a central authority to prove that patients accessing their digital twin is really them. Passports now adays have digital data embedded in them but it requires a specific physical device for identification. Certificates can be also used but they are depended on a central validation system which can become a single point of failure. Moreover, these centralized systems are hard to implement on top of the decentralized sensor monitoring network as part of the digital twins environment.

In terms of authentication and authorization, the current methods rely on traditional password techniques. These techniques are very vulnerable during the first phase of the kill chain which is the reconnaissance where the intruder look for easy hints in the patient social media profiles as an example. So brute force attacks can easily extract passwords intercepted during the reconnaissance phase. Therefore, password techniques can be easily stolen, and it doesn't guarantee the identity of the patient. Password management is always an evolving issue specially if it's used to protect sensitive health data in digital twins. Another method used is cryptographic certificates which are challenging to distribute due to limited sensor power resources [2].

10.3.4 Building cyber resilience in digital twins

To tackle of each of the challenges of digital twins mentioned before and more future ones that will appear with new scenarios and health monitoring applications, cyber resilience should be integrated throughout the whole digital twin process [22]. Cyber resilience can be defined as a measure of how well digital twins can manager a cyber attack or data leakage while continue to operate effectively without causing damages back to the patient. Therefore, digital twins with cyber resilience need to provide data confidentiality, integrity, and availability. This can be achieved by improving and imbedding cyber resilience through the innovation of authentication, data encryption, vulnerability detection techniques, redundancy, and security policies to provide the trustworthy required for healthcare organizations and patients [22]. Different improvements researched will be detailed in the following sections.

10.3.4.1 Stronger IDS

There are two types of IDS. A host-based network detection where agents are installed on sensor devices to monitor the activities and characteristic of a single node for any suspicious activities. The other type is a network-based detection where the network traffic is monitored and analyzed to identify any possible threats.

Three main detection techniques are used by IDS to identify suspicious events:

1. Signature based: the patterns corresponding to known threats are created in this detection technique. This method will create signature patterns that are compared against live network traffic to identify any suspicious events. It looks for predefined indicators such as file hashes, malicious domains, certain bye sequences, and file names. This technique has high accuracy and processing speeds to detect known attacks. It also consumes less energy which can be suitable for sensor monitoring devices that has limited battery power. However, this technique is incapable of detecting unknown new attacks.
2. Anomaly based: the normal behavior profile is developed and updated as the system normal behavior changes over time using statistical methods and machine learning algorithms. This normal behavior baseline is then compared against the current observed network traffic to identify any deviation between them which can result in suspicious events. This type of detection can help detect zero-day new exploits. However, this technique can produce many false positives events and it requires lots of power to perform continuous complex computing.
3. Stateful protocol analysis: baseline of normal activities of a protocol is created and compared with observed network behavior to identify any deviations which can result in suspicious events. This analysis is usually vendor or standard specific that establishes a deep understanding of how the network protocol, transport and application interact and defer from its original state. The primary issue with this analysis is that it requires lots of resources and energy to complete the complex computations and cannot detect attacks that do not violate the standards of a generally acceptable protocol behavior.

Most IDS technologies use multiple techniques, either separately or integrated together to provide more accurate detection. Zero-day attacks are continuously emerging and happening every day due to the addition of new protocols specially from IoT new devices and applications. These new attacks are either variants of already existing attacks or totally new ones which makes it difficult for traditional signature-based detection methods to detect these new attacks [23].

Therefore, enhancing IDS in IoT medical sensors is a continuous issue. For IoT sensors devices, the focus of the current research nowadays is focused on anomaly based to detect zero-day exploits which are more challenging. So many research papers have been done in the machine learning field to come up with the most efficient anomaly behavior detection algorithm. To be effective, these anomaly behavior detection algorithms need to work under low capacity, low power, big data, and fast response.

Machine learning algorithms are constructed through two main phases. The training phase that uses mathematical algorithms fed with regular data to learn the characteristics of the sensor monitoring network. Then it is followed by the detection phase where nonregular data are used to validate detection and classification [23]. A list of the most popular machine learning techniques researched recently for IDS for IoT devices in health care systems is [24]:

- Support Vector Machines (SVMs): a supervised learning technique for non-linear data classification and regression. It uses a margin kernel function to nonlinearly map samples into a higher dimensional space to create a clearer distinction between data. The main feature of SVM that differentiate it from other machine learning techniques is that it maximizes the margin to reduce the numbers of weights that are nonzero to just a few to provide a more useful information out of the nonlinear data input [25].
- Decision Tree: a classification algorithm where each outcome of a monitored sensor security device attribute is represented by a branch. The leaves in the decision tree output the result of the classification as normal or abnormal. The decision tree based on initial values of security attribute features which represents the normal behavior. These attribute features can be actions like changing the sensor settings, action timestamp, number of occurrences of an action, time interval since last occurrence of an action, signal strength indicator, access day, and location of the implanted sensor. Once monitored thresholds are exceeded for any of these features is detected, a warning is triggered that an attack is occurring [26].
- Random Forest Tree: a learning method for data classification and regression using multiple decision trees for training. The output of the random forest for classification in IDS is the profile class selected by the most decision tree. Random forest provides higher accuracy than a single decision tree with shorter time to execute [27].
- K-Nearest Neighbors (KNN): a supervised learning technique for data classification and regression. The KNN classifier assumes that a profile

instance is most like the classifications of other profile instances that are nearby in the distance space. This technique doesn't rely on previous probability in its learning algorithm, so it requires less computation [27].
- Artificial Neural Network (ANN): another learning technique which is based on biological neural networks ANN is based on connected sensors called artificial neurons which models the neurons in a brain and the connections are called edges. Neuron's sensors and edges have weights values that gets adjusted as learning continuous during the lifetime of the sensor.

Most of the used machine learning detection techniques are considered supervised learning techniques which means it requires ongoing human participation to feed the algorithms with training data to teach it on how to respond to a subset of data. However, the data most of the time is unlabeled and training data is not available. So, more research being done under the unsupervised learning category where the machine organizes the data without human intervention into common data clusters or themes with different layers. Deep learning algorithms came into play as a result which introduces sophisticated approach to machine learning and extends the ANN techniques into multilayered approach. This way it can generate more accurate nonlinear transformation of the data without human interventions. Due to current developments in IoT hardware, deep learning can help in detecting complex cyber breaches [28]. The most common unsupervised deep learning detection algorithms are Deep Restricted Boltzmann Machine (DBM), Autoencoder Neural Network (AE) and Generative Adversarial Network (GAN) [23].

10.3.4.2 Stronger intrusion prevention system (IPS)

Another technology is used in conjunction with an IDS is called IPS. An IPS has all the features of an IDS to detect any possible attacks and does one extra step to stop or respond to those attacks. The main response techniques can be divided into:

- Blocking an attack: the IPS will attempt to block the target IP address, user account, source file or any other attributes of the attacker.
- Change the current settings: the IPS will attempt to reconfigure automatically current security controls like blocking a port or applying certain patches to a sensor automatically.
- Deletion: the IPS attempts to change or remove contents of an attack. For example, IPS can simply delete the infected file received by an attacker or it can normalize incoming requests traffic by removing header information in the packets received.

10.3.4.3 Future digital twin authentication methods

There are different types of authentication factors. They can be categorized as follows:

Something you are: using physiological characteristics of the user to authenticate such as retinal scan, fingerprint, face recognition, voice print, and signatures.

Something you know: using secret knowledge that only the user supposed to know and remember such passwords, birth date, birthplace, and passport number.

Something you have: using things that a user might own like smartcards, passport, tokens, or driving license.

Something you do: using behavior characteristics of the user such as daily home routine, traveling routes, bank transactions, and social media activity.

Multi factor authentication is using at least two or more different types of factor combination to complete the authentication process for a more secure way and help in prevent cloud-based user attacks such as account hijacking.

Channel characteristics variation authentication

One type of authentication that can be applied for health sensor monitoring devices is based on behavior pattern of the channel used to communicate between the monitoring sensors for a user. This design is a risk based adaptive authentication method based on the naive Bayesian network algorithm [26].

The channel characteristics variation between monitoring sensors can be measured to model the behavioral patterns of the user by recording the following parameters:

- Received Signal Strength Indicator (RSSI) expressed in decibels (dB) is the power level received by a sensor device.
- Channel gain: due to multipath characteristics of radio signal propagating between sensors, it causes different channel gain values that can be used as a parameter value.
- Doppler measurement detects the sensor while its moving by recording the frequency shift caused by the speed of the transmitting sensor.
- Temporal link signature: the difference in the amount of time for a radio signal to arrive at the destination sensor due to multipath characteristics of radio signal.

These collected parameters are then applied to the naive Bayesian network algorithm to build the behavioral pattern of the user for a given device. After first time authentication, the system keeps monitoring and classifying the user behavioral patterns based on the knowledge obtained on the initialization stage. This process will be used then to authenticate the user for each device pattern [29].

Radio frequency (RF) fingerprinting

The concept of RF fingerprinting is that electronic sensors can be identified and authenticated through their own unique radio frequency emissions. Traditional RF fingerprinting methods can be classified as either transient approach, steady state approach or deep learning approach based on their target signal region. Depending on the input type of the used algorithm, the technique can be either time series based, or image based.

This authentication method can work on its own or to support multi factor authentication which needs a classification algorithm to distinguish between wireless monitoring sensors of the same model and different serial numbers [30].

Biometric authentication

This type of authentication type is based on "Something you are" and it usually offers better authentication than "Something you know" like passwords since its physically and permanently associated with the user. The following methods can be explored to be used and enhanced by researchers for digital twin access for health monitoring applications:

- Fingerprint authentication: the most popular biometric authentication type and most convenient one to use. However, this type is vulnerable to damaged fingerprint, and it can be captured by hackers using the print left by patients when they touch things [31].
- Eye based authentication: One method uses the complex physical patterns of one of both irises of an individual eye. Another one uses the unique patterns on the user's retina blood vessels where a special near infrared imaging scanner is used to capture that. Even though these two methods are reliable biometric authentication methods, but they might get sensitive to changes in lighting conditions which makes it a bet hard to implement in digital twins [32].
- Voice based authentication: this method enables users to access their application using their own speech which is verified against a saved database of recorded voice samples from users. The drawback of this method is that voice recording can be easily duplicated, and it needs a robust noise cancellation mechanism during the recordings to minimize false positives [32].
- Electrocardiogram (ECG) authentication: it uses the electrical signal pattern of the patient heart rhythm to create an authentication profile for the user. This signal is unique for each patient because it is based on the gender, heart mass orientation, conductivity, and order of activation of cardiac muscles [31]. For hackers to bypass this method, they need full access to the body of the user to measure their ECG rate. So, they require special hardware. The drawback of this method that it needs 10 seconds or longer to capture an accurate ECG signal to lower false positives [31].
- Multibiometric authentication: this process combines multiple biometrics methods by feeding the measured values from different sensors into a defined algorithm to achieve higher accuracy and efficiency than

using just one biometric system with better security. One research proposed an algorithm to combine both ECG and fingerprints to achieve high authentication security with less amount of time [32]. This combined method can be applied in digital twins where the user can place both of their thumbs on a single scanner where it can measure both ECG and fingerprints for verifications to ensure privacy of the patient's data [33].

10.3.4.4 Protecting the communication channel for digital twins

IoT sensors may need to use a short-range communication protocol to connect itself with a central device from where data is transferred to the server or cloud. The short-range communication protocols include near field communication (NFC), Bluetooth, IEEE 802.15.4, Wi-fi, ZigBee, and 6LoWPAN.

Body coupled communication (BCC) uses the human body as communication channel. Signals are conveyed over the body instead of through the air in contrast to RF communications where a much larger area is covered. Therefore, communication is possible between devices placed close to the body. This implementation of BANs enables low-cost and low-power consumption than in standard radio systems commonly used for BANs like ZigBee or Bluetooth [34].

This creates possibilities for many applications in the field of authentication and security because of the communication range is limited to the distance within the body which prevents interference from other close by sensors that are not attached to the body. Moreover, this technology consumes less power because the signal between different sensors propagates through the body instead of longer distance through free space [11].

Encryption is used to make sure data is protected even if it's leaked through the communication channel. There are two main types of encryption that can be used in digital twins. One type is asymmetric which uses a public and private key for encrypting and decrypting the information. The other type is symmetric encryption that use the same private key for encrypting and decrypting. Asymmetric keys are not commonly used in IoT network due to the large key size [35]. However, asymmetric algorithms such as Elliptic Curve Cryptography (ECC) public key cryptography is based on algebraic structure of elliptic curves over finite fields. It is commonly used in smart card encryption and considered efficient technique due to its lower key size and hard exponential time challenge for the attacker to break into the system. For example, in ECC a 160-bit key provides the same security as RSA with 1024-bit key. Therefore, ECC requires only lower computation and less memory space. This algorithm can be integrated with blockchain to provide both confidentiality and integrity for digital twins.

Symmetric encryption algorithms on the other hand have smaller key size and are more well suited to device with lower power and storage. The symmetric algorithms can be also used either block or stream methods to generate the cipher text. However, they require more complex commutation, and they are slow to run. There are proposed lightweight encryption algorithms which is based on symmetric key algorithms to use smaller key size and provide efficient security [36].

10.4 Digital twin privacy framework

10.4.1 Lack of privacy and trust challenge

Privacy can be considered one of the most challenging issues that can be faced by digital twins in the health care sector due to dual responsibility on not only the technology and the health care providers from one side but the users and patients themselves on the other side. The users who own their own data can fall under the privacy paradox where actual users' intentions to not share personal information defer from their actual active daily behaviors especially online and on social media platforms such as twitter, Facebook and snapchat for example [37]. The privacy paradox can be captured and tracked under a theory called privacy calculus theory where users perform on their own a risk benefit decision making analysis to see if the benefits outweigh the privacy risk. Then the chance to adopt to digital twins to disclose personal health information would be higher. Key factors that effect a user decision in the privacy calculus theory can be health information sensitivity to the patient, personal appeal to new technologies, any legislative protection that exist, emotional fear, and perceived reputation of the wearable device [38,39].

There are two forms of illegal information collection. One is called interrogation and the other is surveillance. Interrogation is about pressuring patients to release information. So, data from digital twins can be used for marketing and advertising without user consent and without any compensation. Personal data from these digital twins can be used to generate behavioral patterns or predictions to sell to third parties which can cause threats to personal freedom and privacy. In addition, a data leakage can happen if these third parties don't maintain enough protection of this data and are exposed to cyber-attacks and hacking [40].

The other form of information collection is surveillance where digital twins in health can cause segmentation and discrimination between people in two ways. One way where this new technology might not be affordable or available for every individual within the society. The other security related way is that pattens might be identified through surveillance from the data consolidated from different digital twins from different patients so people can be categorized in a certain unacceptable way [41].

Even though some health tracing applications applicable to digital twins might have a large benefit on the society specially when dealing with an epidemic disease like COVID-19 tracing apps but it can expose serious threats. Due to the large use of these kind of applications, they can offer hackers and government unethical surveillance. Moreover, these kinds of tracing application can create an atmosphere of fear for those who are diagnosed with COIVD-19 who can be targeted for fraud and abuse [42].

Another risk or challenge that can be caused by digital twins in health care is identity theft or financial fraud since it can expose both physiological and behavioral profile patterns of the patient. Physiological profile includes fingerprints, eye scans, shapes of a face, blood type, DNA, and urine analysis. Behavioral profile includes hand signatures, voice, and movement patterns.

Finally, building trust in digital twins in the field of healthcare can be challenging because its relatively new. Patients might also have an anxiety and fear against artificial intelligence digital twins to become a dominant force on earth by taking full control of their body.

10.4.2 Privacy by design

Since personal data is always used in digital twins in health care services, then privacy must be taken into considerations from the creation of digital twins to termination. A concept called "Privacy by Design" was developed in 2012 to encourage organizations to integrate privacy design concepts from the beginning of the development phase when dealing with sensitive data such as personal health information. The objective of privacy by design is to balance personal information in a way that is controlled by the owner or the patient. Privacy should also be carefully managed by the health providers by building Fair Information Practices (FIPs) into the design, operations, and management of information systems [43]. The following seven principles are used to implement the objectives of privacy by design:

1. Proactive and preventative: this principle covers the proactive activities that happen before implementing the digital twin technology to prevent any privacy events before they happen. For example, careful risk analysis on the privacy of information based on the specific future scenarios used by the digital twin scenario to create a list of privacy requirements for the initial design phase.
2. Privacy as the default: during the configuration of digital twins, privacy settings must be configured by default and automatically without any user interactions.
3. Privacy embedded into design: privacy must be integrated in the core framework of the design digital twin system for remote health monitoring for example.
4. Positive sum functionality: patients should not be put in a tough situation to decide between functionality and privacy in digital twins. So

digital twins need to be designed in such a way where unnecessary trade offs are made.

5. End-to-end lifecycle protection: data privacy should be imbedded in digitals twins throughout its life cycle from start to finish.

6. Visibility and transparency: all components of how digital twin technologies handle personal data should be visible to both the patient and the health providers. They can use audit logs for example to enable more confidence of the remote health monitoring application. Moreover, patients need to know what data has been collected, how is it being used, and who can view it from the health providers for example.

7. Respect for users' privacy: privacy interest of patients or users should be put forward as the highest priority during the design of digital twins in health care services.

These seven principles for privacy by design were used by researchers to strength it by combing it with other standard requirements like the General Data Protection Regulation (GDPR). For example, some researchers proposed eight privacy design strategies based on GDPR [44]:

1. Minimize: "limiting usage as much as possible by excluding, selecting, stripping, or destroying any storage, collection, retention or operation on personal data, within the constraints of the agreed upon purposes".

2. Hide: "preventing exposure as much as possible by mixing, obfuscating, dissociating, or restricting access to any storage, sharing or operation on personal data, within the constraints of the agreed upon purposes".

3. Separate: "preventing correlation as much as possible by distributing or isolating any storage, collection or operation on personal data, within the constraints of the agreed upon purposes".

4. Abstract: "limiting details as much as possible by summarizing or grouping any storage, collection or operation on personal data, within the constraints of the agreed upon purposes".

5. Inform: "providing as abundant clarity as possible for supplying, explaining, and notifying on storage, collection, retention, sharing, changes, breaches or operation on personal data, in a timely manner, within the constraints of the agreed upon purposes".

6. Control: "providing as abundant means as possible for consenting to, choosing, updating, and retracting from storage, collection, retention, sharing or operation on personal data, in a timely manner, within the constraints of the agreed upon purposes".

7. Enforce: "ensuring as abundant commitment as possible for creating, maintaining, and upholding policies and technical controls regarding storage, collection, retention, sharing, changes, breaches or operation on personal data, in a timely manner, within the constraints of the agreed upon purposes".

8. Demonstrate: "ensuring as abundant evidence as possible for testing, auditing, logging, and reporting on policies and technical controls regarding storage, collection, retention, sharing, changes, breaches or operation on personal data, in a timely manner, within the constraints of the agreed upon purposes".

Another privacy enhancement can be done using intelligent privacy profiles. Each profile will contain different classification levels of the data that can be accessed by certain stakeholders. The profiles will indicate which data can be shared with signed consent from the digital twin and which data can't be shared [39].

Informed consent for personal data use in healthcare is a central requirement for ensuring the protection of personal data. However, current informed consent practices often fail to write clear strict rules that are enforced for the use of personal data to make it easier to use and share.

10.4.3 Enhancing trust with block chain integration

Establishing trust among sensors and other components within digital twins is a challenge that can mitigated using the block chain technology concept which is secure by design to maintain trustworthy data between sensors. Therefore, blockchain can be used to create a secure distributed database between patients and healthcare providers to improve the accuracy of data exchanged to eventually provide a more accurate treatments and prevent attacks on digital twins [45].

Block chain is a peer-to-peer network where it uses either a public or a private distributed ledger to validate and communicate with new sensors. The current researched digital twin blockchain solutions mostly use mainstream sequential chain-structured public or private blockchains, such as Ethereum (ETH) or Hyperledger (HLF). Each node in the block chain network contains information about the previous node or sensor to construct a trusted chain. This information can be a cryptographic hash of the previous sensor, timestamp, and transaction data. Transactions among sensors and other components within digital twins are timestamped and logged in a shared private or public digital ledger. So, to add a new sensor or node into the trusted network of digital twins, the new node must pass a consensus algorithm to solve a resource demanding complex and verifiable cryptographic mathematical equation [46]. Once new nodes solve this consensus algorithm, transactions are grouped to form a new block to append to the shared digital ledger. Some benefits of integrating digital twins with block can be summarized as follows [46]:

- Build a trustworthy digital twin: block chain can help to collect data only from trusted and registered nodes and sensors to prevent any tempering. To enhance this trust, public keys and digital certificates be leveraged in the block chain to register any new sensor to confirm its au-

thenticity. Moreover, patching, and new firmware updates can be done via the trusted blockchain to avoid malware or malicious code injection.
– Enhance digital twin data integrity and confidentiality: block chain cryptographic algorithms as well as the hashing techniques can help secure the creation or access transaction through the life of the digital twin. However, these techniques will require more processing power where further research is needed to optimize power consumption. It can also improve privacy by using public keys to represent sensor profiles for blockchain transactions. At the same time, sensors can use multiple pseudonyms to remain anonymous to strength confidentiality in the network.
– Better Access Control: blockchain can provide a better authentication mechanism based on distributed control which can help save power consumption on one central node. For example, a smart contract-based solution relying on blockchain can be used by the user to send an announcement transaction to the blockchain to record on the distributed ledger. So, the smart contract can handle the access control to any recorded data [47]. An example of a real implementation is Microsoft which recently launched their ION open and public network for providing decentralized identity services running on the Bitcoin blockchain [48].
– Improve digital twin product authenticity.
– Add auditing mechanism for data traceability in digital twins.

10.5 Digital twins compliance with standards and governance

Like any other new technology that appear, it needs proper guidance and standards to be followed so it can be trusted to use within the users and the community in general. Digital Twins in health care integrate different new technologies and components with different purposes. We will touch on the most related standards that can be used as guidance to get digital twins as securely compliant as possible.

National Institute of Standards and Technology (NIST) is one of the most followed practices for Information Security. Currently there is a NIST standard in draft mode NIST-Internal Report (IR) 8356 detailing different challenges facing digital twins' technologies [49]. It recommends that digital twins in general follow a proper cyber security and privacy control framework. A NIST Publication 800-53 titled "Security and Privacy Controls for Info Systems and Organizations" contains a complete list of a cyber security control catalog that covers every aspect of the technology from the physical layer to the application layer [50]. If all these security controls are meet, we can make sure that digital twins application running in health care service are protected and compliant from the framework perspective. Moreover, since digital twins are dealing with sensitive health data for patients, a separate privacy analysis can be done using the NIST privacy framework to make sure all the specific privacy security controls are meet [51].

An important aspect of using digital twins in health care is the collection of sensitive personal data such as DNA details, medical history records and current health status. Therefore, they require strong governance policies or standards locally and globally to ensure privacy, confidentiality, and integrity. For example, in the USA they use the Health Insurance Portability and Accountability Act (HIPAA) created in 1996 to find privacy rules for health care providers, health group plans and business associates including health related tools development. The privacy rule defines personal health data as Protected Health Information (PHI) which includes patients' past, present, or future physical or mental health condition. The HIPAA rule also protects electronic Protected Health Information (e-PHI) that covers all patients' related health data an organization creates, receive, stores or transmits in electronic form [52]. This means organizations need to ensure confidentiality of all e-PHI and secures to make sure it is protected against any cyber attacks for any possible disclosure. In addition, there is the Gramm-Leach-Bliley Act founded in the USA in 1999 to place limits on the reuse of personal data when a company provides it to another company [53]. There is also a specific USA rule for children under the age of 13 called Children's Online Privacy Protection Act (COPPA) which states that websites must protect the confidentiality, security, and integrity of personal information collected from children [54].

In terms of global standards or laws governing personal data, the well known one for data protection is called General Data Protection Regulation (GDPR) of the European Union (EU).

Moreover, over 100 countries have adopted data protection laws [55]. Different countries have something called Tort law which protects citizens against surveillance and privacy affairs.

Another aspect of digital twins is that it uses machine learning and artificial intelligence to make predictive analysis. Therefore, the same principles of ethics that govern the use of artificial intelligence in health in general can apply to digital twins from that aspect. There is an Ethics and Governance of Artificial Intelligence for Health produced by World Health Organization (WHO) [55].

Developers of sensors and equipment's that are being used to produce digital twins should get all the required information security certification to ensure compliant with all the essential cyber security and privacy principles. When manufacturers and developers achieve the highest security standards and certification, it will increase its ability to build a trust with the consumers themselves. One accredited international audit certificate is called International Standard of Information Security Management (ISO) 27001 which requires organizations to establish, implement, maintain and agile improvement of its information security management system [56]. Another certification that is intended for organizations providing information system services is called System and Organization Controls (SOC) Type 2. This standard is acquired after providing extra measures to protect user's data based on organization achieving high level of availability, integrity, confidentiality, privacy, and security [57]. Digital Twins using encryption modules should also follow the 140 series of Federal Information Processing Standards (FIPS) if it's based in the USA that specifies the security requirements for both software and hardware cryptography modules [58]. At the end of the life cycle of the used sensors and equipment's in digital twin, they need to follow proper sanitization guidelines in NIST 800-88 to make sure personal data is destroyed completely and hackers won't be able to retrieve it directly from the physical device itself [59].

The ISO has established advisory groups under the technical committee ISO/IEC JTC 1 to develop standards for digital twins [60]. Another standard worth mentioning is the ISO/IEEE 11073 Personal Health Device standards created in 2008 to help achieve interoperability and integrate personal health sensors devices and monitoring devices such as smart phones [61]. However, this standard lacks the cyber security aspects. So more specific security guidelines for personal health devices are required to recommend the most proper authentication and identity methods for digital twins in health care.

10.6 Conclusion

There is great potential for digital twins to be a key technology in the future of healthcare services. Implantable medical devices are being increasingly deployed to improve diagnosis, monitoring, and therapy for a range of medical conditions anytime and anywhere. This technology will be enabled using IoT real time monitoring devices, AI, augmented and virtual reality, and cloud computing services. Because of the complexity of the integration of these various services, there is a lack of clarity as to what are the vulnerabilities this technology holds and what are the available security solutions researched for each of those challenges. We attempted in this chapter to provide this clarity. We explained this problem by splitting it into cyber security and privacy frameworks. We focused on the various cyber security attacks and what are the latest algorithms research that has been done to identify each possible security hole within the digital twin architecture. We evaluated privacy design issues and what a lack of trust can do to digital twin functionality and quality. We also mapped our evaluations where appropriate to other cybersecurity and privacy guidance and standards.

More future work should be focused on tweaking the current NIST 800-53 to be digital twin focused to make sure that digital twins application running in health care service are protected and compliant. Moreover, ethics that govern the use of digital twins in healthcare should be standardized and government enforced in a similar way like the GDPR standard. Finally, more work should be done in defining more accurate models that will be able to predict sophisticated attacks before they happen without effecting the performance of the system to improve digital twin security in health care.

References

[1] A. El Saddik, F. Laamarti, M. Alja'Afreh, The potential of digital twins, IEEE Instrum. Meas. Mag. 24 (3) (May 2021) 36–41, https://doi.org/10.1109/MIM.2021.9436090.

[2] M. Eckhart, A. Ekelhart, Digital Twins for Cyber-Physical Systems Security: State of the Art and Outlook, in: Security and Quality in Cyber-Physical Systems Engineering, Springer International Publishing, Switzerland, 2019, pp. 383–412, ISBN: 978-3-030-25311-0.

[3] A. Bécue, Y. Fourastier, I. Praça, A. Savarit, C. Baron, B. Gradussofs, E. Pouille, C. Thomas, Cyberfactory#1—securing the industry 4.0 with cyber-ranges and digital twins, in: 2018 14th IEEE International Workshop on Factory Communication Systems (WFCS), 2018, pp. 1–4.

[4] M. Eckhart, A. Ekelhart, Towards security-aware virtual environments for digital twins, in: Proceedings of the 4th ACM Workshop on Cyber-Physical System Security, CPSS '18, ACM, New York, NY, USA, 2018, pp. 61–72.

[5] M. Eckhart, A. Ekelhart, A specification-based state replication approach for digital twins, in: Proceedings of the 2018 Workshop on Cyber-Physical Systems Security and PrivaCy, CPS-SPC '18, ACM, New York, NY, USA, 2018, pp. 36–47.

[6] Salughter and May, Personal data, anonymization and pseudonymization under the GDPR, https://bit.ly/2tEOHTH.

[7] V. Damjanovic-Behrendt, A digital twin-based privacy enhancement mechanism for the automotive industry, in: Proceedings of the 9th International Conference on Intelligent Systems: Theory, Research and Innovation in Applications, 2018.

[8] E.M. Hutchins, M.J. Cloppert, R.M. Amin, Intelligence-Driven Computer Network Defense Informed by Analysis of Adversary Campaigns and Intrusion Kill Chains, Lockheed Martin, Bethesda, MD, 2011.

[9] L. Liu, O. De Vel, Q.L. Han, J. Zhang, Y. Xiang, Detecting and preventing cyber insider threats: a survey, IEEE Commun. Surv. Tutor. 20 (2) (2018) 1397–1417.

[10] R. Langner, To kill a centrifuge: a technical analysis of what Stuxnet's creators tried to achieve, 2013.

[11] C. Li, A. Raghunathan, N.K. Jha, Hijacking an insulin pump: security attacks and defenses for a diabetes therapy system, in: Proc 13th IEEE Int Conf e-Health Networking, Appl. Serv., 2011, pp. 150–156.

[12] S.C. Sethuraman, V. Vijayakumar, S. Walczak, Cyber attacks on healthcare devices using unmanned aerial vehicles, J. Med. Syst. 44 (1) (Jan. 2020) 29.

[13] J. Sen, A survey on wireless sensor network security, Int. J. Commun. Netw. Inf. Secur. 1 (2) (August 2009) 59–82.

[14] A. Perrig, R. Szewczyk, V. Wen, D.E. Culler, J.D. Tygar, SPINS: security protocols for sensor networks, Wirel. Netw. 8 (5) (September 2002) 521–534.

[15] G.R. Mode, P. Calyam, K.A. Hoque, False data injection attacks in Internet of things and deep learning enabled predictive analytics, arXiv preprint, arXiv:1910.01716, 2019.

[16] A.D. Wood, J.A. Stankovic, Denial of service in sensor networks, IEEE Comput. 35 (10) (2002) 54–62.

[17] H. Shafiei, A. Khonsari, H. Derakhshi, P. Mousavi, Detection and mitigation of sinkhole attacks in wireless sensor networks, J. Comput. Syst. Sci. 80 (2014) 644–653.

[18] S. Mohammadi, R.E. Atani, H. Jadidoleslamy, A comparison of link layer attacks on wireless sensor networks, J. Inf. Secur. 2 (2011) 69–84, https://doi.org/10.4236/jis.2011.22007.

[19] A.S. Abu Daia, R.A. Ramadan, M.B. Fayek, Sensor networks attacks classifications and mitigation, Ann. Emerg. Technol. Comput. 2 (4) (October 2018) 28–43, https://doi.org/10.33166/AETiC.2018.04.003.

[20] T. Borgohain, U. Kumar, S. Sanyal, Survey of security and privacy issues of Internet of things, arXiv preprint, arXiv:1501.02211, 2015.

[21] Mirai (malware), https://en.wikipedia.org/wiki/Mirai_(malware).

[22] J. Zhang, L. Li, G. Lin, D. Fang, Y. Tai, J. Huang, Cyber resilience in healthcare digital twin on lung cancer, IEEE Access 8 (2020) 201900–201913.

[23] K. Albulayhi, A.A. Smadi, F.T. Sheldon, R.K. Abercrombie, IoT intrusion detection taxonomy, Ref. Archit. Anal., Sensors 21 (19) (September 2021) 6432.

[24] D. Androcec, N. Vrcek, Machine learning for the Internet of things security: a systematic review, in: ICSOFT 2018, January 2018, pp. 597–604.

[25] C. Ioannou, V. Vassiliou, Network attack classification in IoT using support vector machines, J. Sens. Actuators Netw. 10 (3) (August 2021) 58.

[26] S. Gao, G. Thamilarasu, Machine-learning classifiers for security in connected medical devices, in: 2017 26th International Conference on Computer Communication and Networks (ICCCN). Presented at the 2017 26th International Conference on Computer Communication and Networks (ICCCN), IEEE, Vancouver, BC, Canada, 2017, pp. 1–5.

[27] M. Wu, Z. Song, Y.B. Moon, Detecting cyber physical attacks in cyber manufacturing systems with machine learning methods, J. Intell. Manuf. (2017).

[28] A.A. Diro, N. Chilamkurti, Distributed attack detection scheme using deep learning approach for Internet of things, Future Gener. Comput. Syst. 82 (2018).

[29] M.T. Gebrie, H. Abie, Risk-based adaptive authentication for Internet of things in smart home eHealth, in: ECSA (Companion), 2017, pp. 102–108.

[30] J. Yu, A. Hu, F. Zhou, Y. Xing, Y. Yu, G. Li, L. Peng, Radio frequency fingerprint identification based on denoising autoencoders, in: International Conference on Wireless and Mobile Computing, Networking and Communications (WiMob), October 2019.

[31] J.S. Arteaga-Falconi, H. Al Osman, A. El Saddik, ECG authentication for mobile devices, IEEE Trans. Instrum. Meas. 65 (2016) 591–600, https://doi.org/10.1109/TIM.2015.2503863.

[32] J.S. Arteaga-Falconi, H. Al Osman, A. El Saddik, ECG and fingerprint bimodal authentication, Sustainable Cities and Society 40 (2018) 274–283, 2018.

[33] A. El Saddik, H. Badawi, R.A.M. Velazquez, F. Laamarti, R.G. Diaz, N. Bagaria, J.S. Arteaga-Falconi, Dtwins: a digital twins ecosystem for health and well-being, IEEE COMSOC MMTC Commun. (May 2019).

[34] H. Baldus, S. Corroy, A. Fazzi, K. Klabunde, T. Schenk, Humancentric connectivity enabled by body-coupled communications, IEEE Commun. Mag. 47 (June 2009) 172–178.

[35] M. Usman, Lightweight encryption for the low powered IoT devices, https://arxiv.org/pdf/2012.00193.pdf, Dec 2020.

[36] A. Perrig, R. Szewczyk, V. Wen, D.E. Culler, J.D. Tygar, SPINS: security protocols for sensor networks, Wirel. Netw. 8 (5) (September 2002) 521–534.

[37] E. Hargittai, A. Marwick, What can I really do? Explaining the privacy paradox with online apathy, Int. J. Commun. 10 (2016) 3737–3757.

[38] H. Li, J. Wu, Y. Gao, Y. Shi, Examining individuals' adoption of healthcare wearable devices: an empirical study from privacy calculus perspective, Int. J. Med. Inform. 88 (2016) 8–17.

[39] A. Majumdar, I. Bose, Privacy calculus theory and its applicability for emerging technologies, in: V. Sugumaran, V. Yoon, M.J. Shaw (Eds.), E-Life: Web-Enabled Convergence of Commerce, Work, and Social Life, vol. 258, Springer International Publishing, Cham, Switzerland, 2016, pp. 191–195.

[40] E. Vayena, T. Haeusermann, A. Adjekum, A. Blasimme, Digital health: meeting the ethical and policy challenges, Swiss Medical Weekly 148 (2018) w14571.

[41] K. Bruynseels, S.F. de Sio, J. van den Hoven, Digital twins in health care: ethical implications of an emerging engineering paradigm, Front. Genet. 9 (2018) 31.

[42] R. Raskar, I. Schunemann, R. Barbar, K. Vilcans, J. Gray, P. Vepakomma, . J. Werner, Apps Gone Rogue: Maintaining Personal Privacy in an Epidemic, Whitepaper, PrivateKit, MIT, 2020.

[43] A. Cavoukian, A. Fisher, S. Killen, D.A. Hoffman, Remote home health care technologies: how to ensure privacy? Build it in: privacy by design, Identity Inform. Soc. 3 (2) (2010) 363–378.

[44] M. Colesky, J. Hoepman, C. Hillen, A critical analysis of privacy design strategies, in: 2016 IEEE Security and Privacy Workshops (SPW), San Jose, CA, 2016.

[45] A. Azzaoui, T.W. Kim, V. Loia, J.H. Park, Blockchain-based secure digital twin framework for smart healthy city, Adv. Multimed. Ubiquitous Eng. 107 (113) (2021).

[46] S. Suhail, R. Hussain, R. Jurdak, A. Oracevic, K. Salah, C.S. Hong, Blockchain-based digital twins: research trends, Issues, and Future Challenges. CoRR, arXiv:2103.11585 [abs], 2021.

[47] M.H. Miraz, M. Ali, Integration of blockchain and IoT: an enhanced security perspective, Ann. Emerg. Technol. Comput. 4 (4) (1st October 2020) 52–63, Print ISSN: 2516-0281, Online ISSN: 2516-029X.

[48] I.O.N. Microsoft, Decentralized network based on bitcoin blockchain, https://identity.foundation/ion/.

[49] NISTIR 8356 (Draft), Considerations for digital twin technology and emerging standards, https://csrc.nist.gov/publications/detail/nistir/8356/draft, April 2021.

[50] NIST 800-53, Security and privacy controls for information systems and organizations, https://csrc.nist.gov/publications/detail/sp/800-53/rev-5/final, 2020.

[51] NIST Privacy Framework, A tool for improving privacy through enterprise risk management, Ver. 1.0, https://nvlpubs.nist.gov/nistpubs/CSWP/NIST.CSWP.01162020.pdf, 2020.

[52] Health Insurance Portability and Accountability Act (HIPAA), Guide to privacy and security of electronic health information, https://www.healthit.gov/providers-professionals/guide-privacy-and-security-electronic-health-information, 2015.

[53] L.L. Broome, J.W. Markham, the Gramm–Leach–Bliley Act: an Overview, 2001.

[54] Children's Online Privacy Protection Rule ("COPPA"), 15 U.S.C, 6501–6505, https://www.ftc.gov/enforcement/rules/rulemaking-regulatory-reform-proceedings/childrens-online-privacy-protection-rule, 1998.

[55] World Health Organization. Ethics and Governance of Artificial Intelligence for Health: WHO Guidance; World Health Organization: Geneva, Switzerland, 2021.

[56] ISO/IEC 27001 International Information Security Standard published, bsigroup.com, BSI, November 2005, https://www.bsigroup.com/en-GB/about-bsi/media-centre/press-releases/2005/11/ISOIEC-27001-International-Information-Security-Standard-published/.

[57] System and Organization Controls: SOC Suite of Services, AICPA, https://us.aicpa.org/interestareas/frc/assuranceadvisoryservices/sorhome.

[58] NIST, FIPS general information, https://www.nist.gov/itl/fips-general-information, October 2010.

[59] NIST Special Publication 800-88, Revision 1: Guidelines for Media Sanitization, https://www.nist.gov/publications/nist-special-publication-800-88-revision-1-guidelines-media-sanitization, February 2015.

[60] ISO/IEC JTC 1/SC 41 Internet of things and digital twin, https://www.iso.org/committee/6483279.html, 2017.

[61] H.F. Badawi, F. Laamarti, A. El Saddik, ISO/IEEE 11073 personal health device (X73-PHD) standards compliant systems: a systematic literature review, IEEE Access 7 (2019) 3062–3073.

11

Potential applications of digital twin in medical care

Kaouther Abrougui^a, Hazim Dahir^b,
Ahmed Khattab^c, Jeff Luna^c, Raj Kumar^c, and
Rashika Verma^d

^{a}Voltron Data Inc., Mountain View, CA, United States, ^{b}Cisco Systems, Research Triangle Park, NC, United States, ^{c}Cisco Systems, San Jose, CA, United States, ^{d}Rutgers New Jersey Medical School, Newark, NJ, United States

11.1 Foundations for potential applications of digital twins in medical care

11.1.1 Digital health criteria

The U.S. Food and Drug administration (FDA) has provided guidance for digital health criteria to enable anyone who is thinking, designing, or building software, analytics, artificial intelligence, cloud, cyber security, or medical devices to safely and correctly build solutions with interoperability and security. This acts as both guidance and as direction for future solutions that will incorporate digital twins and the medical and consumer-based devices creating data supporting clinical decision engines [1].

The FDA's digital Health Center of excellence goal is to empower stakeholders to advance health care by fostering responsible and high-quality digital health innovation. Three objectives that the FDA is approaching are: To connect and build partnerships to accelerate digital health advancements; Share knowledge to increase awareness and understanding, while driving synergy and best practices; Innovate regulatory approaches to provide efficient and least burdensome oversight while meeting FDA standards for safe and effective products.

223

The anticipated outcome is to fulfill the objectives set for the by the digital health center of excellence and to advance digital health and scientific evidence while accelerating access to digital health technologies with oversight.

To achieve these outcomes, the latest set of standards became available in 2018, classifying how medical devices can be applied to digital twins for medical care within the scope of Digital Health technologies.

Digital health technologies include categories such as mobile health, health information technology, wearable devices, telehealth/telemedicine, and personalized medicine. Digital health technologies use computing platforms, connectivity, software, and sensors to collect data, computational analysis, and decision making.

Digital health technologies, tools, and sensors are providing an improved view of patient health at any given time to which digital health technologies can offer opportunities to improve health outcomes and enhance clinical efficiencies. The potential for patients to become more proactive in their own health outcomes to prevent disease and promote health outside of traditional healthcare settings is helping to evolve digital health technologies and solutions.

Digital health technologies, applied as a digital twin healthcare solution, have the potential to improve the ability to accurately diagnose and treat disease and to deliver better health outcomes for the patient. Medical applications and software that support clinical decisions are being developed to improve diagnostics, treatments, and long-term outcomes. Digital twins are becoming a true revolution in healthcare to further provide insights and guidance to both doctors and patients.

One of the biggest challenges facing the foundation of digital twins in medical care is consistency in the environments that support data collection, correlation, and decision making. The primary concern is that software may not operate the same in any computing environment and results could be altered. The caution is to ensure that adequate resources are provided to enable software solutions to have consistency across clouds or environments and eliminate differences in computing, security, and networking.

FDA guidance for digital health software capabilities should be designated for specific hardware and medical devices but it should not drive or control the medical device hardware.

In addition, software can run stand alone on mobile platforms such as tablets and smartphones or computers and attached through wireless communications to devices.

The ability for advanced analytics to provide information to medical resources has to protect the identities and data of patients while not exposing any personally identifiable information (PII) or any personal health information (PHI).

11.1.2 Digital health regulatory policies

In the United States, the Food and Drug Administration (FDA) is responsible for regulating and approving much of the digital health technologies that are in clinical use within the country. By and large, digital health is considered a type of "medical device" by the FDA, and therefore, is regulated based on its "intended use in the diagnosis, treatment, cure, mitigation, or prevention of disease or condition". Many of the regulatory policies around approval, and rejection, of digital health technologies center around the technologies' ability to execute their intended function.

Enforcement discretion – FDA doesn't need to regulate low risk medical devices/technologies. While they may still exercise some control over the product, it is not as comprehensive. This creates a significant gray area where consumers, both clinicians and patients, are bombarded with multiple products that may not best serve their healthcare needs.

– Therefore, with AI/ML technologies like digital twins, it's important to really understand what the healthcare needs are and tailor products to deliver on those fronts.

11.1.3 Digital health center for excellence

The United States Food and Drug administration (FDA) has created a Digital Center of Excellence that serves to complement advances in digital health technologies including digital twins. The digital Health Center of Excellence provides services to digital health stakeholders to translate advances into tools that benefit consumers and patients [2].

The FDA is empowering stakeholders of digital health solutions by setting and leading strategic direction in digital health technologies, launching strategic initiatives that advance digital health technologies by building the new capacity to oversee and leverage digital health technologies.

Additional areas of digital health technologies include providing scientific expertise, providing technology and policy advice, promoting and showcasing work, and transparently sharing resources for developers. This aspect of oversight is enabling the stakeholders to build collaboration with scientific research and digital health communities.

One of the most important aspects of facilitating and building strategic partnerships with scientific research communities is to increase velocity for new solutions by taking research methods and using them in clinical settings.

The FDA is also advancing device regulatory policy and international harmonization. This means that the FDA is enabling many solutions to be brought to larger markets in shorter amounts of time. The FDA is also bringing international standards into digital health technology solutions and providing access to internal and external digital health experts.

11.1.4 Network of digital health experts

A network of digital health experts has been assembled by the FDA to provide developers in digital health rapid access to a permanent specialist pool of experts to help provide guidance. This network of digital health experts is comprised of experts available to share knowledge and experiences regarding digital health issues with FDA staff and companies looking to build solutions on an ongoing basis.

The digital health recognized experts are chosen based on their leadership through authorship of peer review publications, committee membership, ongoing speaking engagements, and academic or professional activities. The digital Health Network of experts is truly a partner organization of experts who also must file conflicts of interest and transparency disclosures statements to avoid any potential bias with organizations looking to build digital twin health solutions.

Specific medical fields and topics related to digital health and digital health services security have been applied to identify upcoming healthcare research areas for digital health experts to apply their knowledge. The newest areas that are being explored for new development include: artificial intelligence; physiological sensors and wearables; and medical internet of things devices (MIoT).

The movement to explore these upcoming digital health areas has increased velocity for traditional technology and cloud solution providers to apply their expertise in mobile networks, wireless communication, internet of things, and server security to benefit digital health.

This chapter describes the applications of digital twins in healthcare.

11.2 Applications of digital twin in medical care: state of the art

In their analysis, Bagaria and team [3] introduced the concept of Health 4.0 where they compare the digital transformation in healthcare to the 4th industrial revolution, also referred to as Industry 4.0. The comparison is justified given the similarities between industry 4.0 and any industry that depends on data acquisition, ingestion, and analysis visibility and performance optimization.

Applying similar concepts to healthcare, we can subsequently deduce the following pillars for Health 4.0:

1. Data, Lots of it. The very essence of Industry 4.0, or Health 4.0 for that matter, is the operational visibility through data obtained from sensors at various points. Each step of the data transformation journey from data to information, to knowledge, to wisdom adds value to the initial raw data. It Eventually unlocks business or operational value needed for patient wellness.

2. Digital Twins and Simulations. The ability to model future conditions using current physiological data.
3. Augmented, Virtual, and Extended Reality. Although this is mostly on the visualization side of things, it will become a powerful tool when combined with Digital Twin modeling and analysis.
4. Connected Devices. The Internet of Things (IoT) or Internet of Medical Things (IoMT) facilitates the ingestion of data from various sensors distributed over multiple domains. This field is experiencing massive growth in technology and adoption. Significant research is also happening where synthetic biology and nanotechnology tools that allow the engineering of biological embedded computing devices [4]. This is called Internet of Bio-Nano Things (IoBNT). We can only imagine the amount of knowledge and wisdom we can capture about and possibly influence the inner workings of the human body.
5. Cloud Computing. For most cases, Cloud has been proven to reduce cost, increase flexibility and scalability. As we move towards digital twin applications, we will explore challenges and opportunities of cloud computing in greater detail.
6. Cybersecurity. Wherever patient data exists, security and privacy concerns also exist.
7. System integration. The concept of Connected devices is one thing, connected systems are another. Connected devices are useless if the data they possess is not integrated or shared by multiple systems.
8. Intelligence and autonomous systems. Autonomous systems with AI/ML support that automatically adjust operations based on predetermined policies.

Having a good understanding of the above pillars will pave the way for building smart systems supporting a variety of applications. Personal health or "well-being" is an important and rising area of research for preventative medicine and should be given adequate attention.

11.2.1 Personal health management

Whenever we think of Personal Health or well-being, we have to think of it as a combination of physical, emotional, intellectual, social, and even spiritual. The US Government Department of Health and Human Services, Substance Abuse and Mental Health Services Administration (SAMHSA) published a guide to "Wellness" called "Creating a Healthier Life." In that guide, they even list "Eight Dimensions of wellness", where each dimension builds on another. (See Fig. 11.1.)

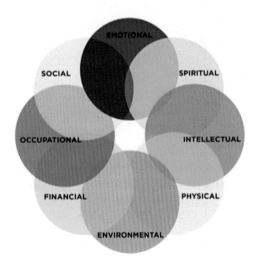

FIGURE 11.1   Eight dimensions of Wellness. *Source: www.samhsa.gov/publications.search for sma16-4958.pdf.*

There is very little doubt that these wellness dimensions are related and affect one another. In the next few sections, we will explore how DT models can help correlate data from the various data sources to assess various wellness factors.

11.2.1.1 Personal health and well-being

Recent advances in biosensors and microsystems, combined with advances in mobile communications have opened new opportunities in "continuous" monitoring for connected health (medical or personal health) [5].

Fig. 11.2 is an excellent representation of the available monitoring capabilities, what biomarkers are used, from what organ, representing what health assessment or health goal. The figure represents a small sample of possibilities and touches two or three of the eight wellness dimensions mentioned earlier.

The Digital Twin models we're attempting to create have a wealth of information available to them and for a variety of assessments. There are a lot of decisions to be made and they each affect the DT, the data used, and the frequency of capture. When used for decisions, the DT models need to be well-defined, have consistent access to data, and specific accuracy thresholds (or ranges) and personas of patients where the data is deemed acceptable. Wearable sensors help us quantify ourselves, document our lives, and the data they produce will allows us to create new realities for ourselves and others.

Let's look at this partial list of decisions depicted in the following table.

Decision	Is it one or more of this?	Or one or more of this?
Purpose	Medical Decisions	Health and Well-being
Monitoring	Continuous (how long?)	Snapshot (how many?)
Sensors	Commercial	Medical Grade
	Active	Passive
	Invasive	Non-invasive
Biomarkers	Most reliable?	Most available?
(and bio-recognition)	(e.g., urine sample is most reliable for Hydration study)	(e.g., sweat sample in continuous monitoring)
Context	In situ	In vivo
Setting / Place	Anywhere (normal activity)	Lab or Medical Facility
Data	Clinically generated data (Molecular)	Demographic or 1 social health data
Communication	Wired	Wireless

Note: The above table is a simple representation of questions to ask in the way of assessing the various data sources to be used in building our model or learning process. Every cell you pick (regardless of the column it sits in) has an implication on your data and how clean and relevant it is.

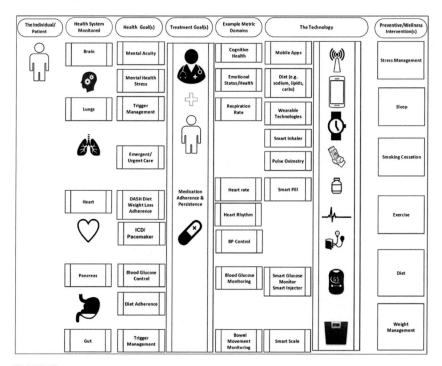

FIGURE 11.2 Possibilities of Personal Health Data and Digital Twin. (*Schwartz, SM, et al. 2020:****)

It is important to understand that correlating various values or dashboards generated by a sensor or a group of sensors are NOT a digital twin.

These dashboards and tools are the *digital phenotype* (*observed values*) [5]. The link between the digital phenotype and the digital twin is a set of algorithms and statistical models that build on a base of scientific knowledge and experiences.

When we start moving beyond observation and into calculation, the main questions we need to answer are:

- How many data?
- Captured within how-long of a duration?
- What purpose?

It's all connected. The "purpose", or the "application" helps us decide the datatype and the duration. The "purpose" also helps us determine the "best" data generator (sensor) and the best biomarker for that purpose.

11.2.1.2 Personal health

Continuous monitoring of patients with chronic diseases and the impact of the treatment can be beneficial and improves the quality of treatments for patients. Few models were presented recently presented for applying Digital Twin concepts into precision medicine. One of them described a reference model integrating data-driven methods such as Machine Learning, and Digital Twins. This model helps track patients' health status in a continuous way, and evaluates the efficiency of medical treatments, as well as decision making processes with regards to treatment's evolution. The main purpose is to support precision medicine techniques in the context of continuous monitoring and data-driven personalized medical treatments [6].

Similarly, another model presented a viewpoint on the necessity to have large scale, deep digital phenotyping to be able to advance in precision medicine. That model discussed how to combine real world with clinical data and omics features in order to identify more precisely the digital twin of a patient and move towards a patient centric medicine [7]. In addition, few other frameworks proposed using strong capabilities of Digital Twin and advanced data processing for identifying the best medicine for a specific disease for a specific patient [8].

There were also few attempts to integrate Deep Learning alongside Machine Learning and Digital Twin to improve the overall health of humans. In this work [9], Digital Twin subtypes that represent an individual person, a designated group, and the AI Software used for the Artificial Neural Network (ANN). The ANN related subtype is a representation of the features of the system as well as its suitability for dealing with specific diseases. As described in the figure below in Fig. 11.3, the health recommendations are built using a digital twin that references digital twins comprising a Personal Digital Twin (PDT), Group Digital Twin (GDT), and System Digital Twin (SDT). In addition, with the presence of machine learning we have

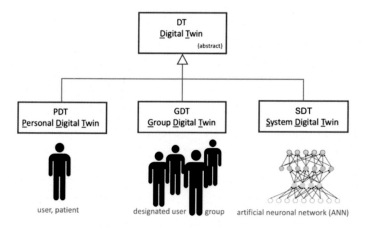

FIGURE 11.3   Digital Twin Subtypes are described by Lutze et al. [9].

the ability to reapply the model to adjust prior conclusions in light of additional knowledge learned.

11.2.2 Precision medicine

It is not uncommon to see "precision medicine" and "personalized medicine" terms used interchangeably. Several definitions exist:

- "The use of new methods of molecular analysis to better manage a patient's disease or predisposition to disease." – Personalized Medicine Coalition.
- "Providing the right treatment to the right patient, at the right dose at the right time." – European Union.
- "The tailoring of medical treatment to the individual characteristics of each patient." – President's Council of Advisors on Science and Technology.
- "Health care that is informed by each person's unique clinical, genetic, and environmental information." – American Medical Association.
- "A form of medicine that uses information about a person's genes, proteins, and environment to prevent, diagnose, and treat disease." – National Cancer Institute, NIH.

11.2.2.1 Personalized medicine

According to the US Food and Drug Administration (FDA) report, the rate of medication found to be not effective on patients with common diseases ranges between 38% and 75% [10]. This is mainly due to the complexity of common diseases. Although different patients can have similar diagnostics, there are thousands of genes involved with different interactions amongst patients. Hence the need of mechanisms for person-

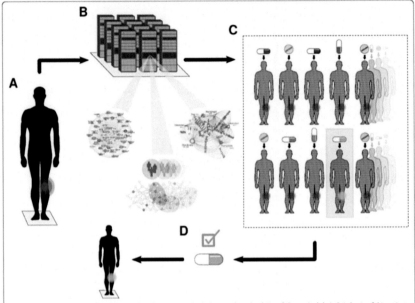

Fig. 1 The digital twin concept for personalized medicine. **a** An individual patient has a local sign of disease (*red*). **b** A digital twin of this patient is constructed in unlimited copies, based on computational network models of thousands of disease-relevant variables. **c** Each twin is computationally treated with one or more of the thousands of drugs. This results in digital cure of one patient (*green*). **d** The drug that has the best effect on the digital twin is selected for treatment of the patient

FIGURE 11.4   The digital twin concept for personalized drug treatment (Björnsson et al., 2019, p. 2 [8]).

alized medicine. This is the objective of the DigiTwin consortium (https://www.digitwins.org) that includes partners from 32 countries from academic, clinical and industrial backgrounds. There are also national initiatives, for example the Swedish Digital Twin Consortium (SDTC) that is trying to develop and build a strategy for personalized medicine.

In Björnsson et al. [8] proposed a solution to personalized medicine by constructing digital twins as models of individual patients that can be treated with different drugs in order to be able to find the optimal drug, and effective treatment. Their strategy consists of initially creating digital twins, which are, unlimited copies of network models with various factors; such as molecular, phenotypic, and environmental that are relevant to mechanisms of a disease in individual patients. Then, leverage computations to treat those digital twins, with different drugs in order to determine the most effective drug. Finally, as a result of these many trials and related outcomes, treat the patient with the most effective drug (Fig. 11.4).

It is worth noting, however, that there is a considerable gap between current state of modern healthcare and the complexity of the strategy for precision medicine. This is due to the fact that modern healthcare diagnostics leverages limited number of biomarkers. Digital and genomic medicine aspire to bridge this gap by collecting, processing, and integrating large amounts of data from wearable devices and sensors and other IoT devices and tools.

Cardiovascular medicine

The most advanced digital twin applications in the healthcare industry are developed for cardiology. The living heart project [11] units leading cardiovascular stakeholders on developing and validating personalized digital human heart models with high accuracy. The purpose of these models is to create a unified foundation for cardiovascular in silico medicine.

11.2.2.2 Drug management

Recently, we've witnessed the development of few models that include digital twin and cloud computing services [12]. This model helps to track the health of elderly and how they are consuming their medication to detect if there is any unreasonable use of drugs.

Based on the features of digital twin, Liu et al. [13] developed a model that combined digital twin and cloud computing service, which could effectively track the health status of the elderly patients and reasonable use of medication.

11.2.2.3 Diseases and treatment

Subramanian et al. [12] proposed a digital twin model for the liver. This model was developed after extensive research and studies to understand various functions of the liver, diseases and effect of drugs. Subramanian [12] was successful in proving that the digital twin model was able to reproduce effectively the normal functions of the liver and the diseases evolution. The model was also effective in showing the impact of treatment. The digital twin was also proved effective when applied to drug development, management, and treatment of other diseases.

Multiple models were created for tissue engineering processing [14]:

- Skeletal tissue engineering: Digital Twin was used in processes related to tissue engineering where multiple digital, model-based strategies were developed. Geris et al. [15] did an extensive study and focused on skeletal tissue engineering. In their use case, they focused on studying the specialization of cartilage cells, which is an important biological process that impacts cells regeneration and bone formation. First, they focused on understanding the molecular regulation of the skeletal tissue engineering process. Second, they came up with cell cultures and design strategies in silico with the purpose of enhancing the experiments behavior in vitro or in vivo. The in-silico screening step permits the identification of the key factors candidates that would reduce the complexity of experiments. They used mathematical models and leveraged results from research done in the laboratory to come up with new treatment that directly benefit the patient. Not only did they leverage pure data driven models, but also, they explored mechanisms following foundational assumptions and using known mechanisms. They provided an approach.

- Lung Tumor Models: In order to identify drug targets and identify biomarkers that can predict diseases, infection, or exposure, that would help with personalized treatment strategies and cure for cancer, there is a need to understand in depth the cells signaling networks and how they are impacted through drug response. In aa, bb used tissue culture techniques to create models that mimicked the tumor microenvironment and presented with different options reflecting the clinical situation. In their use case, Walles et al. [16,17] focused on the lung carcinoma. They combined a 3D in vitro and an in-silico model for dependencies identifications in cells signaling networks. This combination permitted the identification of biomarker profiles and helped with targeted therapies.
- Others: Liver tissue engineering [18–20], Skin Model [1].

11.3 Future applications of digital twin in medical care

11.3.1 Monitoring

Digital Twin can collect health data in real time from different sources including patient record, wearable devices, omics (genomics, proteomics, etc...). All these feeds can help monitor the patient, and different organs in the patient body. This data can help develop ML models to detect symptoms that can lead to illness and give the opportunity to doctors and nurses to intervene in guiding their patients before it is too late. Besides, customized recommendations and guidance can be provided based on patient's habits and social determinants.

This concept is not just limited to patient monitoring in the regular sense but can be extended to cover crisis warning systems. In other words, application in an emergency unit in hospitals to evaluate current state and using artificial intelligence for assessing crisis and incidents state. Again, and as explained throughout the book, that highlights the importance of using data in the Digital Twin realm. Consider a case of major emergencies, such as a tsunami, terrorist attack or an earthquake, it becomes apparent that it is crucial not just to have access to that colossal data, but to have that access in real time, make sure that the data is accurate and augment sources of data for the intelligence controller or decision maker to come up with important decisions that can affect human lives. This has been a main topic of research in the application of Digital Twin in healthcare [21].

Remote monitoring is a complex aspect of this architecture. Again, taking the examples above of a natural or man-made disaster, remote patient monitoring faces many challenges that can be compromised by disruption of communication media and infrastructure. The same applies to having patients in remote areas that are not covered well with communication technologies, such as in the case for mountain climbing or diving situations.

Another major challenge is related to the state of technology based on geography or political situation of the locale itself. This gives a clear advantage to more advanced countries where technology is entwined into the fabric of society.

Concrete implementations of digital twins can already be found for organs such as the heart, for example, by the French software company Dassault Systèmes or by Siemens Healthineers in Germany. Siemens Healthineers has used data collected in a huge database of more than 250 million annotated images, reports, and operational data. The AI-based DT model was trained to weave data about a heart's electrical properties, physical characteristics, and structure into a 3D image. The technology was tested on 100 digital heart twins from patients treated for heart failure in a six-year study. Preliminary results of the comparison between the actual outcome and the predictions the computer made after analyzing DT status seemed promising. French startup Sim&Cure developed a DT system that virtualizes a patient-based aneurysm and surrounding blood vessels. After a patient with aneurysm is prepared for surgery, a DT represented by a 3D model of the aneurysm and surrounding blood vessels is created by processing a 3D rotational angiography image. The personalized DT allows surgeons to perform simulations and helps them gain an accurate understanding of the interactive relationship between the implant and the aneurysm. In less than five minutes, numerous implants can be assessed to optimize the procedure. Preliminary studies have shown promising results [22].

11.3.2 Diagnosis

- Simulate the patient health status based on feeds from different sources.
- Personalize human diagnosis and treatment...

Diagnosis must grapple with the reality that individual differences between patients means many diseases do not present the same way in different people [8]. Artificial Intelligence has the ability to consolidate diagnostic data from many different patients and use it to inform clinicians of the current patient's health status. Digital twins also allow for personalization of diagnosis and treatment by giving clinicians the opportunity to take every aspect of a patient's genetic and lifestyle factors into consideration during medical care.

Digital Twin technology can be particularly useful in the treatment of multi-factorial diseases like Multiple Sclerosis (MS), which is a chronic autoimmune disease that destroys the central nervous system [Voigt]. Not only does MS vary in its presentation from patient to patient, it also varies in severity, length, and response to existing treatments. Leveraging artificial intelligence can give clinicians the ability to diagnose MS at an earlier stage and personalize treatment, therefore resulting in better quality of life outcomes.

11.3.3 Surgery planning: simulation and risk assessment

Digital Twin can help simulate surgery procedure and assess the outcome prior to performing an invasive intervention. This will help with risk assessment and personalization of the surgery in order to increase the success of the intervention and reduce the risks incurred to the patient.

The Living Heart Project (Dassault Systèmes) was leveraged by a Boston hospital to plan surgical procedure and assess its outcome (https://answers.childrenshospital.org/aerospace-tools-repair-hearts/). The Digital Twin even helped with the design and the generation of the cuff between the heart and arteries.

Sim and Cure (https://sim-and-cure.com/patient-care/) is a Digital Twin that helps brain surgeons use simulations in order to treat aneurysms (https://sim-and-cure.com/about-aneurysm/) while increasing the safety of the patient.

With organs virtualization and surgery simulation, invasive clinical intervention can be personalized to the patient, increasing its safety, and reducing the need to have follow up surgeries.

11.3.4 Medical devices

- Test devices on patients.
- Optimize device design.

First, it is important to note that the term device should be considered in the greater context of infrastructure and not just limited to the physical device in isolation. A device work in orchestration with an IT infrastructure, IoT network of different smart devices (by different manufactures and can be using completely different technologies) and different protocols for transmission, authentication, and security.

The main purpose of the medical devices is to collect accurate patient data in real-time. The devices can be stationary connected to a patient, such as vital signs monitor devices used in hospitals, wearable sensors such as watches or fitness bands, or implantable in humans such as a heart pacemaker. They all share the main characteristic about collecting accurate data, with a wide variation of specialization. They are also safety-critical and in certain cases their failure can lead to the patient's life risk. Hence, the importance of monitoring (and maintenance).

Preventive maintenance must be considered as part of the design phase of the medical device. Collecting self-data (i.e., historical performance and maintenance data) thus becomes an important requirement for such devices. Not only that, but implementing AI models for studying collective historical data, from different sources and different runtime environments can improve, tremendously, the accuracy and lifetime of medical devices.

Aside for the importance of the test and monitoring devices, the Digital Twin paradigm extends that aspect to include virtual devices, or more accurately described Digital Twin devices. Similar to the concept of having a patient Digital Twin, a medical device can have a Digital Twin that can be used for testing, improving design, predicting failure conditions and overall, helps in achieving predictable maintenance to ensure quality of service.

Using the same concept, virtual replicas of entire medical facilities can be implemented. This can identify bottlenecks or areas that can use more optimization by running simulations that can cover entire end-to-end use cases. For example, examining the process of emergency rooms and figuring out the average service time for different medical conditions and understanding where delays or additional specialized staff are needed. It becomes apparent that this approach can provide important insights on optimization, training and risk management for medical facilities and staff.

11.3.5 Drug development

- Use Digital Twin to try multiple drugs in order to be able to determine the best one for a specific use case.
- Test new drugs.
- Optimize Dosage.

According to a report from the US Food and Drug Administration (FDA), medication is deemed ineffective for 38–75% of patients with common diseases. [US FDA. Paving the way for personalized medicine: FDA's role in a new era of medical product development. Silver Spring: USDA; 2013. https://www.fdanews.com/ext/resources/files/10/10-28-13-Personalized-Medicine.pdf.]

In a highly complex system, as in drug development, scientific research is key for spring boarding the process and ensuring accuracy to the highest level possible. The process includes collecting research data and identifying patterns and in general, identifying the state of normal (or healthy) and comparing to a large set of population patterns.

Since Digital Twin is at the heart a data-driven paradigm, it has a promising potential in improving the process for drug development in general. Let's consider the case that physicians usually explain as asymptomatic illness. Those cases have a risky side effect of not being identified early on when it comes to diagnosis. It is an area where an application of Digital Twin simulation and studies can be done to try to identify positive patterns that can aid physicians and specialist in identifying potential illness and either proceed with further specialized diagnosis or start a known treatment.

Digital Twins can be also used to assess drug toleration and side effects or potential side effects for a specific population, again, based on employing AI simulation or a large set of accurate collected data. Inclusive data that is.

Digital Twins may also support clinical trial populations matching through use of population attribute screening to better define control and placebo groupings through larger dataset matching of co-morbidities.

Considering all previously mentioned information, Digital Twin can be used for developing a strategy for personalized medicine. The Swedish Digital Twin Consortium (SDTC) aims to achieve that goal with a strategy focusing on 1) constructing unlimited copies of network models of all molecular, phenotypic and environment factors relevant to disease mechanisms in individual patients (i.e., Digital Twins), 2) computationally treating those Digital Twins with thousands of drugs in order to identify the best performing drug and 3) treating the patient with this drug [https://www.sdtc.se].

References

[1] S.H. Mathes, H. Ruffner, U. Graf-Hausner, The use of skin models in drug development, Adv. Drug Deliv. Rev. (2014) 69–70, pp. 81–102.
[2] US FDA Digital Health Center of Excellence Services, https://www.fda.gov/medical-devices/digital-health-center-excellence/digital-health-center-excellence-services, 2018.
[3] Namrata Bagaria, Fedwa Laamarti, Hawazin Badawi, Amani Albraikan, Roberto Martinez, Abdulmotaleb El Saddik, Health 4.0: digital twins for health and well-being, https://doi.org/10.1007/978-3-030-27844-1_7, 2020.
[4] I.F. Akyildiz, M. Pierobon, S. Balasubramaniam, Y. Koucheryavy, The Internet of bio-nano things, IEEE Commun. Mag. 53 (3) (March 2015) 32–40, https://doi.org/10.1109/MCOM.2015.7060516.
[5] S.M. Schwartz, K. Wildenhaus, A. Bucher, B. Byrd, Digital twins and the emerging science of self: implications for digital health experience design and "small" data, Front. Comput. Sci. 2 (2020) 31, https://doi.org/10.3389/fcomp.2020.00031.
[6] L.F. Rivera, M. Jiménez, P. Angara, N.M. Villegas, G. Tamura, H.A. Müller, Towards continuous monitoring in personalized healthcare through digital twins, in: Proceedings of the 29th Annual International Conference on Computer Science and Software Engineering, 2019, pp. 329–335.
[7] G. Fagherazzi, Deep digital phenotyping and digital twins for precision health: time to dig deeper, J. Med. Internet Res. 22 (2020) e16770.
[8] B. Björnsson, C. Borrebaeck, N. Elander, et al., Swedish Digital Twin Consortium, Digital twins to personalize medicine, Gen. Med. 12 (2019) 4.
[9] R. Lutze, Digital twins in eHealth: prospects and challenges focusing on information management, in: Proceedings of the 2019 IEEE International Conference on Engineering, Technology and Innovation (ICE/ITMC), 2019, pp. 1–9.

[10] US Food and Drug Administration. Paving the way for personalized medicine: FDA's role in a new era of medical product development. US Food and Drug Administration, Silver Spring, https://www.fdanews.com/ext/resources/files/10/10-28-13-Personalized-Medicine.pdf, 2013. (Accessed 26 November 2019).

[11] Living Heart Project, SIMULIA – Dassault Systèmes, https://www.3ds.com/products-services/simulia/solutions/life-sciences-healthcare/the-living-heart-project/.

[12] K. Subramanian, Digital twin for drug discovery and development – the virtual liver, J. Indian Inst. Sci. 100 (2020) 653–662.

[13] Y. Liu, L. Zhang, Y. Yang, et al., A novel cloud-based framework for the elderly healthcare services using digital twin, IEEE Access 7 (2019) 49088–49101.

[14] Johannes Möller, Ralf Pörtner, Digital twins for tissue culture techniques—concepts, Expectations, and State of the Art 9 (2021), https://www.mdpi.com/2227-9717/9/3/447.

[15] L. Geris, T. Lambrechts, A. Carlier, I. Papantoniou, The future is digital: in silico tissue engineering, Curr. Opin. Biomed. Eng. 6 (2018) 92–98.

[16] C. Göttlich, L.C. Müller, M. Kunz, F. Schmitt, H. Walles, T. Walles, T. Dandekar, G. Dandekar, S.L. Nietzer, A combined 3D tissue engineered in vitro/in silico lung tumor model for predicting drug effectiveness in specific mutational backgrounds, J. Vis. Exp. (2016) e53885.

[17] A.T. Stratmann, D. Fecher, G. Wangorsch, C. Göttlich, T. Walles, H. Walles, T. Dandekar, G. Dandekar, S.L. Nietzer, Establishment of a human 3D lung cancer model based on a biological tissue matrix combined with a Boolean in silico model, Mol. Oncol. 8 (2014) 351–365.

[18] J.J. García Martínez, K. Bendjelid, Artificial liver support systems: what is new over the last decade?, Ann. Intensive Care (2018) 8.

[19] R. Tandon, S. Froghi, Artificial liver support systems, J. Gastroenterol. Hepatol. 36 (2021) 1164–1179, https://doi.org/10.1111/jgh.15255.

[20] Y.-T. He, Y.-N. Qi, B.-Q. Zhang, J.-B. Li, J. Bao, Bioartificial liver support systems for acute liver failure: a systematic review and meta-analysis of the clinical and preclinical literature, World J. Gastroenterol. 25 (2019) 3634–3648.

[21] V. Augusto, M. Murgier, A. Viallon, A modelling and simulation framework for intelligent control of emergency units in the case of major crisis, in: 2018 Winter Simulation Conference (WSC), 2018, pp. 2495–2506.

[22] Isabel Voigt, Inojosa Hernan, Dillenseger Anja, Haase Rocco, Akgün Katja, Ziemssen Tjalf, Digital twins for multiple sclerosis, Front. Immunol. J. 12 (2021), https://doi.org/10.3389/fimmu.2021.669811.

Digital twins for decision support system for clinicians and hospital to reduce error rate

Raj Kumar[a], Kaouther Abrougui[b], Rashika Verma[d], Jeff Luna[a], Ahmed Khattab[a], and Hazim Dahir[c]

[a]Cisco Systems, San Jose, CA, United States, [b]Voltron Data Inc., Mountain View, CA, United States, [c]Cisco Systems, Research Triangle Park, NC, United States, [d]Rutgers New Jersey Medical School, Newark, NJ, United States

12.1 Introduction to digital twin decision support system for reducing errors in hospitals

Healthcare is a knowledge-based profession and industry. As community and market needs continue to evolve, the healthcare providers who are nimble, learn fast and have consistent frameworks for execution will be able to achieve the best clinical outcomes.

Around the world, healthcare providers are struggling with the challenge of how to minimize and mitigate medical errors during patients' treatment.

Each year in the U.S., over 12 million adults who seek outpatient medical care receive a misdiagnosis, according to a recent study by BMJ Quality & Safety. That translates to about 5% of adults, or 1 out of 20 adult patients. The error rate in diagnosis for a patient by a clinician is between 35–40% in the developing world, but in some cases, it can be greater than 60% once we include near miss and unreported cases.

Developing countries continue to have higher rates of both reported and unreported cases as compared to developed countries. More than these error rates, it is important to focus on the higher incidences of both forced (adverse reaction) and unforced (people and process related) errors in healthcare.

According to reports published by the US Institute of Medicine [1], medical errors claim the lives of ~400,000 people (about the population of Malta) each year due to issues related to data inefficiency. More specifically, it is the inability to access a patient's medical history by clinicians that contributes significantly to this high error rate.

Missed and delayed diagnoses, as well as corrupted health data, also prevent healthcare providers from learning vital patient information that is essential for prompt diagnosis and treatment. Technological advancements around machine learning and artificial intelligence have the potential to significantly reduce these diagnostic errors in the healthcare system.

By connecting the Decision Support Systems (a Data Solution and System capable of collecting the data with context and curation for normalized and harmonized dataset ready for further analysis and insights generation) at provider or payer level to the patients' personal devices, which can capture, store, and notify the health institutes with the real-time health data, we can enable effective health support and reduce mortality rate, cost overrun, and misdiagnosis.

The rise in personal health monitoring devices in the form of mobile applications or built-in sensors that can actively monitor user's vital health parameters such as ECG, BP, heart rate, and blood glucose, can reduce the potential errors of patient data recording. These devices can capture and transfer data anonymously to the cloud and compare it with historical data for symptoms of any illness or notify the appropriate health personnel (doctor, nurse, or health agent) if any new or emergency intervention is required.

Fewer errors mean better performance, cost, efficiency, and improvements in healthcare services where an error can be the difference between life and death.

This is important not only because of the adverse impact on human life and quality of care, but also the negative impact on the bottom line of the providers and hospitals. The situation is getting out of control and is resulting in billions of dollars of fines, litigation costs and many unnecessary deaths. Clinicians and providers are under constant pressure from payers and government to reduce costs and improve quality of care for patients.

The time has come to leverage Digital Twin concept to create a digital diagnostic and prognostic system which can be used as a support and verification mechanism by every clinician to provide just and better care to each patient with the goal of minimizing unavoidable sufferings (medicine side effects etc.) and addressing avoidable sufferings (Communication, follow-up, advise, consult, Next Best Action, and Long-term care considerations etc.) in consistent manner.

To make the decision support system more effective, the system needs to leverage qualitative criteria which should be based on Empathy, Compassion and Gratitude (addressing why for Clinicians) and Due Diligence, Doing the right thing, and thinking anew (addressing how for clinicians), digital twin of decision support system can provide the right nudge and guidance needed to operationalize the transformation journey towards a robust, consistent, and just care for every patient.

Each clinician can work within this broad framework to customize the decision support system to enable himself or herself to provide error free care to their patients. Hospitals can use this framework to provide better care to their patients and reduce accidents from knowledge gaps or human error.

12.2 Why we need the digital twin system to reduce errors in hospitals

> To err is human and to learn is divine

Our understanding of the human brain is constantly changing. Earlier, it was believed that the human brain is a highly sophisticated organ consisting of billions of neurons and capable of recognizing patterns from complex networks, such as languages, without any formal instruction. "Previously, cognitive scientists tried to explain this ability by depicting the brain as a highly optimized computer, but there is now discussion among neuroscientists that this model might not accurately reflect how the brain works." Now, Penn researchers have developed a different model for how the brain interprets patterns from complex networks. This model presents the idea that the brain seeks to represent everything in the simplest way and is constantly trying to juggle accuracy and simplicity in the decision-making process [2].

"This new model is built upon the idea that people make mistakes while trying to make sense of patterns, and these errors are essential to get a glimpse of the bigger picture" [2]. The model takes these errors, as well as various other inputs to get a glimpse of the big picture; to understand how the little parts all work together to contribute to the whole.

Doctors spend several years entrenched in the minute details of health and disease during medical school and residency, and their knowledge base as physicians reflects this. For example, a physician educated in an area of endemic disease will be better at diagnosis and treatment than a physician who had never seen a case of that disease before. Physicians use judgment to diagnose disease and prescribe medications to their patients. The thought process behind this process is more based on previous experience and training received during their medical education, so they are liable to make errors when faced with novel or unorthodox cases, without knowing how or why. For instance, people mistakenly believe that errors "cancel each other out" but they do not. They add up, especially in medicine where errors can lead to unnecessary patient suffering or avoidable healthcare expenditures. In their book, *Noise*, Kahneman and coauthors Olivier Sibony and Cass R. Sunstein expose egregious, undetected errors that a "noise audit" could have diagnosed and avoided creating new problems while trying to solve the existing ones [3].

Physicians determine diagnoses and treatment courses based on probabilities. Often, a physician spends 10–15 minutes collecting a history and physical on patients in an outpatient clinic and will make up his or her mind based on a handful of questions. This process is unique to each doctor, and it has been noticed that the same doctor can produce totally diverse set of conclusions and prescriptions for same chief complaint with minimal variabilities for factors such as race, gender, time of day (i.e., noise). In fact, multiple studies have shown that up to 80% of the information necessary to make a clinical diagnosis can be obtained from the patient history. However, this process assumes that the patient is fully qualified to communicate all symptoms without associated bias and noise, which is often impossible in practice, doubly so if the patient has any speech or language barrier and depends upon an advocate to speak on their behalf.

The American Association of Medical Colleges (AAMC) also reports that the United States is expected to have a significant physician shortage by 2034, with shortfalls between 37,800 to 124,000 physicians. This looming shortage is driven by many factors, primarily the aging US population (i.e., baby boomers) and the aging medical workforce. The COVID-19 pandemic has only exacerbated this issue as demand for medical care has soared, while supply (i.e., medical graduates) remains mismatched [4]. Furthermore, high rates of clinician burnout brought on by the pandemic means more providers are leaving the field than ever before.

AI (Artificial Intelligence) and ML (Machine Learning) technologies have the potential to address one facet of this shortage by giving healthcare providers quick and accurate tools to triage and diagnose disease. Digital Twin for Decision Support System built based on AI and ML technologies can be one such tool. AI programs can image and diagnose patients on a spectrum of "no disease" to "advanced disease" and make it so that patients with serious disease are not left waiting too long for treatment.

In recent times, machine learning – or AI – has come to prominence in making predictions based on vast troves of data. With greater accuracy than any human, AI can predict random events. Doctors making predictive judgments often rely on gut instincts, which are not always accurate.

Wherever prediction exists, assumption does also – and more than you might think. Associating assumptions with the current context is the first step to addressing uncertainty, reducing errors in determining the next

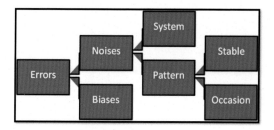

FIGURE 12.1 Noise and biases.

best action for the well-being of a patient and it is an improvement over allowing overconfidence to flourish and noise to accumulate accordingly.

Reducing error is especially important in the healthcare field. Errors can be broken down into three successive, layered categories which contribute to noise in different proportions.

- Error divides into bias and system noise.
- System noise divides into level noise and pattern noise.
- Pattern noise divides into stable pattern noise and occasion noise.

Noise contributes more to error than bias does. Among the various kinds of noise, pattern noise is significantly more prevalent than level noise – usually, by at least double. (See Fig. 12.1.)

Digital twin of Decision Support System (DSS) has lot of potential and new process is being prototyped on almost weekly basis to leverage the digital twin concept in healthcare. DSS can help with reducing error rates in physician's decision making and can help with other inequalities in addition to mitigating or addressing biases and noises in Healthcare Delivery. Some of those inequalities are:

1. Timely access to trusted and reliable information.
2. Access to medical expertise.
3. Access to medical facilities and medical interventions, testing.
4. Access and maturity of public health system.
5. Ability to afford medical interventions (Self-funded vs Government funded).

Other reasons for use of DSS can address access and affordability barriers in healthcare by providing remote patient support and real-time collaboration among clinicians from multiple specialties.

Digital Twin improves clinical reliability and is used to help identify relevant information pertinent to the correct diagnosis or treatment of a patient by mining historical patient records, enabling a clinician to learn from patients with similar symptoms but different socio economic and demography background. Information on rare cases can be brought to the attention of clinicians to help them reliably recognize medical solutions by aggregating and displaying information such as symptoms that might otherwise be easily overlooked.

12.3 What is digital twin for decision support system to reduce errors

Digital twin (DT) is one of the top trending technologies for 2020, especially within the healthcare industry [5]. The concept of this technology refers to a digital replica of a physical object. DT can be very simple and focus on one or two process and technologies or can combine Artificial Intelligence (AI), Data Analytics, IoT (Internet of Things), Virtual and Augmented Reality paired with digital and physical objects [6]. This integration allows real-time data analysis, status monitoring to anticipate and correct problems before they even occur, risk management, cost reduction, and future opportunities prediction.

DSS to reduce error rate will be a virtual replica of the actual system used in hospital, by collecting all data and providing insights to various stakeholders to understand how system used versus intended plan. This will provide visibility and later can help reduce error in diagnosis and prognosis of patient's medical issues and will save hospital from uncertainty, dealing with infections and costly litigations.

12.3.1 Conceptual diagram

This diagram is one of the way hospital systems can leverage power of AI and ML in building robust yet flexible Decision Support System (DSS), which can be used as additional source for validation of hypothesis or further analysis to identify right questions to explore in all aspect of Patient Care. (See Fig. 12.2.)

12.3.1.1 Key components of the DSS are as follows

1. Patient centric digital twin data set

The source for this data can come from lab results, Electronic Medical Records, Electronic Health Records, Patient Activities logs, Patient diet chart, Patients demographic data, current list of medication including list of other type of medicines (Homeopathy, Naturopathy, and holistic/regional therapy), Key data systems of records or engagement used in hospitals to capture patients' centric data regardless of types and functions.

2. Aggregated digital twin data set at hospitals systems

This dataset will be like for like replica of patient central Digital twin data set and will be maintained in Enterprise Data Warehouse of Data Lake at each hospital system. This will be one dataset for all patients using services regardless of payors responsible for payment of services provided

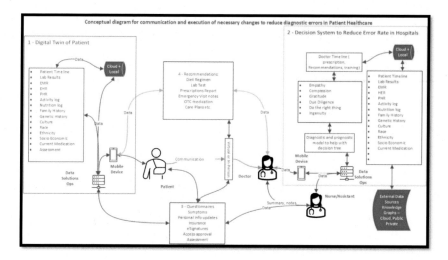

FIGURE 12.2   Conceptual diagrams for decision support system.

by any provider within the hospital systems. This dataset will use Health Exchange Information Integration and interoperability capabilities to constantly update itself to have most up to date capability like Master Patient Index (MPI) capabilities to identify each patient uniquely and help build Patient Timeline to provide critical support to doctors, hospital staff and operation team in rendering services to patients.

This data set will continue to be enhanced as data governance process matures and new data assets become available. This is the best place for hospital systems to leverage its unique differentiation capabilities related to people, process, and population health. Hospital system can use empathy and compassions to provide holistic support to all patients regardless of their social or economic status.

3. Questionnaire dataset

Each hospital System can have a comprehensive set of question-and-answer dataset used by various providers based on specialization. Purpose of this database is to inform and educate providers about interdependencies, adverse reactions, patient experience between various encounters patients have had with other providers either in same hospital system or outside. This will help address gaps, noise and biases related challenges and provide more balanced view of patient medical, emotional, and socioeconomic state to providers and hospital systems.

4. Recommendations dataset

This is single place all recommendations related to the patients are stored and maintained. Some of the examples are Diet regimen, Lab Test ordered, Prescriptions including over the counter, Third party report like behavioral or neurological evaluations and Emergency visit plan etc. This can be great source for providers to follow up and align all future recommendations and referrals to provide better care.

12.3.2 Digital twin for decision support system (DSS)

Health care is a team effort, and shared information supports that effort. Much of the value derived from the health care delivery system results from the effective communication of information from one party to another and the ability of multiple parties to engage in interactive communication of information. DSS can play a significant role in achieving the desired objective of providing care that is based on data and contextualized by keeping patients' current and past situation in mind.

12.3.3 Key components, definitions, challenges, and data sources

The Healthcare data generation is growing so fast that it is doubling every 2–3 months. DSS can be designed with structured in unstructured data. The key is to find the right data source with right context. While detailed description about various types of dataset is outside the scope of this chapter but below you can find some common data sources which can be explored and become part of the DSS as first iteration.

12.3.3.1 Patient health record (PHR)

PHR is an electronic record of an individual's health information. This can generally be found on Patient's mobile devices or personal cloud associated with patient (e.g., iCloud, Google cloud etc.). This is limited to the type of device being used to collect patient's health data.

12.3.3.2 Electronic health records (EHR)

EHR are generated by providers (Hospitals, Clinic, doctors, and Nurses etc.) These records are good way for doctors to cross-check, understand trends, assign proper diagnosis and procedure codes, and generate billing sheets. This data also requires data cleaning and normalization for downstream analysis. Providers own this data and patient can request access to this data for own use or to share with other providers.

12.3.3.3 Electronic medical records (EMR)

EMR are another set of unstructured and structured data points stored in digital format in provider offices or hospital facilities. This data, once curated, is a very good source of information to build patient timeline comprising of medical and treatment history of patients. It is a digital version of the paper charts in the clinician's or hospital office. Patient Timeline is critical component of DSS to understand the context and performs Insilco studies and analysis. Patient Timeline can help providers track data over time, including its association with context. This can help providers identify which patient is due for preventive screening and what precautions must be taken while performing the procedure, what type of medication be offered to patient to mitigate or minimize the precare anxiety and post care suffering.

DSS can be great way to realize full potential of the EMR data because information in EMRs does not travel easily out of the provider office. Curated EMR data as part of Patient Timeline can be very beneficial to reduce error and improve the outcome for patient and providers.

12.3.4 Type of data available and key consideration while building the DSS

The data available to build the DSS is categorized to help with applying the right ML and AI algorithms to produce the right recommendations for a given patient.

- **Structured data** – A dataset with consistent structure; a table with row and column, a relational database is used to manage this type of data.
- **Unstructured data** – this is a free flow of data written and generated by clinicians and medical devices and very cryptic and filled with acronyms and other contextual information. Some examples are Clinical Text written by doctors, X-ray images, MRI, CT scans and signals (EKG) data sets.
- **Observational data** – This dataset is captured over a period by adjacent sub systems as a bye-product and can be useful source to consider additional context to further refine data curation process for AI and ML work.

Healthcare data varies along several dimensions and these dimensions play significant roles in optimizing the AI and ML model, which is important part of any DSS system. Some of these dimensions are:

- Timescale, which is defined as amount of time something takes, for example in case of EKG the timescale is in seconds and based on continuous timeseries of signal, but for diabetes, the Timescale can be in years. Different Timescales, generated at different points in patient journey, offer different values. Because of this knowing the timeline view of patient data will be critical for effective use of data.

It is important to recognize that data generated within the healthcare system may be inaccurate or biased at various points along the delivery chain. These include:

- **Patient** – Selection bias to decide when, where and how to seek guidance to receive care from provider, patients can decide to seek, delay, or not seek help, they may self-medicate by asking friends/family or via internet search.

- **Clinical care** – The data is patchy and there is a lot of missing information. Critical information from other health systems is missing. In general, medical records are often neither complete nor accurate due to lack of information sharing between EMRs.
- **Healthcare providers** – The records captured by the systems or manually over a period which may introduce inaccuracies because of incorrect mapping. Incorrect mapping can happen because of manual data entry, subjective interpretation by different stakeholders.
- Financial incentives in term of bonus or penalties from the payers and government may also skew data collection and recording.
- Finally, data collected may be skewed if the hospital only provides limited specialty care or serves a limited demographic pool.
- **Coding of data** – Diagnosis codes and procedures codes entered medical bills only include information about the services rendered to patients by doctors during the episodes and it does not include information about activities or treatments patients may have received from other healthcare providers or self-medication. Medical codes and the diagnoses they represent are also regularly updated, which makes tracking patient health over time via diagnostic codes difficult.

Data capture misclassification can be characterized into two categories:

1. Exposure related Misclassification – Data captured in EHR, and Pharmacy data does not match and have gap because Patient may have used either free sample of medication received from doctors' office, thus requested prescription late or he did not pick up the prescription in time from the pharmacy.
2. Outcome Related misclassification – This type of misclassification can occur when clinicians are not sure about the prognosis and record the incorrect code into the Clinical notes for expediting the billing process.

One way to mitigate these challenges is to combine and harmonize the data sources, this can be done by leveraging collaboration among providers and subject matter experts from multiple disciplines who understand the context in which the data set was developed, but it can be challenging because of data ownership, missing contextual information and transparency around how data is captured in first place. This can be mitigated by providing clear guidelines and policies with structured and unstructured data. The care should be taken in such a way so that the health data is captured correctly via objective signs and symptoms (fever, weight, heart rate, etc.), lab values, imaging studies, etc. and follow standardized protocols for these across the nation to diagnose a patient.

Another way is to set clean up rules to clean or fill up the missing data by studying the context and adjacent data sets. This will be a downstream analysis either by embedding a data engineer team along with healthcare team or using a simple tool to categorize the data for downstream analysis and mapping by a centralized data science team.

Although there is not one silver bullet to address these challenges, the following steps can be taken to mitigate the data inconsistency related issues to build a robust DSS. These steps are:

1. Require multiple mention of diagnosis code in EHR dataset to ensure that the patient really has a condition. This is very important to improve the explainability and transparency of harmonized data set developed by combining raw data from multiple providers.

2. Require that there be a corresponding procedure code also available along with the diagnosis code in EHR data. This can be achieved by combining qualitative, contextual mapping with prebuilt data mapping tools with predefined rules. This type of hybrid data mapping is still emerging and will continue to be both science and art for a data engineer as he or she tries this to improve the efficacy of harmonized data sets. This can be further complicated if patients decide to refuse sharing of data or care or request a completely different procedure that may not be FDA approved for their condition but had shown promise in some studies. The whole point of a doctor is to make a diagnosis, gives patient the options, and then let them decide what they want to do. DSS can be very powerful tool to aid doctors in analyzing and reviewing all possible care options which can help doctors convince patient in receiving the right care.

3. Require that a medication, if applicable, specific to the decease is also available in the dataset. Best way to handle this is use the predefined data mapping so that the DSS can give choices to providers as top three best options with caveat. This way providers can take into considerations issues specific to patient rather than based on muscle memory. This is emerging field and purpose is to empower providers with as much as possible contextual info to help them take a more complete decision regarding diagnosis and prognosis. This will require some creative thinking on part of doctors and hospital to operationalize the capabilities. For example, there is 100+ antibiotics, each is different. Diagnosing someone with Strep throat is great, but there are probably 20 antibiotics you can use to treat that. It is only the ultra-rare diseases that have only 1 or 2 medications. Most common diseases have many different meds and therapies and again, the doctor must decide the best one. It is not random though, there is a whole flow chart that they have access to that they can use to pick an antibiotic based on type of pathogen, patient's risk factors, etc. So, to that end, having an DSS that consolidates all this information would be helpful. So instead of

the doctor having to look it up, they could just input the information and having the DSS can say "here are the top 3 best options" or something.
4. Leverage Electronic Phenotyping (Strategies used to determine if an outcome really occurred in the patient's timeline) techniques by comparing patient's manual chart with EMR data or putting boundaries on the misclassification of data by constructing a computer simulation that adds random misclassifications to the outcome or exposure dataset and corelating this with the question we are trying to answer using the Decision Support System.

12.3.4.1 Possible data sources for decision support system to reduce errors

There are multiple sources of healthcare data, and this list is increasing every month. In fact, the amount of data generated is so vast that medical knowledge is effectively doubling every 3 months (7). Unfortunately, it is exceedingly difficult to find all the data in one place. Each data source will have some information about what happened to patients in their care timeline, but will rarely, if it all, have a patient's entire health history on record. Hence it is recommended that a robust DSS all available datasets be used to consolidate all available health data to provide a clearer picture of a patient's overall health. Some of those data sources are as follows:

1. Medical Records of the Patient – EMR, EHR, Genetic Tests results.
2. Delivery of Care – Progress notes, clinical notes, lab orders written by doctors, laboratory results etc.
3. Claims Data – Provider Billing Data, Insurance claims reimbursement data.
4. Pharmacy Records – Written prescription, Online prescriptions, timing of prescriptions, payment method used etc.
5. Surveillance Data – This is the database maintained by Government and other agencies to maintain records of reported effects and side-effects. Some of the examples/sources are:
   a. FDA (Food and Drug Administration)
   b. FAERS (FDA Adverse Event Reporting System)
   c. MAUDE (Manufacturer and User Facility Device Experience)
   d. CDC (Center for Disease Control)
   e. Registries run by professional Societies. Examples are:

      i. PRIME
      ii. CancerLinq
      iii. Intelligent Research in Sights (IRIS)

6. Population Health Data – Records data by treatment type, medical condition, and geographic location. Some examples are:

a. AHRQ (Agency for Healthcare Research and Quality) (Agency for Healthcare Research and Quality)
b. National Inpatient Sample (NIS)
c. The Medical Expenditure Sample Survey (MEPS)
d. The National Health and Nutritional Examination Survey (NHANES)

7. Patient Generated Data – Patient records their own health data and maintains records of data and can choose to share with provider at the time of care delivery. Examples include health data obtained from wearable technology like the Apple Watch or the Fitbit.
8. Other data sets: Family History data, Genetics history, Questions and Answers, financial well-being, living will, trust etc.
9. Research Generated Data – Research is systematic analysis, investigation of bio-medical phenomena to find valid and generalized results. All the clinical trial data can be found at www.clinicaltrials.gov. This data is available for reuse and reanalysis once the clinical trial is complete.
10. Clinical trials generate lots and lots of information and generally, researchers only care about a subset of it that they need to prove their hypothesis. There is a lot left over that does not really get used but if this data is curated with right context, then this data can be reused for future analysis, and it can possibly contribute to the overall scientific knowledge on the topic and informed decision-making systems. DSS is best way for hospital to know if this data set usable and help with better decision making.

As there are multiple data sources that can be of potential use while building a DSS system to reduce error and it can be very confusing to determine which data source to use. You can use guidelines or ask the following questions before determining if a data source is mandatory or nice to have requirement to build the DSS system.

Questions to ask when considering Multiple data sources:

a. Is there a well-documented data model available for the data source?
b. What is the source of the data, is it from dependable sources?
c. Is the data freely available, how often is this data updated?
d. How is this data maintained and by whom, are there any known errors documented?
e. Does this data source have the data elements needed to answer pertinent questions concerning reducing error in decision making by hospitals?
f. Is the data set big enough to address noise and biases normally found in smaller set of data?
g. Is the data must be used as it is or a proxy for this data can be used?

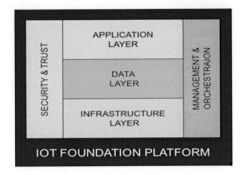

FIGURE 12.3   Digital twin platform.

12.4 Digital twin platform for decision support system to reduce errors

To achieve the objective of reducing error rate on a consistent and transparent basis, a common digital twin platform is necessary, and it can be part of existing or new IOT (Internet of Things) Foundation Platform Framework. This platform needs to be based on sound reference architecture and a foundational and modular component with preintegrated functions and services which in turn will enable the flexible and agile introduction of new data solutions. Unifying services into a common platform layer reduces risk, noncompliance, and operational cost compared to working with fragmented solutions across diverse silos. Unified Platform Approach also helps with operationalizing the Digital Twin system by following framework of Utility, feasibility and monitoring and operability.

The digital twin Platform can also provide centralized visibility and control, even during peak data distribution. This platform will enable highly automated deployment and efficient operations with limited human intervention when deploying and managing services. Finally, consolidated data through a common pipeline is advantageous to avoid multiple implementations of common components.

The Digital twin platform will be composed of technologies having common elements at the overarching Platform level. (See Fig. 12.3.)

This platform will provide common services, technologies, and design principles that are common to all layers within the all-encompassing, "holistic" Digital twin Platform. With the Digital Twin Platform, each contributing technology falls within one of five categories.

12.4.1 Infrastructure layer

Physical and virtual infrastructure elements are located at this layer, including networking functions, compute, storage, virtual machines, and application containers. These elements typically make up the physical topology of the IoT platform ecosystem.

12.4.2 Data layer

This layer involves data that is being generated by elements (sensors, applications, etc.) in the Healthcare environment and typically consumed by applications in higher layers. Additional elements or functions like data persistence, governance, and interfaces are also part of this layer.

12.4.3 Application layer

Applications can take many forms within an IoT Platform. In general, this layer is defined by applications that consume and process data from the Data layer. For example, applications related visualization, data analytics, machine learning or other systems are all part of the Application layer.

12.4.4 Security and trust layer

Any IoT Platform must incorporate strong security and trust mechanisms which span all layers. Security must be strongly designed into the base platform to avoid unintentional attack vectors and compromised elements.

12.4.5 Management and orchestration layer

Each of the previously described layers can be composed of multiple individual technology components and span organizational domains (business, partners, and IT (Information Technology)). A robust, scalable, and flexible management and orchestration layer is critical to deploy and operate the IoT Platform at scale.

While detailed explanations of the all the layers are outside the scope of this chapter but data layer will require deeper understanding in the context of healthcare because of compliance, regulatory and privacy concerns.

12.5 Digital twin system deployment, evaluation and operational consideration

DSS has the potential to provide high performance, fair, transparent, optimize care trajectories, right fit recommendations and insights based on explainable data model, which can help providers improve the process of clinical assertion and decision making.

Although, there is accelerated promise and interest in AI based DSS system in healthcare field but most of them are limited in scope and focus on narrow set of data sets (from one or two providers), narrow definition of research problem, limited to accuracy of the standalone model and potential use by providers to deliver care. Since these models focus on accuracy within narrow set of contexts, which many providers do not find appealing if the context is different for them. Hence there is need for comprehensive evaluation and operational consideration. The list below is not exhaustive but good start for hospitals and providers to start using to operationalize the DSS based on AI and Machine Learning model.

12.5.1 Output action pairing (OAP)

OAP can be first framework can be used to associate the output (Recommendation, Decision guidance, insights etc.) with type of actions taken by the providers for care delivery to patients.

One such framework to align Output with Actions can be based on utility, feasibility, and clinical impact of the output from DSS for the actual use by the providers.

Utility helps us determine the right fit and alignment with patients and providers value system. Feasibility is the readiness of operational backbone or the infrastructure to implement the recommendations from the DSS and finally Clinical impact is the overall effect the recommendation and insight will have on patient care, outcome for patients and care standards as per the requirements from Payors or government agencies.

Other considerations in addition to Utility, feasibility, and clinical impact to design, develop, deploy, evaluate, and operationalize DSS are understanding of Stakeholder involvement and identification of beneficiaries.

Stakeholders need to be chosen carefully and should be from the field with domain knowledge, business decision makers and end users.

Beneficiaries are all about end users and how this will impact the outcome which can be measured and communicated back to decision support system for continuous improvement and mid-course corrections. Beneficiaries can be Payors, Providers, population, or hospitals etc.

12.5.2 DSS deployment considerations

DSS in healthcare will continue to move slower because of its direct impact on human lives thus high stake and regulatory and compliance requirements. Hence there is a need for a holistic system and delivery framework for simple AI and Machine learning based DSS to be usable and effective in real life.

One such framework can be based on Total Product Lifecycle (TPLC). TPLC focuses on end-to-end delivery and operational aspect from ideation to development to packaging to testing and deployment, validation and finally optimization and operationalization. Some of the critical components of TPLC are:

1. Data collection, cleaning, normalization, harmonization, and management.
2. DSS based model training and tuning based on clinical impact, utility, and feasibility.
3. DSS based model validation.

Some Questions to ask prior to start deploying DSS:

1. What are the questions requiring answers based on clinical impact and utility?
2. How the DSS model will help address Stakeholder, beneficiaries and end users care abouts?
3. What type of Training data is available and can be used?
4. What will be the implementation cost for initial set up for development, deployment, testing and validation and implementation?
5. What will be ongoing cost for optimization, support, maintenance, and operationalization?
6. Does the team working on the DSS has necessary digital skills to understand and carry out the tasks?
7. How the DSS system's recommendation and insights will be used by other System of records and engagement in use within Hospitals?
8. Is the DSS system is going to be used as standalone or needs to be scaled and integrated with other systems for compliance and return on investment?
9. What are the audit and reporting requirements, and does it follow Minimum Information for Medical AI Model Reporting? Minimum information for Medical AI model reporting needs to have following minimum requirements met:
   a. Information about Population data used as training data in terms of data sources and cohort selection. This will help with transparency and explainability of AI model based DSS.
   b. Demographic information used in training data for context and comparison purpose during model validation.
   c. AI Model based DSS technical architecture and development process.
   d. Model Optimization and operationalization process logic is easily explainable and independently reproducible.

12.6 Digital twin for decision support system challenges

A significant barrier to the implementation of a digital twin infrastructure is an inherent distrust of not only algorithms, but of the medical field in general. As with any technology, developers need to establish a level of confidence with the general population about the utility, reliability, and security before it can be meaningfully implemented.

Practical uses of Digital twins based on AI and Machine Learning in healthcare do not come without certain challenges as described below. Later we will discuss the best practices to address these challenges. One such challenge is that correlation does not always imply causation.

Digital twins based on machine learning methods, such as neural networks, simply learn associations between the inputs and outputs to find the best fit, regardless of what features or key factors it needs to predict the outcome. As a result, they can sometimes be unreliable, or even dangerous, when used for driving medical decisions. Because Machine Learning is not actually explicitly programmed so machine learning algorithms can use whatever signals to determine correlation are available to achieve the best possible performance in the data set used.

So, we must recognize that the model or function that is created can in many instances come from patterns in the data that are correlations, rather than causative truths about the label, or in other words, about the outcome or the disease. So, if a model to fit the data is based on useless correlations due to either coincidence, or the presence of unforeseen variables, also referred to in machine learning lingo as common response variables, confounding variables, or lurking variables, it can be problematic for translation into clinical care.

For example, there was one group that trained a machine learning model using EHR (Electronic Health Records) data to predict the risk of death for pneumonia patients. The hope was that the model could automatically identify patients at elevated risk who could be treated more aggressively and find those at low risk who could be safely treated at home. And then they built their model using historical EHR data, and it worked well. However, they found later that the model focused on a key correlation to make accurate predictions that was problematic. In this case, they found the model predicted that pneumonia patients who also had asthma were minimal risk, suggesting that asthma is somehow a good prognosis for pneumonia. But this is not accurate as per the medical audience. During reexamination of data it was found model worked perfectly on the data it was used to train but when looked into data more closely it was found that the model was trained on historical data from the hospital which has better outcome for asthma patients but this was because of hospital policy of providing aggressive pneumonia treatment to asthmatic patients, so asthma patients with pneumonia were directly admitted automatically to the intensive care unit, where they received aggressive treatment by protocol, which, over time, improved pneumonia prognosis for patients who also had asthma in this hospital. So, the model did not know that there was a policy that helped the pneumonia patients get better who also had asthma. It just found that there was a correlation in the data than pneumonia patients with asthma had better outcomes than if they did not have asthma.

12.7 Example case studies – DSS

Ophthalmology is one field where artificial intelligence and machine learning are radically improving the ways in which clinicians can diagnose disease. Ophthalmology already takes advantage of many advancements in imaging technology to assess patients' vision and overall health through machines like nonmydriatic retinal cameras, optical coherence tomography, and visual field analyzers, among many others. Traditionally, an ophthalmic visit consists of a series of images from these machines, followed by clinical interpretation of the results into a cohesive diagnosis. This entire process is very labor intensive and requires highly specialized expertise to execute. Couple this reality with the fact that a rapidly aging population will require more vision services in the coming decades and the potential for a serious shortfall in clinical care arises.

Several companies, such as EyeNuk based in California, USA (United States), have developed highly sensitive and specific AI programs to detect vision threatening diseases like diabetic retinopathy which affects upwards of 7.7 million people (about twice the population of Oklahoma) in the USA [7].

The combination of these AI based diagnostic software and digital twins is a concept that is recently being explored. Optimo Medical is working on a new digital twin program called Optimeyes which will give eye surgeons the opportunity to virtually model surgeries via digital clones of the anterior eye segment [8]. The result of these technological developments is a new frontier of personalized medicine where clinicians can visualize and model unique patient characteristics prior to treatment to achieve the best clinical outcomes.

12.8 Conclusion

Digital Twin based on AI and machine learning based model has the potential to provide high performance data-driven medicine, to optimize care trajectories, suggest the right therapy for the right patient, also known as precision medicine, and to improve the process of clinical assertions and decision-making. The best way to understand Digital twins' potential use in healthcare is to think about its application in three separate categories:

1. Biomedical research

2. Translational research
3. Medical practice.

Under biomedical research, Digital Twin can be used in assisting in the automated experiments using simulation, automated data collection, and gene function annotation that we see across a broad spectrum.

For translational research, Digital Twin can be used in assisting in areas such as biomarker discovery, drug target prioritization, and genetic variant annotation.

Finally, under medical practice, Digital Twin based applications are being used for disease diagnosis, treatment selection, patient monitoring, and risk stratification models.

Digital Twin based DSS can help hospitals and providers provide holistic care to its patients irrespective of patient needs. This is great way to incorporate the learnings based on data driven analysis with agility, scale, and reliability. These learnings can help stakeholders in hospitals system to incorporate changes needed to help patients needing desirable outcomes and leverage the power of AI and ML in consistently and continuously.

Master Patient Index (MPI) is one such example with huge potential for hospitals and providers. MPI will help hospital reduce data duplication error and need for manual entry of data every time a patient uses different service from providers funded by different payors.

Some of the domains where MPI can make a huge difference are Revenue Cycle Management, Patient Care Management, Population Health Management, Data Driven Decision Management, seamless transfer of data between Health Care Systems and providing continuity of care etc.

References

[1] J.T. James, A new, evidence-based estimate of patient harms associated with hospital care, J. Patient Saf. 9 (3) (2013) 122–128, https://doi.org/10.1097/pts.0b013e3182948a69.
[2] University of Pennsylvania, Researchers Develop a New Model for How the Brain Processes Complex Information, Medicalxpress.com, Medical Xpress, 2020, https://medicalxpress.com/news/2020-05-brain-complex.html.
[3] D. Kahneman, Olivier Sibony, C.R. Sunstein, Noise: a Flaw in Human Judgment, Little, Brown Spark, 2021.
[4] AAMC, AAMC report reinforces mounting physician shortage, https://www.aamc.org/news-insights/press-releases/aamc-report-reinforces-mounting-physician-shortage, 2021.
[5] R. Saracco, Digital twins: where we are where we go III – IEEE future directions, in: IEEE Future Directions, 2019, https://cmte.ieee.org/futuredirections/2019/07/02/digital-twins-where-we-are-where-we-go-iii/.
[6] S. Julia Jeske, Digital twins in healthcare, https://projekter.aau.dk/projekter/files/360456256/Jeske_MasterThesis.pdf, 2020.
[7] NIH, Diabetic retinopathy data and statistics, National Eye Institute. Nih.gov., https://www.nei.nih.gov/learn-about-eye-health/outreach-campaigns-and-resources/eye-health-data-and-statistics/diabetic-retinopathy-data-and-statistics, 2020.

[8] ANSYS, Optimo medical AG and ansys transform eye surgery to better treat astigmatism, Ansys.com, https://www.ansys.com/news-center/press-releases/02-23-21-optimo-medical-ag-ansys-transform-eye-surgery-to-better-treat-astigmatism, 2021.

Digital twin for cardiology

Jeff Luna[a], GilAnthony Ungab M.D.[b],
Kaouther Abrougui[c], Hazim Dahir[d],
Ahmed Khattab[a], and Raj Kumar[a]

[a]Cisco Systems, San Jose, CA, United States, [b]Lucia Health Guidelines,
San Francisco, CA, United States, [c]Voltron Data Inc., Mountain View, CA,
United States, [d]Cisco Systems, Research Triangle Park, NC, United States

13.1 Introduction to digital twin for cardiology

13.1.1 History

The usage of digital twins, the virtual representation that serves as the real-time digital counterpart to a person, physical object, or process in healthcare has become very prevalent over the last few years as healthcare organizations are seeing benefits for clinical based medicine, evidence-based medicine, research, and medical treatment areas to use digital twin technology to help patient outcomes.

The emergence of digital twin technology in healthcare is a result of healthcare systems using data from clinical measurements and supplying the data to population health tools for patient data analysis and personalized models.

This has not always been the case as most health care systems started adopting electronic medical records in the 1990s and many paper records took years to convert into an electronic medical record format.

Data sharing has also been an issue with both paper records and electronic records. As doctors would try to gain insight from past clinical visits, it was often a consultation or referral with limited information. The advent of electronic medical records was meant to supply insights, condition histories, and testing results to clinical professionals in a fast and efficient manner.

The complexities of data sharing between health care systems and patients placed the burden on the patient to request their medical record, visitation histories, laboratory results, testing results, x-rays to give the medical record to clinical professionals outside of the primary healthcare organization if they wanted the fastest record sharing method. Normal methods to bring healthcare information into new organizations would require a record request to start a request for records and for those records to be published and then sent to another health care entity.

Most health care organizations struggled in the 1990's due to the expense of implementing electronic medical records and most the health care organizations were not building efficient high speed and secure networks to traffic information to requestors of healthcare records.

Health care organizations would compensate by downloading medical records based upon scheduled visitations to ensure that clinicians had proper histories available or preposition data to specialty clinical locations to ensure that health care professionals were able to have data available to make a thorough and comprehensive diagnostic decision. In cases where critical information or case history was needed, it was often unavailable or incomplete due to poor investments from healthcare organizations into properly supporting their networks and record keeping.

In one California based patient practice, due to poor access to medical records at remote locations, patients' medical records were copied and put in the folders to be transported by medical assistants from a central clinical office and delivered to specialist offices. The health care practice professional clinical staff made poor decisions based on incomplete data and relied heavily on the past medical history provided by the patient who may have not had a health care background. It resulted in a medical professional removing a healthy kidney from a patient due to inability to access medical record histories in a surgical location [7] and later surrendered their medical license. Often when patients supply incomplete or incorrect information, it results in repeating costly studies and lab tests that cause care delays or severe errors.

Issues like these have been sensational highlights of poor access and management of centralized medical records in health care organizations in the 2000's. Electronic medical record companies began to provide shared and dedicated outsourced hosting of medical records as a way providing better access and availability for medical professionals and alleviate issues with security and accessibility by healthcare professionals.

One of the outcomes of hosting electronic medical record data in an outsourced fashion was the ability to have more patient data and the potential to share data between healthcare organizations to discover trends, procedures, and predictions for future health care requests.

Many electronic medical record organizations called upon large technology companies to help them sort through data and be able to meet new requirements for privacy of personal identifiable healthcare information. These efforts also alleviated some of the burden on healthcare systems to support regulatory compliance from a data perspective.

Many healthcare organizations were in a better position to start analyzing substantial amounts of anonymized patient data for usage in public health and policy planning through medical record vendors but the need for granular information drove many healthcare organizations to look at extracting information for specific uses or funding areas including reduction in heart diseases.

13.1.2 Focus

The call to tackle chronic illness and disease in populations has been the focus of Health care organizations. Major chronic illnesses and diseases such as lung, heart, cancer, and diabetes were top choices for health care organizations in the 1990s and 2000s to analyze and find better ways to manage patient care due to the massive costs and efforts needed to combat these diseases.

Population health tools helped health care organizations better plan and prepare for operations to apply their resources where needed and have improved population health outcomes at a lower cost to patients, providers, and insurers. The new models for care coordination and high-risk care management including unplanned acute episodic care have shown promising results and encourage health care providers to build care coordination models outside of the health care facilities [8].

The shift towards using population health information care coordination model into early intervention and prevention of chronic disease alleviates the burden on health care systems to build and mobilize the delivery of resources.

The focus on cardiology in American patients increased in the last few years due heart disease being a major chronic disease in the United States. According to the Centers for Disease control and prevention, heart disease is the leading cause of death for women men and people of most racial and ethnic groups in the United States.

13.1.3 Facts

One person dies every 36 seconds from cardiovascular disease which equates to about 660,000 people (about half the population of Hawaii) dying each year or one in every four deaths. Heart disease is one of the most expensive healthcare costs in the United States accounting for $363 billion (about $1,100 per person in the U.S) of health care services, medications, and lost productivity due to death.

The most common type of heart disease is coronary artery disease which kills over 360,000 people (about half the population of Vermont) each year. About 18 million adults over 20 years old have coronary artery disease. This type of heart disease is now claiming more lives at younger ages. Two in ten deaths from coronary artery disease happen in adults less than 65 years old.

Deaths from heart attacks occur every 40 seconds in the United States. 75% of heart attack patients experience a first-time heart attack and the remaining 25% of patients have already had a heart attack sometime in the past. This accounts for over 800,000 people (about half the population of Nebraska) in United States. 20% of all heart attacks are considered silent heart attacks where the patient does not know they have had a heart attack, but damage has been caused [9].

Cardiologists consider heart disease a lifestyle illness due to the major risk factors being high blood pressure, high blood cholesterol, and smoking. Other increased factors for heart disease include diabetes, obesity, poor diet, physical inactivity, and excessive alcohol use.

13.2 Digital twins to challenge heart disease

Heart disease is a significant burden on patients because of the cost of intervention, the number of specialty clinical disciplines, complexity of care coordination, a high number of invasive procedures, long recovery times, elevated risk of death due to the process of heart disease being hard to reverse.

The advent of digital twin technologies in cardiology to combat and cure heart disease is a much-needed technology that can supply an approach to create awareness for healthier lifestyles. Patient centered activities like diet, exercise, weight reduction, and early screening/detection of issues could become more impactful if it were done in comparison with a digital twin for cardiology to compare results. The gains in long-term health outcomes can become more impactful.

From April 2020 through April 2021 the top fitness apps of Mi fit, Strava, MyFitnessPal, and Peloton increased subscriptions by 30% or more. Many subscribers have turned to connected devices to keep moving and track their fitness. The afore-mentioned apps supply users the ability to not only track their fitness but to understand ranking and positioning against cohorts of users.

It may seem surprising that elements of fitness apps like leaderboards are large draw for users, but just like in the world of video games where top ranked players expose their score and supply their initials or gamer identity tags, the elements and mechanics between games and exercises show self-determination theory, competence, autonomy, and empathy. Various journals have published information about gamification in healthcare context through which promotion of physical activity by the advancement through levels of completion in exchange for earning points, scores, and ranking proves capacity or ability [10].

Gamification of exercise applications like Peloton and Strava helps users in a community focus on live class formats to understand their improvement as individuals as compared to other individuals or to larger groupings of fitness seekers. Fitness enthusiasts have increased their participation with these apps and user based subscriptions have seen phenomenal demand as consumers sought outside activities to supplement their homebound lifestyles.

Connected devices have supplied fitness tracking apps insights and information to users to compare their results to others. Downloads of Strava surged 115% globally with Brazil, the United Kingdom, and the United States leading the demand.

Fitness enthusiasts have used applications like Strava to build affinity, rankings, and overall wellness from an aspect of a digital twin with over 80 million users (about twice the population of California). The fitness digital twin has existed to remind users to take part and compare their results to others as a virtual training partner or as a guide in unfamiliar training circuits or racecourses. The results from fitness digital twins usually help fitness enthusiasts become more competitive and strive to reach new levels of performance.

Patients with cardiovascular disease have received help from using applications like Strava to supply insights into fitness and use the fitness digital twins as a virtual training partner for cardiac rehabilitation through monitoring and recreational or competitive activities used to increase physical ability. The reports from the Strava platform have enabled patients to better communicate to clinical professionals their fitness routines and heart rates using monitoring devices like smart watches [11].

Many health care professionals do record patient supplied fitness data in a health care patient record to support the improvements in cardiac rehabilitation. Updates from third party applications like Strava are not automatically uploaded into a healthcare record because the amount of data provided to healthcare systems is considered as nonclinical or too detailed for a healthcare provider to review. In some cases, device generated data would overwhelm electronic medical record systems with the amount of data to be stored, processed, and analyzed. Clinicians also question the accuracy of the data due to errors in devices, incorrect placement of devices (blood pressure cuffs, heart rate monitors, blood oxygen sensors), low quality sensors, or incomplete transmission of data from devices to patient healthcare records.

Another challenge to digital twins for cardiology is the prevention of cardiovascular disease through analysis of health risk factors and results from early cardiovascular disease testing. The data from risk factors and testing scores may be translated into cardiology digital twins to uncover trends of cardiovascular disease progress in asymptomatic individuals to prevent future health issues or to create a care plan for early intervention [12].

Most healthcare systems struggle to create allies of patients and use strict protocols for intervention based upon using data from past patients due to the structure and complexity of data analysis services and clinical decision engines.

13.2.1 Opportunities

There are several opportunities for digital twins for cardiology starting with the benefits of using nonclinical data into a digital twin format to alert health care professionals when intervention is needed. The idea of using data to call attention to patients in most need based upon reported information was controversial. Scholars, Doctors, and healthcare professionals felt that only clinical data could be trusted, and that artificial intelligence and machine learning would become more efficient than traditional diagnostic tools and drop many clinical routines [2].

The ongoing debate of the long-term ramifications of digitalization of healthcare has been argued as the power of healthcare will gradually shift from doctors to patients. Dr. Eric Topol's 2016 publication "The Patient Will See You Now" uses the argument of devices such as smartphones will play a significant role to help patients supply information to health care organizations and that information will be democratized [13].

The democratization of healthcare is characterized by two major factors; the distribution of data and the ability to generate and apply insights at scale. Democratization of healthcare is meant to enable access to data, technology, and ability for patients and care givers to take charge of their health care outcomes [3]. Less time will be spent on routine tasks and more time will be spent by the clinicians in the area that is most concerning to patients.

Computers will replace physicians for diagnostic tasks and citizen scientists will give rise to citizen medicine that will lead to innovative approaches to cure illness through data science methods for understanding outcomes of past patients.

The possibilities of assembling patient data in the form of a digital twin could greatly enhance the capabilities for both patients and clinicians to gain insights through evidence-based medicine to help decide possible outcomes for prevention, treatment, and ongoing care.

The success of a digital twin for cardiology will be based on the systems that will support it and organizations that will support clinical healthcare staff to learn and train them to use these systems and build trust over time. Atrial fibrillation (afib), an irregular and rapid heart rate (arrhythmia) can lead to blood clots in the heart which can raise the risk of stroke, heart failure, and heart related complications. Afib clinical prediction scoring is a splendid example of how decision-based medicine can use scored information by machine learning and artificial intelligence methods to help clinical support decision making to prevent advancement of disease or death.

One of the tools used by clinical staff in emergency rooms and intensive care units is CHA_2DS_2-VASc scoring. CHA_2DS_2-VASc is used to guide physicians on blood thinning treatments to prevent or lower the risk of stroke. The score is based on your past medical history and can very accurately calculate your risk of stroke per year if you were diagnosed with afib [4]. The scoring of the conditions has been advanced to use the past medical history and ingestion of electrocardiogram readings where machine learning is applied to advanced clinical insights. In a digital twin framework, this enables doctors to use the data evidence for prompt and actionable diagnostics at the point of care.

Past methods for diagnosing Atrial fibrillation, a cue knows as **HAS-BLED** (Hypertension; Abnormal renal and liver; Stroke, Bleeding; Labile INR; Elderly – Age 65+, Drugs or Alcohol) does not have the same compelling data but it is used to decide if you are too elevated risk for bleeding if blood thinners are prescribed [5]. Each letter corresponds to a chronic disease state like H=Hypertension and they are weighted in value either 1 or 2 points and if you have scored >2 then you have a substantial risk for stroke per year, and a score of >3 is considerable risk but does not necessarily mean that an anticoagulant can be given as some risk factors may be changed [6].

Artificial intelligence and machine learning can be more exact at reading/diagnosing Afib on the electrocardiogram than the Emergency room or primary care physician supplying the insights to aid clinical professionals in a digital twin model*. CHA_2DS_2-VASc and **HAS-BLED** scores have been based on large randomized clinical trial and can accurately supply the clinician a percent risk for stroke per year and if the patient is at a higher risk to have a bleeding side effect complication if a blood thinner is prescribed. Companies like Lucia Health Guidelines have helped clinicians in emergency rooms to treat all atrial fibrillation patients to save lives and prevent strokes.

13.2.2 Digital twin structures for cardiology

The structures that will enable clinical staff to gain insights into patient's well-being and long-term health care outcomes will be based upon complex systems that will ingest healthcare data from multiple resources including an exchange of information from the traditional electronic medical record system. These structures will enable any noncardiologist clinician to supply care at the highest level or at the level of a specialist when a structure for digital twins is assembled.

Health care organizations will need to partner with organizations that have ability to rapidly build infrastructures for high-speed networking that enable external nonclinical devices to report information through trusted and secure methods. Healthcare systems will also need to have the ability ingest clinical data consisting of both digital outputs (numerical values from testing and measurement devices) and analog outputs (forms with recorded values, electrocardiogram rhythm strips, MRI images, high-definition ultrasound video).

These forms of data will need to be secured, stored, then passed through to the systems standing for a digital twin for cardiology. Healthcare data will then travel through a lifecycle where it will need to be normalized, classified, sorted, and processed.

The processing will expose information that may be unstructured to a structured process. The structured processes will allow clinicians, researchers, public health officials to understand what changes will need to be made for the best outcome based upon a series of decisions by clinicians. The mechanism that will connect this information back to patients will supply feedback or augmentation based on what patients are doing when they are not in a clinical setting.

This approach creates a structure for cardiology where the patient is the center of the process. The patient will become the provider of data and the receiver of advice; through a blend of guidance from clinical decision engines (AI and Healthcare professionals) and a range of outcomes from past patients. Further enhancements to enable chat bots to translate patient and clinician conversations to show risk of stroke or risk of bleeding could also help find patients who need more care. This may also help patients ask their primary care physicians questions if they should be on blood thinners or to help prevent errors that can be lifesaving.

In a 2017 Duke/UCLA University study focused on 95,000 stroke patients at more than 1000 hospitals due to untreated atrial fibrillation. 84% of patients did not receive guideline-recommended anticoagulation therapy preceding a stroke event. The study UCLA would also report that if a patient goes untreated in the ER for atrial fibrillation (afib), they are 2.7 times more likely to stroke, die, represent withing 1 year [19]. An estimated 6.6 million patients (about twice the population of Nevada) will be present to the emergency room with atrial fibrillation conditions in 2022. Many of those patients would like to see a cardiac specialist (cardiac electrophysiologist) to treat them because the side effect of no treatment is a high rate of death. However, health care reform is making it difficult for patients to access cardiac specialists (cardiac electrophysiologists) especially in rural areas.

The solution may be to use digital twins and artificial intelligence with machine learning to read electrocardiographs and build a clinical decision support tool to guide emergency room physicians. Software from companies like Lucia Health Guidelines has provided this type of service to build a better set of predictive analytics and new intervention strategies to help patients and clinicians to the best outcome. This may pose an interesting avenue for clinical intervention where patients and advocates of patients, in the form of chatbots standing for a digital twin for cardiology, may have a discussion with doctors to prevent a patient's own stroke or death. Some intervention techniques could prompt clinicians to start blood thinners promptly because modeled data shows a better outcome by preventing a stroke to save lives.

The patient and the digital twin for cardiology will supply guidance in the management of their own health. This may also be the opportunity to find other sources of data, such as research and clinical trials data, and supply patients a choice to apply those layers of data to their own digital twin for cardiology instance. The digital twin for cardiology instance could sort through the newly added layers of data and look for new ways to further improve long term cardiac health care options and present options to patients and clinicians with natural language translation and sentiment analysis.

Those new ways to improve long term cardiac healthcare options could be screened and approved by health care professionals as a practical choice based upon rank and scoring through their clinical experiences or the outcomes of other cardiac patients.

13.2.3 Bring your own data (BYOD)

Streamlining the clinical data produced by devices carries complexities ranging from lack of standardization of data results to the lack of support to use advanced technologies such as computer vision and artificial intelligence to analyze analog and digital data for cardiac care patients.

Many healthcare organizations are vendor dependent and include a range of vendors that they use for testing and clinical diagnostics. Many diagnostic tools in clinical settings have become more sophisticated and their abilities to supply rich data such as 4K resolution ultrasounds, high-definition electrocardiography which produce high quality data that can be helpful for clinicians to add to their patient profiles. In the case of electrocardiogram (EKG) machines, some hospitals are better funded than others and can buy the newer EKG machines. High resolution digital imaging information can also be provided to digital twin for cardiology systems by taking pictures through mobile devices and tablets of EKG strips produced during studies. Leveraging mobile devices to ingest information into machine learning can enable a more democratized approach to prevent disparities between well-funded and under-funded hospitals while supplying prompt diagnostics and treatments [20].

A system of digital twins for cardiology could supply these sophisticated devices a way to ingest more depth and layers of data that could enhance a digital twin for cardiology to use more data to help clinicians take advantage of new data. This means that many health care vendors would need to allow enriched data and metadata results to systems outside of their systems and into a digital twin model. This may also create controversy as the digital twin model may interpret the results of 4k resolution ultrasound faster than a clinical professional and inform the patient.

Patients would be given very prompt information instead of today's system where results are packaged and shipped to a doctor. The doctor must take time to analyze the results and then arrange time to review the results with the patient possibly in the clinical office. In many clinical settings in the United States, this process could take weeks especially if a patient did not have access to specialty medical professionals. A digital twin for cardiology system may only take minutes to classify and interpret results to a patient. The results from healthcare digital twin solutions can empower the patient to ask about their own care and often will tell friends and family to act as an advocate for the patient to the primary care physician or care teams.

Combined with information coming from a patient's own health devices including blood oxygen levels, heart rate, and more advanced sensors that could predict atrial fibrillation. We may now have layers of data within a context or framework of information that supply stronger diagnostic capability and better real time feedback to cardiac patients known as evidence-based medicine. Consumer devices like smart watches are already alerting patients to speak to their physicians when they detect possible atrial fibrillation conditions often intervening before a health condition becomes critical.

13.2.4 Timely data sharing

The development of the method for systematic reviews and meta-analysis from patient care records has supplied direction for development of evidence-based medicine by looking at outcomes in health care practices. It has been widely debated how much a healthcare record can predict outcomes of long-term health because of the best research evidence with clinical experience and patient values [1]. Evidence based medicine uses the principle of information hierarchies to obtain information from the patient through clinical testing layered with date from randomized clinical trials. Each layer of data is reassembled in a hierarchy where patient records are blended with clinical data and evaluated for quantitative results more than clinician's experience and expert opinion.

Digital twins in cardiology could use a set of quantitative and qualitative events from patients that are recorded and layered into a digital representation that can be used for outcome analysis, care planning and coordination, integration into clinical trials and treatments. Future solutions and devices will lead to better datasets that will supply better predictive analytics. This supplies assurance that digital twins for cardiology will guarantee better predictors of health outcomes in the future.

The importance of this data can help tackle some of the most severe acute and chronic conditions for disease prevention and health promotion. Cardiac device monitoring manufacturers are recognizing the value of clinical informatics being used in a digital twin configuration to help patients and clinician to ingest meta data that would have normally not been easy to interpret by a human, but easy for an advanced system to supply diagnostic information.

The advances in computational sciences have created areas of interest including bioinformatics and genetic studies. Layering bioinformatic and genetic data into a digital twin model further insights can be given about patience and the range of clinical outcomes that are available. As these advances have been able to capture data, artificial intelligence and machine learning patterns will have to be created to allow clinicians to have better insights and outcomes of unstructured data.

This may enable clinicians to recognize undiscovered or unknown issues the traditional methods of clinical study may have never discovered, or a trial-and-error method will discover over time. Many innovative solutions like Lucia Health Guidelines will be able to ingest genetic test data in minutes, possibly supplanting point in time EKG studies. This may help clinical decision support tools to recognize risk of stroke and to supply guidance for treatment based on the predictive analytics of the better data set collected by digital twin for cardiology solutions.

Most cardiac diagnostic testing information produced had been delivered in paper format and was oftentimes coded into electronic medical records as summaries of full studies, only within one healthcare practice or organization. Islands of data from various cardiac reports and studies have isolated important data making it difficult for patients and clinicians to gain insights and to share with health care clinicians outside of the primary organization, thus hindering care coordination and oftentimes lead to delays in treatment. Since this form of documentation was adequate to look at case history for clinical reviews, it was not adequate for a process of digitization and clinical solution insights.

Patient health outcomes based on digital twin for cardiology modeling and analysis patterns from multiple patients can create pathways for clinicians to gain insight from a digital twin representation of the patient and close matches to the patient's digital twin. This matching process may boost the confidence of a clinician at the point of care to follow guidelines or to examine the treatment outcomes from other patients and clinicians. The amount of prompt data sharing could become a guide for clinical practitioners for better disease identification and diagnosis at the point of care than to have a patient not be treated for a specific coronary disease.

Many cardiovascular diagnostic tests such as Electro cardiography, echocardiography, blood pressure determination, blood biomarkers, risk factors, and ventricle dysfunction can be quantified and inserted into a record for complete picture of overall health. The digital twin for cardiology could help create a more robust layer for data structures. In addition, consumer devices such as pedometers, wrist worn health devices, pulse oxygenation and blood oxygenation sensors can supply valuable data to give key insights into cardiovascular health.

13.2.5 Opportunities

In the United States, most patients who have been diagnosed with cardiovascular disease are given considerable amounts of prescription drugs to alleviate symptoms that may sometimes cause complications while waiting for scheduling of life saving procedures.

Often these prescribed medications do cause side effects which give a false impression to patients that they're on a track for health, yet lifestyle issues persist contributing to poor cardiovascular health.

Studies from the journal of American College of cardiology show that in case studies, performance measures of patients enrolled in programs to provide insights in the quality of outpatient cardiac care show that compliance with performance measures was variable, even after accounting for exclusion criteria suggesting the importance of the opportunity to improve the quality of care [14].

In this study physicians underwent a series of education training sessions to help participants understand what data was to be collected and that the data would be used to change the electronic medical record and data collection systems to comprehensively capture other requisite data elements for program participation.

Data quality checks and analysis were performed, and clinical settings and patients were informed of data updates based upon cardiac performance measures and REITs reported at the patient level. The outcomes of the study looked at performance measures a patient with atrial fibrillation, heart failure, and coronary artery disease. As the patients were admitted to the study based upon eligibility because they were not using a beta blocker after myocardial infarction in a patient with hypotension or personal contraindication for each measure.

The study further suggested that the unit of assessment for patients for beta blocker therapy as well as questioning about smoking patterns, smoking cessation, symptom, and activity assessment showed less than an 86% compliance rate. As this study was performed from July 2008 through June 2009, no personalized devices were used.

The argument that personalized devices could help supply better quality information to a health care record to capture and improve L patient performance measures and compliance could be an opportunity for digital twins and cardiology to help supply models to patients to improve healthcare outcomes.

In 2019 the New England Journal of Medicine published an article about large scale assessment of smartwatches to show atrial fibrillation. The study method required participants without atrial fibrillation to use a smartphone and a smart watch with irregular pulse notification algorithm to find possible atrial fibrillation if detected a telemedicine visit was started an electrocardiography patch was mailed to the participant to be worn for up to seven days. Added surveys were administered to participants upon the notification of an irregular pulse at the end of the study. The main goal was to estimate the proportion of notified participants with atrial fibrillation (AFIB) shown on an ECG patch and the positive predictive value of irregular pulse intervals [15].

400,000 participants were studied over an 8-month span and over 2000 participants received notification of irregular pulse. 450 participants return the ECG patches having data that could be analyzed which on average was 13 days (about 2 weeks) after the first notification of atrial fibrillation. Atrial fibrillation was present in 34% of the detected and returned patches with a 97.5% confidence interval and participants aged 65 and over. (See Fig. 13.1.)

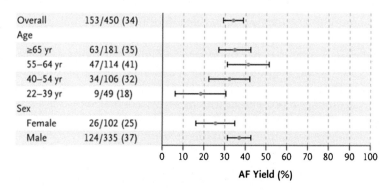

FIGURE 13.1   Yield of atrial fibrillation on ECG patch monitoring.

Surprisingly 50 participants were notified of each real fibrillation ages 54 years and younger with most atrial fibrillation events happened outside of a clinical setting.

These results could be an indicator for a proposal to use consumer device data into a health care record to help discover serious health events through evidence and data presented in a digital twin for cardiology instance that clinicians would examine or be alerted to. As technology companies are trying to diagnose atrial fibrillation, the larger issue is that primary care locations will prescribe aspirin due to those clinicians may not know how to treat you or understanding advanced protocols to prevent stroke by prescribing oral anticoagulants [21].

13.3 Digital twin for cardiology futures

The digital twin of cardiology in the future could take multiple approaches to not only ingest data but to create representations of disease that are easily understood and interpreted by the patient. This approach may help patients better understand their condition through immersive visualizations, augmented reality, and virtual reality experiences.

These technologies do exist today to help cardiac surgeons with ramp up to surgeries but may play a beneficial role in helping patients understand how severe their condition is or how causal factors have changed their outcomes for cardiac disease.

This approach has a historical reference as medical imaging system pioneers envisioned those representations of the heart as the living organ to be represented in a two- or three-dimensions to give clinicians the best vantage point for diagnostics of heart conditions. Radiology pioneers formed ideas that the representation of the heart could show the progression of disease and the capabilities of treatments to define better outcomes and for patients.

The intention was focused on an image of the heart, clinicians should realize the value of having patients view the representations of the heart in a digital twin format to better understand what surgical procedures could be performed, what the number of interventions could be performed, and supporting the planning of postsurgical care. The opportunity for a digital twin for cardiology representation along with data that supports statistical outcomes could help guide patients and clinicians to the best outcomes.

The challenge for cardiac care patients is timing as many patients have acute illness and have noticeably short intensive periods of illness combined with other causal factors to increase mortality rates. Innovative solutions that are enhancing prompt diagnostics and actionable insights are based upon common EKG readings along with historical clinical data and the ability for clinicians to use advanced artificial intelligence algorithms to diagnose undiscovered issues at the point of care. This also helps with discharge planning for patients on guideline therapies and with a digital twin can ensure prompt and safe care coordination of patients to long term care practitioners.

Population health solutions have become more sophisticated at predictive outcomes based on past patient outcomes, research data, and clinical reviews for accuracy. Population health tools have also become exceptionally good at continuously updating and synchronizing information across different devices and clinical records. We will see more bi-directional synchronization of patient records to build more complete analysis and interpretation that clinicians will become more comfortable with to gain real time insights.

The digital representation of various health states of patients can also help larger populations of patients to seek trends or health index markers for interventions.

Care providers, healthcare payors, and patients can gain insights and use those insights to drive new results in areas like clinical priorities, individualized medicine, and population health outcomes.

Currently we are seeing EMR applications draw in population health information to supply guidance to clinicians to build case prioritization and to coordinate care teams. This type of structured data from medical records with historical information is being combined and results extracted into new data structures to form digital representations of patients that can be iterated or tested against data from clinical and research trials.

13.3.1 New software by doctors for doctors

Several cardiologists have come together to build software companies that create health care guidelines to help synchronize and coordinate patient data along with care protocols. The biggest discovery is that coordination of care teams does save patients' lives.

The application of digital twins allows care teams to take on more complicated cases and have better decision-making outcomes based upon modeled data in a digital twin. These efforts can lead to faster intervention on digital twins for cardiology outcomes and then be applied to patients.

The added scenarios include creating health outcomes and looking for the most practical options that would lead clinicians to choose the best outcome based upon artificial intelligence decision-making, clinical experience, and current patient needs.

As evidence-based medicine becomes more prevalent and contributes to more case studies with positive patient outcomes it will become more trusted by more clinicians.

13.3.2 Personalization of evidence based medicine

The personalization of healthcare outcomes can become a better guide for clinicians to serve patients with complications or advanced cases through interpretation of evidence-based medicine. The desire to collect information from past clinical examinations and studies such as EKG, EEG, and patient centered collected data from health devices, wrist worn monitoring (heart BPM/Afib), or worn devices (Pulse Oximeters) could help add more data that can be reflected into digital twin and analyzed for health care outcomes.

In a study, researchers looked at different vital signs that can be measured including pulse/heart rate, heart rate variability, respiration, blood oxygenation, and arterial fibrillation. As this information is collected by monitoring devices and both professional clinical settings and nonclinical settings, there is sometimes a presented bias that clinical setting data is more correct and data collected from consumer sensors due to the sophistication of sensors, approval as an FDA (Food and Drug Administration) device, and the collection of the data by medical professionals in medical setting [16].

There is a larger argument to bring other information in they could help show trends over time that can enhance the data collected for a cardiology based digital twin, and not just a point in time indicator but as an ongoing indicator.

In the future, patients will be able to add to their digital twin latest information such as blood pressure, body temperature, glucose levels, hydration levels, and fitness information via wearable devices that can be connected into health care records.

As researchers are developing digital twin technologies, enriched data sources would help better predict future trends and help discover gaps in data. As you think about cardiology and the constant rhythm of a heartbeat to detect recovery times of heart rates postexercise can be critical in figuring out heart fitness. The state of exercise conditioning collected at point in time treadmill testing with sensors is cumbersome for patients to complete. As a result, it would be easier to collect data from wearable devices to ensure that patients are practicing their care routines as part of their recovery or as part of their preoperation routines to make them a better candidate for surgery.

Cardiologists will also have to understand that the accuracy of information coming from wearable devices is near parity with clinical medical grade devices. Patients should be allowed to supply information to doctors via their device and into their digital twin for artificial intelligence review, discovery, and analysis supplying data to systems for further insights.

The usage of artificial intelligence on data within a digital twin context could point out exceptions or anomalies in data to better assist cardiologists to perform further insights that could lead to lifesaving interventions.

How a doctor would act upon this would be to inform patients of positive trends and activities that promote heart health. It may also help cardiology patients that have had ongoing issues or other chronic diseases to take better care of themselves and have strict outcomes.

Currently many patients only receive information as provided by cardiology studies which must be scheduled and performed in a clinical setting.

Patients who have little to no access to specialist care or live in rural areas have the most difficult time with these care routines.

In several research articles focused on patient centered data driven cardiology, challenges surface when no reliable evidence can be obtained from inferior quality data. It is up to data scientists and software engineers to collect, gather, ingest, and process other forms of cardiac data that can be helpful to build a larger data sample size that could be interpreted through population health [1].

13.4 Conclusion

Cardiologists and clinical specialists are exploring and using digital twin data structures created from tasks that define healthcare procedures and surgical approaches that connect as a simulation for future healthcare procedures on patients (patent filing) [17]. This allows cardiologists to model the tasks and items associated for individual patients to look for the best outcomes while avoiding risks. This may help cardiologists take the best approach for patients that may create early interventions and may save lives, while building equality in treatments and patient outcomes.

Digital twins' data structures will be used as a method to help patients discover risk behaviors and change those behaviors based upon data modeling of similar patients in a digital twin format. Those patients who adhere to their health plan may not need future interventions or surgical interventions. Data from digital twins' could make patients understand their future results and make predictions of how well coordinated care planning will result in the best long term outcomes and reduce admissions into healcare acute facilities. It is up to healthcare systems to help patients discover plausible outcomes if they could ease the supply and delivery of clinical data and range of clinical decisions. Patients now can see the range of possible results for their journey through a digital twin experience. The digital twin for cardiology could become a true partner and advocate for the patient in real life to strive towards healthier living.

The advancement in deep learning medical systems and procedures will only be able to get more data insights in the digital twin world and provide more personalized medicine results that can help both clinicians and patients become more focused with care management protocols and supply relief from expensive and intrusive medical intervention (patent filing) [18].

The opportunity to create personalized care road maps through digital twins and gain real time insights through data synchronization from patients to medical records has massive potential to guide health care providers to the best outcomes for their patients. This may also give an opportunity for cardiac patients to use monitoring devices in new ways to help or steer behaviors that are not conducive to long term healthy outcomes.

Acknowledgments

GilAnthony Ungab, M.D.– Contributing Medical Author, CMO & Cofounder – Lucia Health Guidelines (Luciaguidelines.Com).

Dr. Ungab is a private practice Cardiac Electrophysiologist and former Medical Director of Cardiac Electrophysiology at Sharp Chula Vista Medical Center in San Diego, CA. He served as a member on R&D committees at Boston Scientific, St Jude, and Biotronik. Dr. Ungab was also a cofounder of Geneva Healthcare Solutions, a remote monitoring company (Acquired by BioTelemetry Inc.). He is passionate about advancing medical technology to improve patient care and the healthcare system. His latest healthcare start up is Lucia Health Guidelines a point of care, AI/Machine learning, clinical decision support tool.

Lucia Health

https://www.luciaguidelines.com/about/

https://onlinelibrary.wiley.com/doi/full/10.1002/emp2.12534

https://medcitynews.com/2021/08/meet-the-two-picks-from-jj-and-village-capitals-culturally-competent-care-accelerator/

References

[1] Antonio Luiz Ribeiro, Gláucia Maria, Moraes de Oliveira, Toward a patient-centered, data-driven cardiology, Arq. Bras. Cardiol. 112 (4) (2019) 371–373, https://abccardiol.org/en/article/toward-a-patient-centered-data-driven-cardiology/.

[2] Jorge Corral-Acero, Francesca Margara, Maciej Marciniak, Cristobal Rodero, Filip Loncaric, Yingjing Feng, Andrew Gilbert, Joao Filipe Fernandes, Hassaan Bukhari, Ali Wajdan, Manuel Martinez, Mariana Santos, Mehrdad Shamohammdi, Hongxing Luo, Philip Westphal, Paul Leeson, Paolo Diachille, Viatcheslav Gurev, Manuel Mayr, Pablo Lamata, The 'digital twin' to enable the vision of precision cardiology, Eur. Heart J. 41 (2020), https://doi.org/10.1093/eurheartj/ehaa159.

[3] Stanford Medicine, Health trends report, https://med.stanford.edu/content/dam/sm/school/documents/Health-Trends-Report/Stanford-Medicine-Health-Trends-Report-2018.pdf, 2018.

[4] Margaret Wei, Duc Do, Pok-Tin Tang, Kevin Liu, Michael Merjanian, Lynnell McCullough, Richelle Cooper, Noel Boyle, Optimal disposition for atrial fibrillation patients presenting to the emergency departments, J. Am. Coll. Cardiol. 71 (11 Supplement) (2018) A509, https://www.jacc.org/doi/abs/10.1016/S0735-1097.

[5] HAS-BLED definition, Use. Score. https://en.wikipedia.org/wiki/HAS-BLED.

[6] Kathryn Kiser, Oral Anticoagulation Therapy: Cases and Clinical Correlation, Springer, 2017, p. 20, ISBN 9783319546438.

[7] Veronica Rocha, Surgeon removed the wrong kidney is put on probation, Los Angeles Times (2014), https://www.latimes.com/local/lanow/la-me-ln-surgeon-removed-wrong-kidney-20141204-story.html.

[8] Janice L. Clarke, Scott Bourn, Alexis Skoufalos, Eric H. Beck, Daniel J. Castillo, An innovative approach to health care delivery for patients with chronic conditions, https://doi.org/10.1089/pop.2016.0076, 2017.

[9] Heart disease facts. United States Centers for Disease Control and Prevention, https://www.cdc.gov/heartdisease/facts.htm, 2021.

[10] Sakchai Muangsrinoon. Poonpong Boonbrahm, Game elements from literature review of gamification in healthcare context, J. Technol. Sci. Educ. 9 (1) (2019) 20–31, https://doi.org/10.3926/jotse.556.

[11] H. Moldovan Oana, Balasz Deak, Alin Bian, Diana Gurzau, Florina Frangu, Alexandru Martis, Bogdan Caloian, Horatiu Comsa, Gabriel Cismaru, Dana Pop, How do I track cardiac rehabilitation in my patient with ischemic heart disease using Strava, Balneo Res. J. 10 (2019) 114–117, https://doi.org/10.12680/balneo.2019.248.

[12] Jay N. Cohn, Lynn Hoke, Wayne Whitwam Paul A. Sommers, Anne L. Taylor, Daniel Duprez, Renee Roessler, Natalia Florea, Screening for early detection of cardiovascular disease in asymptomatic individuals, PMID: 1456432.

[13] Eric Topoll, The Patient Will See You Now: the Future of Medicine Is in Your Hands, BN-10: 0465040020, 2016.

[14] Paul S. Chan, William J. Oetgen, Donna Buchanan, Kristi Mitchell, Fran F. Fiocchi, Fengming Tang, Philip G. Jones, Tracie Breeding, Duane Thrutchley, John S. Rumsfeld, John A. Spertus, Cardiac performance measure compliance in outpatients: the American college of cardiology and national cardiovascular data registry's PINNACLE (practice innovation and clinical excellence) program, J. Am. Coll. Cardiol. (ISSN 0735-1097) 56 (1) (2010) 8–14, https://doi.org/10.1016/j.jacc.2010.03.043.

[15] M.V. Perez, K.W. Mahaffey, H. Hedlin, J.S. Rumsfeld, A. Garcia, T. Ferris, V. Balasubramanian, A.M. Russo, A. Rajmane, L. Cheung, G. Hung, J. Lee, P. Kowey, N. Talati, D. Nag, S.E. Gummidipundi, A. Beatty, M.T. Hills, S. Desai, C.B. Granger, M. Desai, M.P. Turakhia, Apple heart study investigators. Large-scale assessment of a smartwatch to identify atrial fibrillation, N. Engl. J. Med. 381 (20) (2019) 1909–1917, https://doi.org/10.1056/NEJMoa1901183.

[16] Cardiology Magazine, Cover story | implicit bias: recognizing the unconscious barriers to quality care and diversity in medicine ("who's who at heart house – American college of cardiology"), https://www.acc.org/latest-in-cardiology/articles/2020/01/01/24/42/cover-story-implicit-bias-recognizing-the-unconscious-barriers-to-quality-care-and-diversity-in-medicine, January 2020.

[17] Inventor-Marcia Peterson, Current assignee-general electric co., https://patents.google.com/patent/US20190087544A1/en, 2017.

[18] Inventors-Jiang Hsieh, Gopal Avinash, Saad Sirohey, Current assignee-general electric co., https://patents.google.com/patent/US10438354B2/en?q=digital+twin;+healthcare&oq=digital+twin;+healthcare, 2016.

[19] Y. Xian, L. O'Brien, E.C. Liang, et al., Association of preceding antithrombotic treatment with acute ischemic stroke severity and in-hospital outcomes among patients with atrial fibrillation, JAMA 317 (10) (2017) 1057–1067, https://doi.org/10.1001/jama.2017.137.

[20] Kim Schwab, Dacloc Nguyen, GilAnthony Ungab, Gregory Feld, Alan S. Maisel, Martin Than, Laura Joyce, W. Frank Peacock, Artificial intelligence machine learning for the detection and treatment of atrial fibrillation guidelines in the emergency department setting (AIM HIGHER): assessing a machine learning clinical decision support tool to detect and treat non-valvular atrial fibrillation in the emergency department, https://doi.org/10.1002/emp2.12534, August 2021.

[21] Jonathan C. Hsu, Thomas M. Maddox, Kevin Kennedy, David F. Katz, Lucas N. Marzec, Steven A. Lubitz, Anil K. Gehi, Mintu P. Turakhia, Gregory M. Marcus, Aspirin instead of oral anticoagulant prescription in atrial fibrillation patients at risk for stroke, J. Am. Coll. Cardiol. 67 (25) (2016) 2913–2923, https://doi.org/10.1016/j.jacc.2016.03.581.

Applications of Digital Twins to migraine

Ali Mohammad Saghiri[a],
Kamran Gholizadeh HamlAbadi[b,c], and
Monireh Vahdati[b,c]

[a]Computer Engineering Department, AmirKabir University of Technology, Tehran, Iran, [b]Young Researchers and Elite Club, Qazvin Branch, Islamic Azad University, Qazvin, Iran, [c]Faculty of Computer and IT Engineering, Qazvin Branch, Islamic Azad University, Qazvin, Iran

14.1 Introduction

Digital Twin (DT) technology was introduced in [1]. Because of using a collection of technologies including Artificial Intelligence (AI), Internet of Things (IoT), Augmented Reality (AR), Virtual Reality (VR), and cloud computing, DT has a high potential for managing complex and unpredictable systems [2]. A main reason for appearing this technology is that many of systems and organisms are going to be complex and unpredictable because of using many elements that are injected to them during digital transformation era and hence managing such complex structure is not easy task for humans. This problem will be more challenging in healthcare systems. For example, consider a patient that has complex disease such as migraine. The patient may use pulse maker in his heart, smart glass for managing his eye disease, and some prosthesis in his brain. All of mentioned parts: heart, eye, and brain play important role in controlling migraine disease [3]. In this situation a main question that arises is that how a physician can manage migraine treatment considering many elements that are involved in the environment of this patient. It seems that DT is able to provide promising solutions. In the rest of this section, migraine disease, applications of cutting-edge technologies for managing diseases, and their problems are introduced. Then, the DT and its capabilities to manage complex and unpredictable systems are explained.

Migraine is a well-known neurological disease. This disease leads to appearing a variety of symptoms including pulsating headaches and some sensory disturbances such as aura, we call them migraine attacks [3]. These attacks are affecting more than one billion people around the world every day. Migraine is classified as a well-known disabling disease in the world [4]. These attacks have no unique characteristics and the symptoms may change from one patient to another one. Another strange fact about these attacks for a specific person is that the symptoms are time-varying and during life of that person new symptoms may be detected [5], [6]. These attacks may be going to chronic and some of them may lads to harmful consequences such as losing vision [7]. Appearing the products such as smartphones, smart glasses, and wearable devices that are the results of digital transformation will have had substantial changes in different dimensions of migraine diseases. With the aid of smartphone, the attacks can be predicted such as some reported in [8], [9].

Digital transformation is going to change traditional healthcare systems rapidly [10]. During this transformation, emerging technologies such as Blockchain [80], AI [79], IoT, VR, AR, Quantum Computing, DT play important roles. In our last research, we used Blockchain technology in order to demonstrate how emerging technologies can manage, detect, and predict some ailments, such as Covid-19 [82]. We are going to use these technologies to organize smart healthcare systems in DT.

Technologies including AI, IoT, AR, VR, simulation techniques, and cloud computing are widely used to address challenges of migraine disease [12], [13], [14], [15]. In this domain, a main problem is that there is no unique concept to bring together all of mentioned efforts to manage migraine disease. This problem may be solved during organizing a DT for a patient. A DT is able to unify all mentioned solution to bring a unit digital replica of a human. This replica is able to resolve complexities at the core of DT of a human.

In this chapter, the applications of DT to migraine disease are studied. The organization of this chapter is given as follows. Migraine disease is explained in section 14.2. Section 14.3 is dedicated to fundamental concepts of DTs. In section 14.4, the applications of cutting-edge technologies to migraine disease and also their drawback is also highlighted to better justify the key role of DT. Challenges and corresponding solutions based on DT are also suggested in section 14.5. A discussion is given in section 14.6 that covers different perspectives. The conclusion is given in section 14.7.

14.2 Migraine disease

In this section, some basic concepts about the Migraine disease are given.

14.2.1 Definitions and complexities related to treatment processes

Migraine is a well-known neurological disease that causes a variety of symptoms. This disease leads to attacks that porolongs over hours to days. The phases of a migraine attack can be identified as premonitory, aura, headache, and postdrome [5]. Headache is the most notably recognizable, stereotyped, and quantifiable feature of a migraine attack. Migraine attacks may get worse with physical activities, lights, sounds, or smells which lead to high complexity in treatment processes [5]. These attacks are affecting more than one billion people around the world every day. Migraine is classified as a well-known disabling disease in the world [4]. In the rest of this section, the following items about migraine diseases are studied to better understand the complexity in the nature of this disease and its treatment processes.

- Migraine classification, symptoms, and diagnosis process
- A study on attack triggers and their complexity
- Treatment processes in migraine.

All above items are described by three subsections in the rest of this section.

14.2.2 Classification, symptoms, and diagnosis process

A wide range of symptoms is reported for migraine. These symptoms can be classified into several types as explained below:

- Migraine with aura (Complicated migraine): this type of migraine appears by headache along with some visual and sensory symptoms. Visual symptoms may include seeing spots, sparkles, and lines. In some situations vision lose may occur. Sensory symptoms may include some pains with known and unknown organs. International Headache Society (IHS) reported a guideline for diagnosing migraine [6]. In this guideline, the reported symptoms can be summarized as below:
  - ◇ Minimum of five headaches, during which at least two episodes may be appeared with an aura
  - ◇ Homonymous visual symptoms
  - ◇ Unilateral sensory
  - ◇ At least one symptom gradually increases with time

- Migraine without aura (common migraine): This type of migraine appears by headache along with one of the following symptoms [6].
  ◇ Pain on one side of head
  ◇ Pulsing or throbbing pain
  ◇ Sensitivity to light (photophobia)
  ◇ Sensitivity to sound
  ◇ Nausea and/or vomiting
  ◇ Pain or discomfort that is made worse by physical activity.
- Migraine without head pain: this type of migrane leads to aura symptom but not the headache.
- Hemiplegic migraine [16]: this type of migraine leads to temporary paralysis or neurological or sensory changes on one side of the body of the patient. Sometimes this type of migraine attack leads to head pain and sometimes it doesn't.
- Retinal migraine (ocular migraine) [17]: this type of migraine results in temporary, partial, or complete loss of vision. This attack may affect one of the eyes, along with pain behind the eye that may propagate to the rest of the head.
- Chronic migraine [7]: this type of migraine attack occurs at least 15 days per month. The strange fact about this type is that the symptoms and level of pain may change frequently.
- Migraine with brainstem aura [18]: this type of migraine leads to vertigo, slurred speech, double vision, or loss of balance. These symptoms may appear before the headache. In attacks of this type, the back of the head may be the origin of the headache. These symptoms occur suddenly.
- Status migrainosus [19]: This type of migraine attack can last longer than 72 hours. Certain medications, or medication withdrawal, may lead to this type of migraine attack.

14.2.3 Attack triggers and their complexity

Many factors may result in triggering migraine attacks that some of them are summarized in Table 14.1 [3]. It is obvious that these wide range of factors and triggers may result in high complexity in treatment process for patients especially for those that suffer from other diseases. This complexity increases by the age and risk factors of the patients.

14.2.4 Treatment processes in migraine

Several goals are considered by migraine therapist. Some of these goals are given below;

- Reduce the severity and duration of the migraine attack [20].

TABLE 14.1 Factors and triggers in migraine disease [3].

Factors and Triggers	Diversity scope
Food	• Chocolate
	• Citrus fruits
	• Sucralose, gluten
	• Fatty and fried food
	• Food colorings
	• Tea, coffee
	• Beverages contain caffeine
	• Alcoholic beverage
	• Dairy products
Daily Life Pattern	• Weather changes
	• Fasting and exhaustion
	• Changes in sleep pattern
	• Emotional stress
Genetics	• Polyglutamine
	• Repeats
	• Frameshift
	• Variation
Physiological and biochemical factors & Genetics	• Estrogen level
	• High insulin level
	• Free iron deposition
	• Monosodium glutamate
	• Metabolic derangements
	• Increased oxidative stress
	• Changes in ovarian hormone secretion
	• Disruptive neural networks in head
	• Mitochondrial enzyme dysfunction
	• Level of nitric oxide
	• Concentration of heavy metal
	• Deviated enzyme level

- Restoring functioning ability. Reducing the use of rescue medications and promoting overall management with no or minimal side effects [21].
- Preventive treatment by managing diet, lifestyle, and magnetic stimulation [22].

According to [3], the existing treatments include acetaminophen, triptans, nonsteroidal antiinflammatory drugs, nonopioid analgesics, NSAID-triptan combinations, and antiemetics. Opioids should be strictly managed due to their addiction risk. Prescriptions including naproxen, ibuprofen, acetaminophen, and aspirin form the first line of treatment for migraine disease. Patients with chronic migraine require prophylactic treatment, although migraine patients with a low frequency of symptoms may be managed with effective acute therapy. The main risk during treatment is the overuse of medications that should be considered during migraine progression. The challenges of treatments in migraine disease will be summarized in the rest of this chapter.

14.3 Digital Twins technology: definitions, required technologies and applications

A DT refers to a replica of something real. In the literature, several definitions for a DT are reported focusing to map the behavior of the physical thing into its virtual copy. The basic features of DTs are explained as below [1], [2], [23]:

1. DT is going to describe "complex" systems that will be emerged as result of industry 4 revolution in different domains including healthcare. This complexity may be arisen in some diseases such as cancer and migraine. DT is able to help doctors to face with some complex diseases and organs utilizing capabilities of DT.
2. A significant feature of a DT is to resolve the "unpredictability" of phenomena and behaviors that may appear in the real world. Many of diseases may lead to unpredictability challenges because the number of treatments, medications, risk factors for patients that increase day by day.

There hasn't been a complete description for DT to show all its dimensions and also corresponding technologies. Historically, the first formal definition was released by Dr. Michael Grieves in 2002 as "digital equivalent of a physical system" [1]. Later on, researchers inspected and completed the definition of this new technology. Some of well-known definitions are given below;

- IBM definition of DT [24] focuses on real time data in a lifecycle of a product.
- In [25], a semantic insight to a model for DT paradigm is highlighted.
- In [26], A DT concept is considered as an improvement of current modeling and simulation toward real-time sample of the thing.
- An integrated multiphysics, multiscale, probabilistic simulation of a system that uses the best available physical models, sensor updates, to mirror the life of its corresponding twin [27].
- A coupled model of a real machine that operates in the cloud platform and simulates the health condition with an integrated knowledge from both data driven analytical algorithms as well as other available physical knowledge [28].
- A model that fully describes a potential or actual physical manufactured system from the micro atomic level to the macro geometrical level.

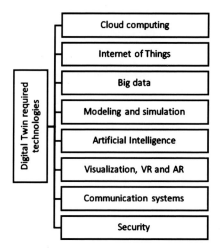

FIGURE 14.1   Required technologies of Digital Twin.

Every information that could be obtained from inspecting a physical manufactured product can be obtained from the DT [29].

- An analysis in [30] extracted a 8-dimensional model to show behaviors and the context of a DT. These dimensions are categorized as: integration breadth, connectivity modes, update frequency, CPS intelligence, simulation capabilities, digital model richness, human interaction, and product lifecycle.

14.3.1 Required technologies

DT initiation needs technical support of several trending technologies. Some of the recent technologies are named in [2], [31], [32] as enabling technologies. (See Fig. 14.1.)

Eight technologies that have close relationship with DT are extracted and explained as below;

1. **Cloud computing:** In this technology, cloud refers to a network or a range of broadband networks such as the Internet that the user is not aware of what is happening behind the scenes. In fact, a cloud is a vast series of computers that are interconnected and functioning as a single ecosystem [33]. In DT design, cloud is a powerful platform providing reliable and sustainable servers and computing resources along with large storage units. Cloud computing also provides fast data processing and presents analytical results to the DT managing center.
2. **Internet of things (IoT):** The Internet of Things (IoT) is a set of equipments, objects, and everything that has a unique identifier and the ability to communicate and transmit data over the network [34], [78]. DT is

realized after linking physical and virtual worlds. It needs several connections and a lot of real time data transfer which could be established on IoT schemes [35].

3. **Big data:** Due to emergence of sensors and smart devices, the volume of data is increasing exponentially during recent years. This issue leads to the emerging a new concept, namely big data. Since DT is a data-driven technology, it strongly relies on data structures. Therefore, big data analyses incorporation into DT will extract some valuable information, provide insights into effective management and result in more precise decisions [36].

4. **Modeling and simulation:** A model represents a system structure with the aim of analysing its performance. Modeling is a complicated process but the outcomes give a clear opinion of how the whole system works and carries on. Besides, simulation is the process of studying a model to recognize and derive system's behavior. It is actually a tool to evaluate the built model and extract operations of the objects. Hence, modeling and simulation come in succession. In a DT structure, modeling and simulation are parts of making a complete picture of the physical object, adding up real physical phenomena to its virtual copy [37].

5. **Artificial Intelligence (AI):** There are various definitions of artificial intelligence that can encompass a wide range of intelligent actions from any agent or object. Currently, deep learning, statistical methods and computational intelligence are some of the most popular approaches in this field. In recent years, the new trend is to develop systems that utilize general artificial intelligence [81]. These systems are not designed for specific purposes and can be used to solve a wide range of previously unknown problems [38], [39]. AI methods have been popular in industries and sciences including DT application. This is because the complexity of DT applications is increasing and we must design intelligent agents that are able to make decisions in the context of DT applications.

6. **Visualization, VR and AR:** Virtual Reality (VR) and Augmented Reality (AR) have been used widely in various industries during last years [40], [41]. In VR, a virtual environment is created for the user's operations, but in AR, the virtual environment is merged with the real environment to provide a single environment for the users. DT applications utilize a wide range of capabilities in AR, and VR. This is because a digitalized model of an entity in multiscale simulations that is deployed at the heart of DT requires a perfect digitalized model that reflect visual and geometrical characteristics.

7. **Communication systems:** A bi-directional data transmission system with ultra-low latency is an essential infrastructure for several industries. A DT includes connection of several physical components to the

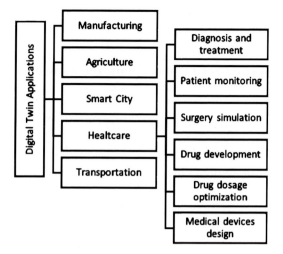

FIGURE 14.2   Applications of Digital Twins.

virtual system. It needs a smart communication system with particular features of high quality of service (QoS), ultra-high data rate, very low latency, simultaneous data transfer between data sources and ultra-low power consumption.

8. **Security:** Information protection and system security against cyber-attacks have been regarded an important issue in recent years. A DT introduces simulation, monitoring, optimization and prediction of a physical system and so is vulnerable to attacks.

14.3.2 Applications of Digital Twin

There is no literature about the usage of DT in migraine disease but a wide range of fields in healthcare have utilized this technology. In [42], [43], [44], [45], [46], [47], [48], [83], the applications of DT in some fields such as cancer disease, nutrition science, diet calculation, cardiovascular disease, and fitness are reported. From another point of view, DT can be used to develop new drugs, instruments, and also simulate critical surgeries [49], [50]. Fig. 14.2 shows general applications of DT technology with a special focus on healthcare which will be discussed in the rest of this chapter [23], [49], [50], [51], [52], [53], [54], [55].

In addition to the above specific solutions, the hospital construction process may also use DT as explained follows. Many DT based tools and methods are proposed in the literature to manage the complex processes in factories such as construction, production line simulation, and optimizing product lifecycle [27], [56], [57]. All of these methods can be used in hospital construction. A live model of the hospital can be used for regulatory and policy determination.

14.4 Applications of Digital Twins Technology to migraine disease

The use of DT in the management of migraine disease will lead to revolutionizing clinical processes and hospital management. This will be done by enhancing medical care with tracking and advancing modeling of the human body. DT of a human may be used to organize modern treatment for migraine considering a wide range of capabilities that are required for personalized treatment. In the rest of this section, challenges of migraine disease are summarized and then the solutions based on DT are highlighted.

14.4.1 Challenges of migraine disease and the importance of personalized medicine

This subsection is dedicated to summarize the substantial challenges that appear in migraine disease and focus on the importance of personalized medicine. In migraine disease, as it was previously mentioned, diagnosing and treatment are very difficult tasks because of the high diversity of attack triggers and treatments among patients. From the patient's point of view, the duration of an attack, and attack triggers are different in every person, and these parameters may also change over time. From the therapist's point of view, treatment for each person according to the risk factors of that person during his lifetime is different from others. In this subsection, the challenges that arise during the management of migraine disease are summarized as bellows;

- **Trigger detection:** As it was shown in Table 14.1, the number of triggers is high and finding specific trigger in each person is a complex decision for therapists [3].
- **Lifestyle selection:** Finding appropriate life style may be used to prevent migrane attacks [3]. This task is challenging in practice.
- **Drug interaction:** As an example, opioids are used in some type of migrane disease. In [58], some reasons are given to avoid opioids in headache and migraine treatment because of drug Interaction and serotonin toxicity with opioid. In other words, drug interactions are considered as a challenging problem in migraine treatment processes.
- **Attack prevention:** Preventive cares play important role in the management of migraine treatment processes [59]. From the patient's point of view, because of a large number of triggers, patients usually are not successful in this process. From therapist's point of view, many as-

sumptions along with many rules should be considered during online monitoring and control of a patient. This can be done by IoT and DT.

- **Attack phase detection:** Migraine attacks include several phases that can be identified as premonitory, aura, headache, and postdrome [3]. Detecting the phase may be used to conduct different treatment processes. Phase detection because of its varying nature is not easy task and new technologies are required for solving this problem [13].
- **Dosage adjustment:** Determining the dosage of medications plays important role in the management of migraine disease. In [60], it was shown that inappropriate dosage may lead to migraine attacks. In addition, this issue becomes more challenging when some parameters such as gender, age, gene, and environmental conditions should be considered.
- **Classification of headache:** Classifying a headache as a symptom of migraine is not possible in different situations. For example, headache can be a symptom of COVID-19 but patients and doctors might not be able to definitely detect the origin of a headache. Some guidelines are reported by HIS that were summarized in [6] although these guidelines may not work in all situations. Cutting-edge technologies based on AI and IoT have shown promising results in this regard that we will talk about them in the rest of this chapter.
- **Medication (Drug) overuse prevention:** Drug overuse may occur in a patient because of negative effects of migraine disease on the memory and mind of that patient [60].
- **Finding Diet for different purposes considering attack triggers:** Finding specific diet for a patient considering all constraints and triggers of that person is a challenging task for therapists [61], [62], [63].
- **Organizing personalized treatment:** Many factors from genetics and personal environment of a patient strongly affect on treatment processes [64]. Therefore, migraine diseases should be supported by technologies such as DT that provide personalized medicine [65].

All of above challenges should be addressed considering the personal information of each patient. Therefore, personalized medicine is very important for managing migraine diseases. It should be noted that by noticeable increase of information for each patient during the digital transformation era finding appropriate treatment without appropriate model and high computation might not be possible.

14.5 Digital Twin solutions for migraine disease

As it was previously mentioned, DT utilizes a wide range of technologies including IoT, AI, AR, and VR to make live digital models which can mimic the behavior of a physical system in an online fashion. This model learns and updates itself continuously using real-time data from sensors of the physical system. The physical system can be either asset that is involved in healthcare systems or the human body and its inner parts. In this section, at first, we summarize the applicabilities of cutting-edge technologies to solve the challenges of migraine disease. Then the problem of these solutions is highlighted. Finally, the possible solutions of DT for migraine disease are suggested.

14.5.1 Applicability of cutting-edge technologies for migraine disease

In fact, many cutting-edge technologies especially those that play a key role in DT have been applied for solving different challenges in the management of migraine disease. Table 14.2 summarizes some of these recent studies.

14.5.2 Problem of existing solutions

The main problem of the existing solutions is that there is no unique infrastructure to unify existing solutions to combat migraine attacks. Since migraine is very complex and its complexity increases day by day, utilizing diverse and separated solutions may not efficiently provide well treatments processes. For example combining migraine headache with sinus headache may lead to challenging problems for doctors to detect the type of headaches and also treatment plans [71]. As another example, combining migraine headaches with thrombocytopenias with brain damages may lead to another challenging situation to detect the origin of headache [72], [73]. In both situations, utilizing a preconstructed DT of a patient may provide numerous benefits from diagnosis to treatment. We are able to examine many treatment plans on DT and also predict the future situation of patients.

14.5.3 Possible solutions of Digital Twins technology for migraine disease

There is no literature about the application of DT in migraine. Therefore, we try to extract possible solutions by analyzing the existing solutions in the relevant field and also the characteristics of challenges in migraine treatment. As it was previously mentioned and also according to [23], [49], [50], [56], [51], [52], [53], [54], [55], DT technology has been used in other diseases for the following purposes. For each purpose, some solutions based on DT are considered.

- **Diagnosis and treatment:** Because of the increasing risk factors in Migraine disease in recent years, traditional diagnosis and treatments might not be efficient. DTs of patients can be used to suggest new treat-

TABLE 14.2 Applications of cutting-edge technologies for migraine diseases.

Reference	Challenge	Solution	Technology
[13]	Classification of Migraine Stages	Self-Constructing Neural Fuzzy Inference Network (SONFIN) is adopted as the classifier	Artificial Intelligence
[14]	Prediction of medication overuse	Support vector machines and Random Optimization	Artificial Intelligence
[66]	Classification of headache disorders	XGBoost classifiers	Artificial Intelligence
[67]	Examine the migraine patients with maximum and minimum pain levels	ANOVA test and neural networks	Artificial Intelligence Statistics
[68]	Distinguish between migraineurs and healthy individuals	Deep learning	Artificial Intelligence
[15]	Migraine Occurrence Prevention	Linear Discriminant Analysis (LDA)	Internet of Things and Machine Learning
[12]	Trigger detection predict the onset of migraine	Deep learning	Cloud computing, Artificial Intelligence, and IoT
[69]	Understanding how symptoms impact the health of a patient	Visual migraine simulation in a virtual environment utilizing a VR head-mounted display	Virtual Reality
[70]	Mimicking migraine symptoms and presence	Head-mounted display (HMD) as a high immersive condition and a handheld display as a low immersive condition	Augmented Reality
[11]	Classification of Migraine Stages	Classification of Migraine Stages based on analyzing EEG	Artificial Intelligence

ments and diagnoses. We will be able to see the result of any treatment plan in every patient before applying it in practice.

• **Patient monitoring:** Patient monitoring may result in complex challenges because of the large complexity of symptoms and factors in migraine disease. This complexity can be resolved by DT. In DT, with the aid of IoT, AR, and VR, the monitoring can be done so that more information will be produced to understand the nature of problems that each patient suffers because of his characteristics.

• **Surgery simulation:** Migraine patients have more sensitivity than other patients in surgical operations. A DT of a patient can be designated to play the role of a real patient for surgery simulation. In this regard, the DT can be used to design a plan for surgery, conduct surgery, and produce an online estimation of the consequence of each action during real surgery.

- **Drug development:** Sensitivity to drugs is a critical problem in migraine patients as described before. Producing a DT for a patient can lead to studying new drugs' effects in an entirely virtual fashion. From another point of view, constructing a million of DTs can lead to studying each drug's effect on a large population, such as vaccines evaluation procedures.
- **Drug dosage optimization:** Determining the appropriate dosage of drugs is an important problem in migraine patients as described earlier. Determining drug dosage for each patient requires considering many elements, including age, sex, and some genetic factors. Providing dosage for each patient can be handled by a DT efficiently.
- **Medical devices design and management:** The medical devices that are used in the body of patients can produce some attacks in migraine patients. This issue should be considered during manufacturing to the maintenance of that device in the body of patients. The creation of medical devices such as dental implants can be conducted by DTs. DTs can also be used to manage a device from creation to life cycle management.

According to Table 14.2, many of DT enabler technologies such as AI, IoT, AR, and VR are utilized to solve some problems of migraine treatments. For each problem, solutions based on DT can be considered as explained below;

- **Trigger detection:** The number of migraine triggers is high. Another surprising fact in migraine disease is that the triggers of attacks in each patient may be different from other patients. This problem might be solved in a fully personalized fashion utilizing a DT for each patient and with AI at the heart of DT
- **Lifestyle selection:** Finding an appropriate lifecycle may lead to multidisciplinary problems requiring many information pieces, from genetics to habits. A DT can be conducted in this regard that collects required pieces of information using IoT. Then DT processes data using an AI engine to provide an appropriate lifestyle that considers the triggers of migraine attacks.
- **Drug interaction:** Recently, the mortality rate of drug interaction has increased because of the increasing use of drugs. This problem in migraine patients with many sensitivities is complex to manage. This problem can be handled by organizing a DT for a patient and following up with the patient considering the challenges of drug interaction when the DT is simultaneously shared with the patient, doctor, and healthcare system.

- **Attack prevention:** A DT can communicate with its real patient to better manage lifestyle, treatment plan, and environmental activities to prevent migraine attacks.
- **Attack phase detection:** Migraine attacks phase detection in an online fashion is not an easy task with existing devices. This issue might be complicated when the treatment plan in each phase is different. With the aid of AI and IoT, it seems that DT can provide an infrastructure to solve this problem.
- **Classification of headache:** The number of headaches and their triggers increase day by day. As explained earlier, this problem in migraine disease is a little more complicated. This problem may be addressed by producing a DT for a patient and then executing learning algorithms that use information from many sources such as MRI, EEG, and fMRI to classify the types of headaches.
- **Medication (Drug) overuse prevention**: This problem was a common problem for migraine patients. Utilizing a DT for a patient, we can make the patient or his doctor aware of the drug overuse.
- **Finding diet:** Calculating the appropriate diet for each patient is not an easy task. This task becomes more challenging when the patient has a migraine. This problem can be solved by creating a DT of the patient.
- **Organizing personalized treatment:** A DT of a patient can evolve after receiving new information about that patient. This DT can be converted to a mature model for suggesting personalized treatment.

It seems that solutions that are reported in Table 14.2 can be unified in a DT solution. Existing solutions are separately developed and therefore they can not work together for solving migraine disease problems. In the rest of this section, we focus on some potential applications that have not been matured enough yet.

- **Prostheses lifecycle monitoring and optimization:** According to the definitions given in [1], [74], and [75], DT may be used to analyze the effect of prostheses on the human body and migraine triggers during the lifecycle of that prostheses.
- **Software-as-a-medical device:** the complex models and also multimedia technologies associated with DT can be used as a service for many purposes. The DTs of brain and other relevant organs of patients can be used as a service for finding new drugs by scientists for finding new drugs and treatment for migraine [49].
- **Virtual organs for migraine treatment:** some organs such as the heart, brain, and lung are very vital, and therefore constructing a DT for them can be very useful in a wide range of treatment processes [76]. A virtual brain can be very useful for online monitoring and treatment of migraine diseases. This virtual entity can be used to unify all detection and treatment mechanisms that are based on EEG, MRI, and fMRI such as those reported in [11].

- **Genomic medicine:** many genetic disorders are the origin of diseases such as migraine. DT has a high potential to find a treatment based on genetic information [77].
- **Migraine Healthcare Services:** lifecycle of migraine patients can be followed by DT. Utilizing DT and its AI-based systems, we are able to predict some attacks and also suggest a Diet to prevent attacks. In [53], an example for elderly patients treatment is reported that can be customized to migraine patients.

14.6 Discussion

Because of the high complexity, unpredictability, and uncertainty in the nature of migraine disease, traditional treatment may not be efficient any longer especially after the rising digital transformation that leads to a drastic increase in the risk factors of this disease. In this section, benefits, future dynamics, development hurdles of utilizing DT in migraine diseases, and managerial implications are discussed in more details.

Benefits: Utilizing DT in migraine research has the following benefits:

- The existing approach for managing migraine diseases suffering from heterogeneity in the infrastructure. In other words, interoperability in communities that fight migraine diseases is low and each community utilize one or limited set of technology to solve the problems associated with these diseases. These problems can be mitigated using DT technology.
- Personalized healthcare which includes personalized medicine and treatments can be reached by organizing DTs for patients.
- Drug sensitivity and resistance can be studied with high accuracy by evaluating drugs on DTs of patients.
- Multiscale simulation to follow a hypothesis with high accuracy can be done.
- Detecting the optimal treatment pathway and optimal treatment monitoring may be conducted.
- Better understanding of the nature of migraine attacks in each person can be obtained.
- DT is able to unify diverse researchers in various fields in AI, IoT, AR, VR, and statistics on migraine disease.
- DT models can be shared with different doctors and organizations that focus on other diseases of a patient.
- Training patient with AR and VR based on DT can be used to better manage attacks for a patient.

Future dynamics: As a future dynamic of research in this area, we may refer to the possible evolution of some parts of DT to bring extraordinary capabilities to manage migraine diseases. In this regard, we may refer to the following.

- The evolution of AI to artificial superintelligence and perfect cognitive systems may result in finding a new treatment that humans are not able to consider because of the limitations of human's mind and abilities.
- The evolution of IoT to nanoscale and molecular scale may be used to monitor and manage every part of the human body. These types of IoT enable us to monitor every organ of the body. This feature can lead to predicting the aging and growth pattern of each organ of the body. Consequently, a DT of a patient can be used to predict the future of disease in a patient considering status of that patient.

Development hurdles: During transition from existing solutions to DT based solutions, we can point out the following:

- Lack of biosensors and nanosensors that may be required to organize the DT of each part of the body.
- Lack of appropriate cognitive systems and intelligent systems to organize learning and cognitive processes related to human body.
- Lack of data for constructing multiscale models and dependent data structures. Patients' data are not gathered in an exemplary fashion and we may require several experiments to prepare a perfect model for each part of the human body.
- Security and privacy preserved solutions are costly and also complicated because of the huge volume of data, participants, and technologies.
- Infrastructure of the IoT that is required in DT is very costly and also high-tech.
- Real-time communication must be a support to create a connection between the DT and its related entity.
- There is no standard for structuring data and data flow for organizing DT. This leads to low interoperability. Data sharing strategies based on peer-to-peer systems and Blockchain technology may facilitate this problem.
- We may require new professions to model different parts of body based on a corporation with a wide range of experts including clinical oncologists, pathologists, and radiologists.

Open research: several open research directions can be followed which some of them are explained below:

- DT based solutions could identify high-risk migraine attacks. This capability may be used to prevent harmful migraine attacks.
- In order to design a fully featured DT for migraine care, many players such as scientists and clinicians should work together. During this cooperation, people across many disciplines, including biology, neuroscience, and computing, try to extract insights from DT to define better treatment plans.

- DT will bring a novel approach for personalized treatment. Doctors are able to evaluate many treatment plans on the DT of a patient to find an appropriate one.

Managerial implications: The number of diseases and also health threats increases day by day. In addition, after rising industry 4 the intrinsic complexities of diseases will be rocketed because of the drastic increase of new environmental paradigm that affects human health. In this situation, finding a treatment for migraine will lead to very challenging problem. Therefore, DT technology can assist physicians to find an appropriate treatment.

14.7 Conclusion

Migraine disease is a complex disease. The complexity of this disease has been drastically grown in recent years because of increasing the risk factors during the digital transformation era. Numerous solutions based on technologies such as AI and IoT were reported in the literature. A survey on these solutions was given in this chapter. We highlighted problems of existing solutions. Then, some solutions based on DTs were proposed. A primary benefit of DTs is to unify the research and development in migraine disease to provide personalized healthcare. In addition, DTs support novel solutions for traditional treatment that have not been reported in the literature yet. Several issues such as development hurdles, open research, and managerial implicants that may arise during the development of DTs were also addressed in the discussion part to support initial steps regarding utilizing DTs in migraine disease. Since there is no literature about the applications of DTs in migraine treatment, this paper opens a new horizon in this field.

Acknowledgment

This Chapter is dedicated to my late mother Dr. H. Afsar Lajevardi. I will never forget her kindness and support. The authors would like to thank the editor and anonymous reviewers whose helpful comments improved the quality of this paper.

References

[1] M. Grieves, J. Vickers, Digital twin: mitigating unpredictable, undesirable emergent behavior in complex systems, in: Transdisciplinary Perspectives on Complex Systems, Springer, 2017, pp. 85–113.

[2] A. El Saddik, Digital twins: the convergence of multimedia technologies, IEEE Multimed. 25 (2018) 87–92, https://doi.org/10.1109/MMUL.2018.023121167.

[3] J. Khan, L.I. Al Asoom, A. Al Sunni, N. Rafique, R. Latif, S. Al Saif, N.B. Almandil, D. Almohazey, S. AbdulAzeez, J.F. Borgio, Genetics, pathophysiology, diagnosis, treatment, management, and prevention of migraine, Biomed. Pharmacother. 139 (2021) 111557.

[4] WHO, Headache disorders, https://www.who.int/news-room/fact-sheets/detail/headache-disorders, 2016. (Accessed 30 November 2021).

[5] A. Charles, The evolution of a migraine attack–a review of recent evidence, Headache 53 (2013) 413–419.

[6] M. Arnold, Headache classification committee of the international headache society (ihs) the international classification of headache disorders, Cephalalgia 38 (2018) 1–211.

[7] A. May, L.H. Schulte, Chronic migraine: risk factors, mechanisms and treatment, Nat. Rev. Neurol. 12 (2016) 455–464.

[8] H. Koskimäki, H. Mönttinen, P. Siirtola, H.L. Huttunen, R. Halonen, J. Röning, Early detection of migraine attacks based on wearable sensors: experiences of data collection using empatica e4, in: Proceedings of the 2017 ACM International Joint Conference on Pervasive and Ubiquitous Computing and Proceedings of the 2017 ACM International Symposium on Wearable Computers, 2017, pp. 506–511.

[9] Z. Cao, C.T. Lin, K.L. Lai, L.W. Ko, J.T. King, K.K. Liao, J.L. Fuh, S.J. Wang, Extraction of ssveps-based inherent fuzzy entropy using a wearable headband eeg in migraine patients, IEEE Trans. Fuzzy Syst. 28 (2019) 14–27.

[10] S.J. Berman, Digital transformation: opportunities to create new business models, Strategy & Leadership (2012).

[11] S.B. Akben, D. Tuncel, A. Alkan, Classification of multi-channel eeg signals for migraine detection, Biomed. Res. 27 (2016) 743–748.

[12] S. Mohan, A. Mukherjee, Migrainecloud, in: SoutheastCon 2018, 2018, pp. 1–7.

[13] Z.H. Cao, L.W. Ko, K.L. Lai, S.B. Huang, S.J. Wang, C.T. Lin, Classification of migraine stages based on resting-state eeg power, in: 2015 International Joint Conference on Neural Networks (IJCNN), 2015, pp. 1–5.

[14] P. Ferroni, F.M. Zanzotto, N. Scarpato, A. Spila, L. Fofi, G. Egeo, A. Rullo, R. Palmirotta, P. Barbanti, F. Guadagni, Machine learning approach to predict medication overuse in migraine patients, Comput. Struct. Biotechnol. J. 18 (2020) 1487–1496, https://doi.org/10.1016/j.csbj.2020.06.006.

[15] R.J. Day, H. Salehi, M. Javadi, Iot environmental analyzer using sensors and machine learning for migraine occurrence prevention, in: 2019 18th IEEE International Conference on Machine Learning and Applications (ICMLA), 2019, pp. 1460–1465.

[16] D. Pietrobon, Familial hemiplegic migraine, Neurotherapeutics 4 (2007) 274–284.

[17] E. Doyle, B. Vote, A. Casswell, Retinal migraine: caught in the act, Br. J. Ophthalmol. 88 (2004) 301–302.

[18] N. Yamani, M.A. Chalmer, J. Olesen, Migraine with brainstem aura: defining the core syndrome, Brain 142 (2019) 3868–3875.

[19] D.A. Marcus, Treatment of status migrainosus, Expert Opin. Pharmacother. 2 (2001) 549–555.

[20] M.J. Marmura, S.D. Silberstein, T.J. Schwedt, The acute treatment of migraine in adults: the american headache society evidence assessment of migraine pharmacotherapies, Headache 55 (2015) 3–20.

[21] S.D. Silberstein, et al., Practice parameter: evidence-based guidelines for migraine headache (an evidence-based review): report of the quality standards subcommittee of the American academy of neurology, Neurology 55 (2000) 754–762.

[22] F. Puledda, P.J. Goadsby, An update on non-pharmacological neuromodulation for the acute and preventive treatment of migraine, Headache 57 (2017) 685–691.

[23] S. Gochhait, A. Bende, Leveraging digital twin technology in the healthcare industry–a machine learning based approach, Eur. J. Molec. Clin. Med. 7 (2020) 2547–2557.

[24] K.L. Lueth, How the world's 250 digital twins compare? Same, same but different, https://iot-analytics.com/how-the-worlds-250-digital-twins-compare/, 2020. (Accessed 11 April 2022).

[25] C. Boje, A. Guerriero, S. Kubicki, Y. Rezgui, Towards a semantic construction digital twin: directions for future research, Autom. Constr. 114 (2020) 103179.

[26] S. Boschert, R. Rosen, Digital twin—the simulation aspect, in: Mechatronic Futures, Springer, 2016, pp. 59–74.

[27] E. Glaessgen, D. Stargel, The digital twin paradigm for future NASA and us air force vehicles, in: 53rd AIAA/ASME/ASCE/AHS/ASC Structures, Structural Dynamics and Materials Conference 20th AIAA/ASME/AHS Adaptive Structures Conference 14th AIAA, 2012, p. 1818.

[28] J. Lee, E. Lapira, B. Bagheri, H.a. Kao, Recent advances and trends in predictive manufacturing systems in big data environment, Manuf. Lett. 1 (2013) 38–41.

[29] M. Grieves, J. Vickers, Digital twin: mitigating unpredictable, undesirable emergent behavior in complex systems, in: Transdisciplinary Perspectives on Complex Systems, Springer, 2017, pp. 85–113.

[30] R. Stark, C. Fresemann, K. Lindow, Development and operation of digital twins for technical systems and services, CIRP Ann. 68 (2019) 129–132.

[31] A. Fuller, Z. Fan, C. Day, C. Barlow, Digital twin: enabling technologies, challenges and open research, IEEE Access 8 (2020) 108952–108971.

[32] Q. Qi, F. Tao, T. Hu, N. Anwer, A. Liu, Y. Wei, L. Wang, A. Nee, Enabling technologies and tools for digital twin, J. Manuf. Syst. 58 (2021) 3–21.

[33] Q. Zhang, L. Cheng, R. Boutaba, Cloud computing: state-of-the-art and research challenges, J. Internet Serv. Appl. 1 (2010) 7–18.

[34] F. Wortmann, K. Flüchter, Internet of things, Bus. Inf. Syst. Eng. 57 (3) (2015).

[35] A.M. Madni, C.C. Madni, S.D. Lucero, Leveraging digital twin technology in model-based systems engineering, Systems 7 (2019) 7.

[36] K.Y.H. Lim, P. Zheng, C.H. Chen, A state-of-the-art survey of digital twin: techniques, engineering product lifecycle management and business innovation perspectives, J. Intell. Manuf. 31 (2020) 1313–1337.

[37] R. Ganguli, S. Adhikari, The digital twin of discrete dynamic systems: initial approaches and future challenges, Appl. Math. Model. 77 (2020) 1110–1128.

[38] J. Duncan, R.J. Seitz, J. Kolodny, D. Bor, H. Herzog, A. Ahmed, F.N. Newell, H. Emslie, A neural basis for general intelligence, Science 289 (2000) 457–460.

[39] J.E. Laird, A. Newell, P.S. Rosenbloom, Soar: an architecture for general intelligence, Artif. Intell. 33 (1987) 1–64.

[40] A.C. Boud, D.J. Haniff, C. Baber, S. Steiner, Virtual reality and augmented reality as a training tool for assembly tasks, in: 1999 IEEE International Conference on Information Visualization (Cat. No. PR00210), IEEE, 1999, pp. 32–36.

[41] B. Furht, Handbook of Augmented Reality, Springer Science & Business Media, 2011.

[42] N.K. Chakshu, I. Sazonov, P. Nithiarasu, Towards enabling a cardiovascular digital twin for human systemic circulation using inverse analysis, Biomech. Model. Mechanobiol. 20 (2021) 449–465.

[43] K. Gkouskou, I. Vlastos, P. Karkalousos, D. Chaniotis, D. Sanoudou, A.G. Eliopoulos, The "virtual digital twins" concept in precision nutrition, Adv. Nutr. 11 (2020) 1405–1413.

[44] P. Shamanna, M. Dharmalingam, R. Sahay, J. Mohammed, M. Mohamed, T. Poon, N. Kleinman, M. Thajudeen, Retrospective study of glycemic variability, bmi, and blood pressure in diabetes patients in the digital twin precision treatment program, Sci. Rep. 11 (2021) 1–9.

[45] H. Elayan, M. Aloqaily, M. Guizani, Digital twin for intelligent context-aware iot healthcare systems, IEEE Int. Things J. 8 (2021) 16749–16757.

[46] R.G. Alves, G. Souza, R.F. Maia, A.L.H. Tran, C. Kamienski, J.P. Soininen, P.T. Aquino, F. Lima, A digital twin for smart farming, in: 2019 IEEE Global Humanitarian Technology Conference (GHTC), IEEE, 2019, pp. 1–4.

[47] T. Ruohomäki, E. Airaksinen, P. Huuska, O. Kesäniemi, M. Martikka, J. Suomisto, Smart city platform enabling digital twin, in: 2018 International Conference on Intelligent Systems (IS), IEEE, 2018, pp. 155–161.

[48] B.R. Barricelli, E. Casiraghi, J. Gliozzo, A. Petrini, S. Valtolina, Human digital twin for fitness management, IEEE Access 8 (2020) 26637–26664.

[49] J. Corral-Acero, F. Margara, M. Marciniak, C. Rodero, F. Loncaric, Y. Feng, A. Gilbert, J.F. Fernandes, H.A. Bukhari, A. Wajdan, et al., The 'digital twin' to enable the vision of precision cardiology, Eur. Heart J. 41 (2020) 4556–4564.

[50] S.M. Schwartz, K. Wildenhaus, A. Bucher, B. Byrd, Digital twins and the emerging science of self: implications for digital health experience design and "small" data, Front. Comput. Sci. 31 (2020).

[51] A. Al-Ali, R. Gupta, T. Zaman Batool, T. Landolsi, F. Aloul, A. Al Nabulsi, Digital twin conceptual model within the context of Internet of things, Future Internet 12 (2020) 163.

[52] H. Laaki, Y. Miche, K. Tammi, Prototyping a digital twin for real time remote control over mobile networks: application of remote surgery, IEEE Access 7 (2019) 20325–20336.

[53] Y. Liu, L. Zhang, Y. Yang, L. Zhou, L. Ren, F. Wang, R. Liu, Z. Pang, M.J. Deen, A novel cloud-based framework for the elderly healthcare services using digital twin, IEEE Access 7 (2019) 49088–49101.

[54] A. Schmidt, H. Helgers, F.L. Vetter, A. Juckers, J. Strube, Digital twin of mrna-based Sars-Covid-19 vaccine manufacturing towards autonomous operation for improvements in speed, scale, robustness, flexibility and real-time release testing, Processes 9 (2021), https://doi.org/10.3390/pr9050748.

[55] B.R. Barricelli, E. Casiraghi, D. Fogli, A survey on digital twin: definitions, characteristics, applications, and design implications, IEEE Access 7 (2019) 167653–167671, https://doi.org/10.1109/ACCESS.2019.2953499.

[56] C. Cimino, E. Negri, L. Fumagalli, Review of digital twin applications in manufacturing, Comput. Ind. 113 (2019) 103130.

[57] W. Kritzinger, M. Karner, G. Traar, J. Henjes, W. Sihn, Digital twin in manufacturing: a categorical literature review and classification, IFAC-PapersOnLine 51 (2018) 1016–1022.

[58] H. Ansari, L. Kouti, Drug interaction and serotonin toxicity with opioid use: another reason to avoid opioids in headache and migraine treatment, Curr. Pain Headache Rep. 20 (2016) 50.

[59] S.D. Silberstein, Emerging target-based paradigms to prevent and treat migraine, Clin. Pharmacol. Ther. 93 (2013) 78–85, https://doi.org/10.1038/clpt.2012.198.

[60] M.E. Bigal, R.B. Lipton, Overuse of acute migraine medications and migraine chronification, Curr. Pain Headache Rep. 13 (2009) 301–307.

[61] P. Gazerani, Migraine and diet, Nutrients 12 (2020), https://doi.org/10.3390/nu12061658.

[62] J.L. Medina, S. Diamond, The role of diet in migraine, Headache 18 (1978) 31–34.

[63] J. Millichap, M.M. Yee, The diet factor in pediatric and adolescent migraine, Pediatr. Neurol. 28 (2003) 9–15, https://doi.org/10.1016/S0887-8994(02)00466-6.

[64] L.M. Pomes, M. Guglielmetti, E. Bertamino, M. Simmaco, M. Borro, P. Martelletti, Optimising migraine treatment: from drug-drug interactions to personalized medicine, J. Headaches Pain 20 (2019) 1–12.

[65] Y. Feng, X. Chen, J. Zhao, Create the individualized digital twin for noninvasive precise pulmonary healthcare, Significances Bioengineering & Biosciences 1 (2018).

[66] J. Kwon, H. Lee, S. Cho, C.S. Chung, M.J. Lee, H. Park, Machine learning-based automated classification of headache disorders using patient-reported questionnaires, Sci. Rep. 10 (2020) 1–8.

[67] E. Sayyari, M. Farzi, R.R. Estakhrooeieh, F. Samiee, M.B. Shamsollahi, Migraine analysis through eeg signals with classification approach, in: 2012 11th International Conference on Information Science, Signal Processing and Their Applications (ISSPA), IEEE, 2012, pp. 859–863.

[68] H. Yang, J. Zhang, Q. Liu, Y. Wang, Multimodal mri-based classification of migraine: using deep learning convolutional neural network, Biomed. Eng. Online 17 (2018) 1–14.

[69] S. Misztal, G. Carbonell, L. Zander, J. Schild, Simulating illness: experiencing visual migraine impairments in virtual reality, in: 2020 IEEE 8th International Conference on Serious Games and Applications for Health (SeGAH), IEEE, 2020, pp. 1–8.

[70] H. Doh, Augmented Reality and Presence in Health Communication and Their Influence on the Empathy of Healthcare Professionals, Temple University, 2021.

[71] R.K. Cady, C.P. Schreiber, Sinus headache or migraine?: Considerations in making a differential diagnosis, Neurology 58 (2002) S10–S14.

[72] A. Benedick, A. Zeharia, T.E. Markus, Comparison of thrombocyte count between pediatric patients with migraine or tension-type headache: a retrospective cohort study, J. Child Neurol. 34 (2019) 824–829.

[73] H. Damasio, D. Beck, Migraine, thrombocytopenia, and serotonin metabolism, Lancet 311 (1978) 240–242.

[74] M. Zhang, F. Sui, A. Liu, F. Tao, A. Nee, Digital twin driven smart product design framework, in: Digital Twin Driven Smart Design, Elsevier, 2020, pp. 3–32.

[75] R. Söderberg, K. Wärmefjord, J.S. Carlson, L. Lindkvist, Toward a digital twin for real-time geometry assurance in individualized production, CIRP Ann. 66 (2017) 137–140.

[76] P. Healthcare, Cardiac 3d chamber quantifications driven by advanced automation, https://www.usa.philips.com/healthcare/resources/feature-detail/ultrasound-heartmodel, 2021. (Accessed 7 November 2021).

[77] K.S. Leifler, Digital twins – an aid to tailor medication to individual patients, https://liu.se/en/news-item/digital-tvillingar-hjalpmedel-for-skraddarsydd-medicinering-, 2019. (Accessed 7 November 2021).

[78] K.G. HamlAbadi, A.M. Saghiri, M. Vahdati, M. Dehghan TakhtFooladi, M.R. Meybodi, A framework for cognitive recommender systems in the Internet of things (iot), in: 2017 IEEE 4th International Conference on Knowledge-Based Engineering and Innovation (KBEI), 2017, pp. 0971–0976.

[79] A.M. Saghiri, M. Vahdati, K. Gholizadeh, M.R. Meybodi, M. Dehghan, H. Rashidi, A framework for cognitive Internet of things based on blockchain, in: 2018 4th International Conference on Web Research (ICWR), 2018, pp. 138–143.

[80] M. Vahdati, K. Gholizadeh HamlAbadi, A.M. Saghiri, H. Rashidi, A self-organized framework for insurance based on Internet of things and blockchain, in: 2018 IEEE 6th International Conference on Future Internet of Things and Cloud (FiCloud), 2018, pp. 169–175.

[81] A.M. Saghiri, K.G. HamlAbadi, M. Vahdati, The Internet of Things, Artificial Intelligence, and Blockchain: Implementation Perspectives, Springer, Singapore, Singapore, 2020, pp. 15–54.

[82] M. Vahdati, K. Gholizadeh HamlAbadi, A.M. Saghiri, IoT-Based Healthcare Monitoring Using Blockchain, Springer Singapore, Singapore, 2021, pp. 141–170.

[83] K.G. HamlAbadi, M. Vahdati, A.M. Saghiri, A. Forestiero, Digital twins in cancer: state-of-the-art and open research, in: 2021 IEEE/ACM Conference on Connected Health: Applications, Systems and Engineering Technologies (CHASE), 2021, pp. 199–204.

Digital twins for nutrition

Monireh Vahdati[a,b], Ali Mohammad Saghiri[c], and Kamran Gholizadeh HamlAbadi[a,b]

[a]Young Researchers and Elite Club, Qazvin Branch, Islamic Azad University, Qazvin, Iran, [b]Faculty of Computer and IT Engineering, Qazvin Branch, Islamic Azad University, Qazvin, Iran, [c]Computer Engineering Department, AmirKabir University of Technology, Tehran, Iran

15.1 Introduction

In recent years, the number of smartphone applications and wearables that focus on nutrition and physical exercise to improve lifestyle has exploded enhance significantly. While many of these apps collect a lot of data from users, such as food plans, water intake, step counts, sleep patterns, and the like, seldom do they use it. Rather of collecting these data separately, integrating them might provide helpful customized insights that could be utilized to adapt dietary recommendations [13]. To put it another way, the creation of a data base on previous virtual nutritional adviser recommendations might be used to provide personalized recommendations.

The use of Artificial Intelligence (AI), Internet of Things (IoT) sensors, and big data analytics are making Digital Twins (DTs) more advanced for the complex requirements of future applications [14]. The findings of this chapter will help us obtain a better knowledge of individual users' dietary habits, laying the basis for supplying tailored meals to them. To comprehend the metabolic impairment in people's bodies, which is unique to each other, a DT, driven by AI and the IoT, was created. The platform uses data from body sensors to monitor and analyze the body's health signals so that therapies may be tailored to the individual's needs and a better eating plan can be suggested.

This part begins with a short description of nutrition concepts, followed by unique explanations of advanced technology in nutrition, customized nutrition of food, and DTs in nutrition, and lastly a summary of the paper's contributions.

15.1.1 Nutrition concepts

Nutrition is an important aspect of overall health and well-being. Better nutrition is linked to better physical health, maternal health, stronger immune systems, a decreased risk of noncommunicable illnesses (such as diabetes and cardiovascular disease), and longer life expectancy. Fig. 15.1 depicts risks posed by all forms of nutrition. The following are the facts that investigated the dangers presented by all types of diet [39]:

- **Health diet:** A health diet helps to prevent malnutrition in all kinds, as well as noncommunicable diseases (NCDs) such as diabetes, heart disease, stroke, and cancer. However, dietary trends have changed on account of increasing manufacturing of processed foods, rapid urbanization, and changing lifestyles. Individuals are eating more foods that are heavy in calories, fats, free sugars, and salt/sodium, and many people are not eating enough fruit, vegetables, and other dietary fiber such whole grains [36].
- **Obesity and overweight:** Obesity and overweight are characterized as abnormal or excessive fat buildup that may be harmful to one's health. Obesity and overweight are caused by an energy imbalance between calories ingested and calories expended. The fundamental cause of obesity and overweight is an energy imbalance between calories consumed and calories expended [40].
- **Malnutrition:** Deficiencies, excesses, or imbalances in a person's energy and/or nutritional consumption are referred to as malnutrition. The phrase malnutrition refers to three distinct situations [38].
- **Salt reduction:** People all across the globe are eating more energy-dense meals that are heavy in saturated fats, trans fats, sugars, and salt. Salt is the most common source of sodium, and consuming too much of it has been linked to hypertension, heart disease, and stroke [37].
- **Sugars and dental caries:** Dental caries is the most common NCD worldwide and is a serious public health concern. Dental caries may be avoided by eliminating free sugars in the diet [35].

15.1.2 Advanced technology in nutrition

Several studies have been conducted using modern technology such as the IoT, Blockchain, and Machine Learning to improve nutrition precisions [11], therapies [20], supply chains [2], food safety [4] and the like. To be more specific, an IoT-based m-health service that encourages users to develop healthy eating habits [31]. Saghiri et al. [23] and HamlAbadi et al. [8] presented a cognitive recommender system based on IoT and Blockchain

FIGURE 15.1   Risk posed by all forms of nutrition.

to deliver food suggestions for consumers based on previous health evidence such cardiovascular diseases, obesity, and even overweighting. Furthermore, in light of the blockchain's smart contract potential [29], [22], choosing nutritious meal packages for patients must be transparent [32]. Raphaeli and Singer [20] machine learning-based screening technologies are used to provide a customized nutritional therapy for malnutrition. Vahdati et al. [28] illustrates a broad set of conceptual models based on IoT, Blockchain, and cognitive systems that may be used to track people's health and diet.

15.1.3 Personalized nutrition of food

Individualization is the goal of personalized nutrition [18], thus suggestions and nutritional counseling should not be based on average nutritional intake standards applied to gender and age groups of people who are divided by their degree of physical activity. As we go from stratified to individualized and high-accuracy nutrition, more and more indications or traits must be used to get the desired result. One or more factors, such as age, gender, or health condition, may be used to stratify a population. This technique enables for the creation of a product that fulfills the demands of the customer, i.e. customized foodstuff (ethnicity, cultural preferences, geographical and environmental features, and lifestyle), while also lowering decision-making time and cost [17], [30].

15.1.4 Digital twins in nutrition

Digital twins are being redefined as digital replicas of living and nonliving organisms that allow data to flow freely between the real and virtual worlds [5]. A data-driven digital platform that can forecast an individual's best diet and lifestyle recommendations to human DT [33], will provide this wonderful result. The system will be used to gather data from DT populations in order to aid in the prediction, prevention, detection, identification of early illnesses, and the creation of personalized goods and services [9]. A DT of a food product is a simulation model that is a virtual replica of the product. A simulation model is a logical and mathematical description of a food product that is used to create desired traits and attributes via a computerized experiment. The DT takes into account a wide range of variables, including chemical composition, functional and technical qualities, and organoleptic indications [17].

Technological engineers may examine the nutritional, biological, and energy value, as well as other features of a food product, using its DT before mass manufacturing. The virtual simulation model will enable technologists to react in real time to changes in the physical and chemical composition of raw materials used, as well as the replacement of main or auxiliary raw materials, and to adjust the recipe accordingly to obtain a product with the specified chemical composition and guaranteed quality [17]. The notion of the DT has recently arisen as a tool for more adaptable process operational management in the context of industrial digitalization and the advent of the IoT [33].

15.1.5 Contribution of the paper

We discuss current studies on nutrition and DT in this study and then present some open research in this area. The following are the contributions of this paper:

- A nutrition survey in DT in order to locate connected work on diseases as well as open research opportunities in the use of nutrition.
- A statistical analysis was used to provide a review of DT and nutrition.
- A DT framework for nutrition is presented in this chapter. With the use of DT, it can not only give personalized food diets for patients based on their Electronic Health Records (EHR), but it can also anticipate and diagnose nutrition.
- A case study of "hair loss illnesses" was provided to demonstrate the applicability of the suggested model.
- The development barriers of employing DT in nutrition-related diseases will be addressed, followed by several open research routes, in order to illustrate the potential of DT.

The rest of the chapter is organized as follows: Section 15.2 contains related works. The research methodology is described in Section 15.3. Section 15.4 is about documentation on DT and nutrition. The ecosystem of the DT of nutrition was given in Section 15.5. A case study is included in Section 15.6. Section 15.7 is devoted to Discussion, which includes benefits, challenges, and future directions. Section 15.8 contains the conclusions.

15.2 Related work

Nutritional and lifestyle modifications are still at the heart of illness prevention and good aging. The influence of genetic, immunological, behavioral, and metabolic factors on individual reactions to foods is becoming clearer, opening the way for the stratification of dietary recommendations. It is required to consider a large number of various variables and factors. This is almost difficult in a laboratory scientific experiment and/or requires a significant amount of time and effort [17]. However, this might be used to define a virtual DT a digital version of oneself that can be used to advise diet in a customized way. Such a strategy has the potential to change obesity treatment and serve as a backbone for healthy aging [7].

It is also the case to create a DT for usage in neural network technology to personalize dietary items for persons who have a hereditary propensity to diabetes. Precision nutrition recommendations based on continuous glucose monitor (CGM), food consumption data, and machine learning algorithms to offer suggestions for specific patients to avoid meals that cause blood glucose rises and replace them with those that do not. Physicians who have access to daily CGM data were able to adjust medication dosages and keep track of patient conditions [24]. When creating an information system for personalizing food products, it has been demonstrated that the main development direction is the modeling of a DT of the product and the consumer, as well as the determination of the technologies that form the foundation of the personalized food model to create an accurate, properly functioning system, protein, etc. should be included in the generalized mathematical model, but also knowledge of the dependencies of functional, technological, structural, and mechanical characteristics that determine the conversion of the optimal formulation in the technological processing process into a stable system [17].

It is feasible to create a DT of the food using mathematical modeling approaches and a simulation model of the planned product. With the aid of new technology and a DT, it is now feasible to include a wide variety of qualitative and quantitative indicators of meat products when developing recipes for new food items with complicated compositions and features, allowing nondrug illness prevention. Physical scientific investigations save time, money, and resources when they are conducted in the virtual world [17]. As a consequence, the system can do not only food personalization systems, but also medical ones [30].

Food Processing Industries, on the other hand, will inevitably adopt digital technologies to ensure product safety and quality, reduce costs in the face of low profit margins, shorten lead times, and ensure timely delivery of an increasing number of products despite production dead times and uncertainties. A digital twin is a computer representation of a manufacturing process that may be used to design, monitor, and improve its performance [12]. With the introduction of multiscale, multiphase, and multiphysics techniques, food process modeling has progressed. In order to improve insights and optimize designs and processes, more extensive numerical tools and software platforms have arisen.

The notion of the DT has recently developed as a tool for more adaptable process operational management in the context of industrial digitalization and the advent of the IoT. The DT is a virtual replica of a real-world process activity that is linked to the actual world through sensor data and powerful big data analytics. While all ingredients are available for constructing DTs, with various kinds of models playing a vital role, effective installation and operation would need a multidisciplinary approach. The initial agri-food applications must yet be proven [33].

15.3 Research methodology

The goal of this study is to assess the present state of DT research in nutrition by reviewing the literature and identifying current trends. It also investigates the challenges of DT, as well as outstanding open research problems and potential future directions of nutrition. A thorough review of the literature was conducted in order to achieve this goal. Journal articles, conference papers, and edited volumes were among the sources examined. Since the DT in nutrition is relatively novel, there is a scarcity of sources for a comprehensive evaluation of the topic. By searching scholarly databases for the terms "DT in nutrition" relevant literature was identified. The following scholarly databases were searched: In this section, the article found from the research method is presented conceptually and statistically. The following scholarly databases were searched:

- ACM Digital Library
- Google Scholar
- IEEE Xplore digital library
- John Wiley & Sons
- MDPI
- Science Direct
- Scopus
- Springer Link
- Taylor & Francis.

15.4 Documentation on DT and nutrition

A total of eight papers were reviewed. Each article was thoroughly examined and classified into two distinct groups. The year of publication, the digital library classification of each article, and the distribution of papers by article type.

According to Fig. 15.2, since this field is relatively novel, the publication in this area starts in 2019. At the beginning of this trend, the number of research stood at two papers. Then, it experiences a sharp rise to five for the following years, and until October 2021, it decreases to only one paper.

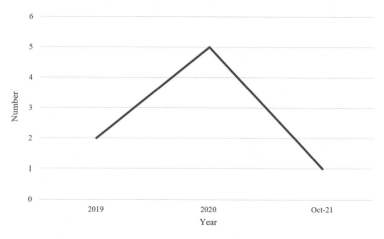

FIGURE 15.2   Distribution paper by year.

As shown in Fig. 15.3, much of the work being done on DT in nutrition is being extracted from IEEE Xplore and Elsevier which have published four out of eight articles. All of the papers, except two, were published in conference by IEEE Xplore. The remaining four papers are published by four different publishers. The result shows that, the majority of the reviewed papers were focused on journal accounting for 75%. In terms of conferences, it has the lowest proportion accounting for only 25% of all eight publications.

15.5 Ecosystem of the digital twin for nutrition

In this section, we use an ecosystem of the DT for health and well-being which was conducted by [6]. Since our work in nutrition could greatly match with this ecosystem, we will propose our model based on the mentioned ecosystem. The brief of the ecosystem being discussed is as follows.

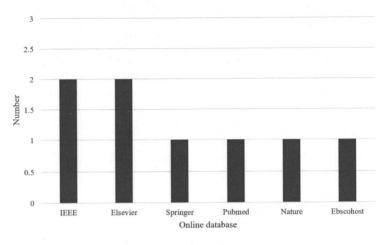

FIGURE 15.3   Distributed paper by online database.

The *'DTwins'* ecosystem which was proposed based on [6] included three layers: data source, AI-Interface Engine and Multi model interaction. The main purpose of this model is to provide a perception of prohibitive healthcare. The prohibitive health DT utilize some sensors from real twin health and AI component to analysis those data and make some recommendations. As a result, by incorporation of DT technology with other technologies, communication between all components will be trustable.

This section explains the steps of making a DT of human and food to provide individual nutrition based on their health status. In general, this model consists of three main components including data source, AI interface, and multi model interaction

15.5.1 Data source

Due to the complexity and long-term nature of nutrition, a particularly large and multidimensional amount of data sources must be collected from real world and organized for the construction of Digital World. These data sources are important not only to create a DT for humans but also to build digital food. Because this system provides appropriate nutrition for each person based on their health circumstances, such information from different types including medical history and clinical and paraclinical data are key factors to personalize food construction to decline diseases progression in the different organs and malnutrition [16].

As for medical history, since nutritional abnormalities are often associated with certain disease states, it is essential to identify underlying medical conditions. Such situations include information regarding previous acute and chronic illness, hospitalizations, operations, social and cultural background, especially as related to diet therapy which is assessed through the medical record and the conversation between physician and patient.

Regarding to clinical and para-clinical data, clinical data are of great importance for diagnosing and monitoring of every individual's health problem. Lab data ranging from standard laboratory to state of the art immunological or neurobiological parameters can complete the fundamental quest of real human evidence for individually improved treatment decisions and balanced diet risk assessment.

A suitable diet is good for both physical and mental health. So, analysis of food is of great importance on account of creating digital food to recommend the proper nutrition program to real human. It can reduce the risk and severity of obesity, heart disease, diabetes, hypertension, depression, and cancer. There are several essential factors for a balanced diet such as carbs, protein, fat, fiber, vitamins & minerals, water, and calories [16], [10], [21]. These factors could be gathered from real food to construct appropriate nutrition regarding real human health condition.

Wireless sensors gather data from real human and real food in order to monitor process conditions. These sensors use a variety of techniques that exploit different parts of the human and food. Furthermore, sensors as well as the actuators (like processing equipment, controllers, machines, robots) should be smart, meaning that communication to the devices is both ways over the wireless network, thereby allowing not only to capture data, but also activation, programming and operation of the sensors and actuators over the network [33].

The Internet is a real-time accessible platform with tools to manage communication, data analysis and interface applications. When it incorporates communication with not only people but also sensors and actuators, it becomes the IoT [27], [34], [3]. An IoT platform operates mainly in the cloud, making use of ease of connectivity and abundant availability of low-cost storage and computing infrastructure, but can also easily connect to local computer hubs for data storage and processing [3]. All the exchanged data must be stored in a data storage system, accessible by the DT [33]. It is thereby ideally suited to implement and manage advanced big data analytical and data sharing tools [7]. For data analysis, different tools are available such as data mining, machine learning, decision support, AI, cloud computing and deep learning [33].

15.5.2 AI interface

With the help of data mining and analysis techniques diet habit and electric health record (EHR) will be extracted. This knowledge of providing appropriate recommendation and predicting digital food in order to improve health diet and reduce diseases such as diabetes and obesities which are obvious result of malnutrition should be taken into account. Machine learning and deep learning techniques facilitate intelligent decision support systems, recognize diet patterns, and predict the accurate personalized food for individuals. Some of the components are described in detail as below.

The data collected from data sources coming from wide variety of individuals' diet habit and health record history needs to be analyzed in order to have ideal nutrition with high level of data accuracy. In the following, the analyzed data is classified into variety of categories such as human health status and food's nutrient content. These medical data can be used to generate the digital food which reduces health problems. It is also noticeable that the contribution [15] between the digital food and digital human plays a vital role to recommend the best real food to real human. The standardization and security of data provided should be taken into consideration.

The results from Data Analytics module are to provide various recommendations by the use of AI methods to the real human through the interaction between DT and digital food. The DT monitors the current health state of the real human and collects data from the data source consistently in real time.

15.5.3 Multimodal interaction (MMI)

The proposed approach of this section is to create a model of the "DT" of the foodstuff based on digital human information providing useful personalized nutrition to evaluate how such diet keeps the real human in a good shape and mitigates the probable diseases at the same time. This process contains several stages. The first stage involves optimization of the nutritional and biological value of the designed product. The second stage is related to designing the food product's structural forms. But even if the recipe of a food product is optimally selected in the first stage, it does not guarantee its transformation into a stable system with the required structural, mechanical, functional and technological parameters during the process [17]. Finally, some several essential factors for a balanced diet are needed to create suitable food based on real human health problems. Fig. 15.4 depicts the concept and elements of a digital twin of human and food, with focus on the individualize of food in order to improve the nutrition program. These factors could create suitable food based on real human health problems.

To meet these expectations for food applications, a DT requires the following elements in place [33]:

- Sensor networks that measure essential variables and properties of the product, process, actuators, inputs, outputs and environment in real time.
- A platform to connect the sensors, actuators with cloud-based data storage and high-performance computing, big data analytics, and applica-

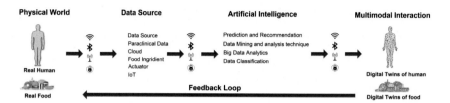

FIGURE 15.4   Ecosystem of the Digital Twin for nutrition.

tions are to be used by relevant users within and across enterprises, which is conveniently provided by the IoT. The IoT platform provides the necessary framework and tools for the integrated sensor communication, data storage, data analysis and decision support that links real human and DT [33].

• A DT simulation platform with the help of computer models that use the data from the platform as inputs performs computations for testing, designing, optimizing and controlling and provides outputs for improved data analysis, process performance and product quality, and therefore provides decision support.

Based on data from real human sensors, each person's health status is extracted in detail. This data is analyzed using a machine learning algorithm and general knowledge of a person's physical condition is obtained. At the same time, the status of important food health data is extracted through sensors embedded in food. At the end of this cycle the digital food which was constructed from the constant communication with DT recommends the best real food for real human to enhance human well-being. It is also noticeable that in the whole process of creating the appropriate food for human the cybersecurity plays an essential role to make sure that under no circumstances the third party has access to other's data but from privacy point of view, information should be shared with other DTs and digital foods.

15.6 Case study: hair loss

In order to investigate the proposed model of producing a personalized diet, hair loss diseases have been studied. The aim of this study was to produce appropriate food based on the treatment of hair loss. In order to produce a great source of personalized diet for hair loss patient, we proposed three algorithms known as DT of human, DT of food, and personalized digital food for individuals who suffer from hair loss diseases. The following step could provide personalized nutrition for people having alopecia areata.

Algorithm 1: Microservice of DTs of human (DTH-Services).

Input: Individual data are collected by the sensors and paraclinical
   parameters
Output: Create DTs of human.
Data: EHR: Electronic health records
1 **Begin**
   /* Medical history, medical test, and paraclinical data   */
2 All personal data are received through the sensors and EHR;
3 Personal data are accumulated in data source;
4 AI interface layer extracted relevant knowledge from all human
   data from data source;
5 The knowledge obtained is modeled using multimedia models
6 DT for human are created;
7 **End.**

Algorithm 2: Microservice of DTs of food (DTF-Services).

Input: Materials of food data are collected by the sensors.
Output: reate DTs of food.
1 **Begin**
   /* Vitamins, protein, sugar, water, etc.   */
2 All food data are received from sensors;
3 Food data are accumulated in data source;
4 AI interface layer extracted relevant knowledge from all food data
   from data source;
5 The knowledge obtained is modeled using multimedia models
6 DT for food are created;
7 **End.**

Algorithm 1 shows how to build a DT from a real human. In this algorithm, in the first phase, real-time data of each person is gathered through
sensors and the paraclinical data accumulate in the data source. The obtained data is then analyzed with the help of machine learning algorithms
and provides knowledge, this knowledge is modeled and the digital twin
of humans is constructed.

Algorithm 2 shows how to construct DT of food. In this algorithm,
each data of food including fruits, vegetables, meat, fish, etc., are received
through sensors and then stored in the data source. This data is analyzed using machine learning algorithms and appropriate knowledge is
acquired from each food. Eventually, this knowledge is modeled with the
help of Multimodal Interface, and digital food is created.

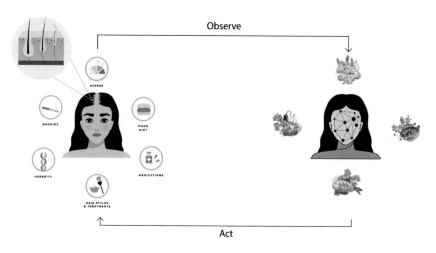

FIGURE 15.5   Case Study: Hair loss ailments.

In Algorithm 3, with the help of calling DT of human service and DT of food service, appropriate food is produced for each person who suffers from hair loss ailments. In this algorithm, the existence of basic nutritional factors which play a crucial role in hair loss is evaluated in DT of humans. In the absence of nutrients, the booster digital food is constructed for each person. In the next stage, the booster digital food materials are tested in DT of humans in order to evaluate hairless treatment. If the desired tests pass, booster real food will be made and prescribed to the real human.

Fig. 15.5 shows a schematic of the real world and digital world for the treatment of hair loss diseases using nutrition. The digital world receives real-time data on the cause of hair loss ailments such as poor diet, medications, smoking, etc. from real world, a DT (DT of human and DT of food) is created. A DT of human can interact with their either real or DT' counterparts, and act on some data to real world.

15.7 Discussion

One of the main purposes while creating a functional and/or specialized diet is to provide the nutrients missing in the diet, as well as to reduce the risk of developing and preventing a particular disease. In the rest of this section, challenges, benefits, and future dynamics utilizing DT in nutrition are discussed in more details.

Algorithm 3: Personalized digital food for individuals who suffer from hair loss diseases.

Input: Personal medical data
Output: Provide personalized nutrition for hair loss patient.
Data:
    DTH-Services: DT microservice of human
    DTF-Services: DT microservice of food
    Hair loss ailments risks: Lack of nutrition materials such as vitamins D, and C, B, iron, omega 3, etc.
    HLN: Hair loss nutrition
    BDTF: Booster DT of food.
    BRF: Booster Real food

1   **Begin**
      /* DT of human is call in order to extract basic information about human */
2   Call DTH-Services;
3   **while** *hair loss ailments risks in DTH* **do**
         /* Evaluate lack of neutrinos including vitamins, proteins, zinc, etc. */
4   Evaluate hair loss ailments risks in DTH
5   **if** *hair loss ailments evaluation detects the lack of nutrition* **then**
6   insert details of materials to HLN factors
7   **else**
8   output is "Hair loss ailments are not caused by lack of nutrition"

      /* DT of food is call in order to extract basic information about food. */
9   Call DTF-Services;
10   **while** *basic factors of nutrition in DTF* **do**
         /* Evaluated HLN factors in DTF Including vitamins, protein, zinc, etc. */
11   Evaluated HLN factors in DTF;
12   **if** *materials of HLN detects in DTF* **then**
13   prescript the food to DTH
14   **else**
15   inject HLN materials to DTF, which is called BDTF

16   **while** *BDTF in DTH* **do**
         /* Evaluate hair loss treatment approximately one month, this variable could manner by physicions. */
17   Evaluate hair loss treatment in whole of time;
18   **if** *BDTF could treat the hair loss ailments* **then**
19   produce the real food of BDTF which is called BRF
20   **else**
21   based on communication between DT of human and DT of food go to stage 10
           /* until stop the hair lost in DT of human this stage should be repeated. */

22   The real human uses BRF;
23   Update DTH based on real human;
24   **End.**

Benefits: Utilizing DT in nutrition has the following benefits:

- Smart and optimized model-based algorithms for more accurate and reliable process control.
- Augmented reality monitoring of the process, where the DT acts as a true virtual duplication of the real process providing better nutrition habit with high accuracy.
- Nutrition adapted to the needs of people will help reducing the risks for those who already have diseases and will meet the needs of those who would like to make their diet more appropriate to individual needs [17].
- Recommending a suitable diet for individual is challenging due to the complexity of the data which is gathered by sensors and actuators. DTs help to understand possible diseases which might occur for individuals on account of inappropriate nutrition. Based on the constantly new data collected through continuous monitoring and provided by the people from the real world, the DT generates new knowledge. This information in turn flows into the people's further diet programs, which is thus continuously improved not only suitable for their diets but also to advice suitable medication intake.

Challenges: Providing the suitable nutrition habit is faced with the challenge of how to optimally use the large amount and diverse data being generated. To render successful frameworks for such tasks, tools from different areas of development are currently being explored. IoT technologies facilitate the transfer of data from and to different sensors, computers, and machines [4], [27], [34]. Cloud Computing offers ways to store, share and work with the data more effectively [25]. Data Mining and AI allow to process data in a smart and efficient way [41], [26], [19]. Apart from the technological challenges of collecting and processing data, digitalized food production also poses other questions concerned with data sharing, such as transparency, trust, and ownership. Since data security is very important to avoid data gaps that could potentially be used for hacker attacks to the detriment of patients, it is also necessary to ensure the protection of privacy, which becomes more and more difficult with the increasing functionality of techniques. To this end, there are initiatives such as blockchain, which is a decentralized database of records in the form of encrypted blocks of all transactions or digital events executed and shared among participating parties [1].

Future Direction: Several future directions can be followed of which some are explained:

- DT could identify high-risk populations in order to allow nutritionist to evaluate different monitoring practices to recommend the best solution to their problems.
- To design a fully featured DT for nutrition, all participants like health professionals, farmers, clinicians, and people should work together.

During this cooperation, people across several disciplines, including nutrition specialists, physicians, manufactures, and computing, attempt to extract insights from DT to define appropriate personalized food diets product.

- DT will bring a novel approach to personalized foods. Health professionals can evaluate several treatment plans of the DT of a patient to find an appropriate food diet. An approach that brings together all capabilities that are resulted via DT and conventional treatment may be required.
- Although DT has a great potential for appropriate nutrition, a minority of research was conducted on DT and nutrition, and more studies can be conducted in the future. Few studies will be conducted on nutrition risk factors such as malnutrition, salt reduction, and sugar and dental caries as open research.
- The DT of the food product will allow producer of food industry enterprises to respond quickly to changes in the properties and types of raw ingredients, to respond to changes in consumer preferences. In addition it creates products with a predetermined chemical composition, nutritional value, and functional orientation to a particular category of consumers and even recategorizes based on new restitutions [17].
- Nutrients have several problems, such as high complexity, unpredictability, and uncertainty. DT can solve these problems.
- DT considers ethnicity, cultural preferences, and environmental factors to create a personalized foodstuff.

15.8 Conclusion

We have discussed the concept of the digital twin, its background, and history. We Learned that when developing products for personalized nutrition, metabolism of nutrients should be taken into account. The concept of the digital twin, its background, and its history has been discussed. We Learned that when developing products for personalized nutrition, the metabolism of nutrients should be taken into account. A Digital Twin ecosystem is composed of a physical world and a digital world, which are interconnected in a loop so as to provide data and information to recommend real food to enhance human wellbeing. We also learned that it is necessary to consider the whole variety of different variables and factors. It is essential to choose the optimal data from both real human and real food to make the best model of digital food based on digital human. The personalize final model should treat not only health problems such as obesities and diabetes, but also recommend the best nutritional program. Thanks to the use of multimodal interaction and AI interface, it is possible to obtain a "digital twin" of the foodstuff. Advanced technologies with the help of a digital twin make it possible to take into account the whole range of clinical, para clinical and even diet habit of every individual to develop a specific food with a complex composition and characteristics, which allow nondrug prevention of diseases. Testing in the virtual world saves time, money, and resources for physical scientific experiments. Using the case study of hair lost, which are showed the possibility of creating a digital twin model of the personalized foodstuff in order to treatment the hair lost ailment. The chapter concludes with a discussion of the benefits, challenges, and future directions of the digital twin in nutrition.

Clearly the lessons learned

- The causes of diseases like hair loss include vitamin deficiency, poor nutrition, smoking, and so forth. DT can be used to make treatments personalized based on individual characteristics of health, rather than testing them on real people or animal models. Although the solution may appear to be expensive, it is not only more accurate and in line with the characteristics of each individual, but it ensures that living organisms are not endangered.
- By using DT, it is possible to recommend a more accurate diet that will help strengthen the immune system to minimize the risk of diseases such as cancer, obesity, heart disease, etc.
- The primary difference between DT and other simulation technologies is that with DT, all changes made in the real world can be applied directly to the digital version so that this method is much more realistic and can be used to accurately predict, diagnose, and treat diseases caused by poor nutrition.

Acknowledgment

Last but not least, I am dedicating this chapter to my late father Mohammad Vahdati gone forever away from our loving eyes and who left a void never to be filled ever. Though your life was short, I will make sure your memory lives on as long as I shall live. I love you all and miss you all beyond words. The author would greatly appreciate Eng. Raika Monazzam for providing some graphical contents, and Eng. Rouzbeh Ghalandarzadeh for some reviews and grammatical checks.

References

[1] F. Antonucci, S. Figorilli, C. Costa, F. Pallottino, L. Raso, P. Menesatti, A review on blockchain applications in the agri-food sector, J. Sci. Food Agric. 99 (2019) 6129–6138.

[2] K. Behnke, M. Janssen, Boundary conditions for traceability in food supply chains using blockchain technology, Int. J. Inf. Manag. 52 (2020) 101969.

[3] M. Ben-Daya, E. Hassini, Z. Bahroun, Internet of things and supply chain management: a literature review, Int. J. Prod. Res. 57 (2019) 4719–4742.

[4] Y. Bouzembrak, M. Klüche, A. Gavai, H.J. Marvin, Internet of things in food safety: literature review and a bibliometric analysis, Trends Food Sci. Technol. 94 (2019) 54–64.

[5] A. El Saddik, Digital twins: the convergence of multimedia technologies, IEEE Multimed. 25 (2018) 87–92.

[6] A. El Saddik, H. Badawi, R.A.M. Velazquez, F. Laamarti, R.G. Diaz, N. Bagaria, J.S. Arteaga-Falconi, Dtwins: a digital twins ecosystem for health and well-being, IEEE COMSOC MMTC Commun. Front. 14 (2019) 39–43.

[7] K. Gkouskou, I. Vlastos, P. Karkalousos, D. Chaniotis, D. Sanoudou, A.G. Eliopoulos, The "virtual digital twins" concept in precision nutrition, Adv. Nutr. 11 (2020) 1405–1413.

[8] K.G. HamlAbadi, A.M. Saghiri, M. Vahdati, M.D. TakhtFooladi, M.R. Meybodi, A framework for cognitive recommender systems in the Internet of things (iot), in: 2017 IEEE 4th International Conference on Knowledge-Based Engineering and Innovation (KBEI), IEEE, 2017, pp. 0971–0976.

[9] K.G. HamlAbadi, M. Vahdati, A.M. Saghiri, A. Forestiero, Digital twins in cancer: state-of-the-art and open research, in: 2021 IEEE/ACM Conference on Connected Health: Applications, Systems and Engineering Technologies (CHASE), 2021, pp. 199–204.

[10] M. Hayder, M. Trzaskowski, L. Ruzik, Preliminary studies of the impact of food components on nutritional properties of nanoparticles, Food Chem. (2021) 131391.

[11] D. Kirk, C. Catal, B. Tekinerdogan, Precision nutrition: a systematic literature review, Comput. Biol. Med. (2021) 104365.

[12] A. Koulouris, N. Misailidis, D. Petrides, Applications of process and digital twin models for production simulation and scheduling in the manufacturing of food ingredients and products, Food Bioprod. Process. 126 (2021) 317–333.

[13] H. Krcmar, M. Böhm, Digital twin: towards a rich nutrition relevant user profile (fa5 module 1), https://www.enable-cluster.de/en/forschung/der-virtuelle-ernaehrungsberater/digital-twin, 2021. (Accessed 18 October 2021).

[14] F. Laamarti, H.F. Badawi, Y. Ding, F. Arafsha, B. Hafidh, A.E. Saddik, An iso/IEEE 11073 standardized digital twin framework for health and well-being in smart cities, IEEE Access 8 (2020) 105950–105961, https://doi.org/10.1109/ACCESS.2020.2999871.

[15] F. Laamarti, A. El Saddik, Multimedia for social good: green energy donation for healthier societies, IEEE Access 6 (2018) 43252–43261.

[16] A. Maqbool, I.E. Olsen, V. Stallings, et al., Clinical Assessment of Nutritional Status. Nutrition in Pediatrics, 4th edition, BC Decker Inc, Canada, 2008, pp. 5–13.

[17] M. Nikitina, I. Chernukha, Personalized nutrition and "digital twins" of food, Potravinarstvo 14 (2020).

[18] J.M. Ordovas, L.R. Ferguson, E.S. Tai, J.C. Mathers, Personalised nutrition and health, BMJ 361 (2018).

[19] D.I. Patrício, R. Rieder, Computer vision and artificial intelligence in precision agriculture for grain crops: a systematic review, Comput. Electron. Agric. 153 (2018) 69–81.

[20] O. Raphaeli, P. Singer, Towards personalized nutritional treatment for malnutrition using machine learning-based screening tools, 2021.

[21] A.C. Rizzo, T.B. Goldberg, C.C. Silva, C.S. Kurokawa, H.R. Nunes, J.E. Corrente, Metabolic syndrome risk factors in overweight, obese, and extremely obese Brazilian adolescents, Nutr. J. 12 (2013) 1–7.

[22] A.M. Saghiri, K.G. HamlAbadi, M. Vahdati, The Internet of things, artificial intelligence, and blockchain: implementation perspectives, in: Advanced Applications of Blockchain Technology, Springer, 2020, pp. 15–54.

[23] A.M. Saghiri, M. Vahdati, K. Gholizadeh, M.R. Meybodi, M. Dehghan, H. Rashidi, A framework for cognitive Internet of things based on blockchain, in: 2018 4th International Conference on Web Research (ICWR), IEEE, 2018, pp. 138–143.

[24] P. Shamanna, B. Saboo, S. Damodharan, J. Mohammed, M. Mohamed, T. Poon, N. Kleinman, M. Thajudeen, Reducing hba1c in type 2 diabetes using digital twin technology-enabled precision nutrition: a retrospective analysis, Diabetes Therapy 11 (2020) 2703–2714.

[25] A. Singh, S. Kumari, H. Malekpoor, N. Mishra, Big data cloud computing framework for low carbon supplier selection in the beef supply chain, J. Clean. Prod. 202 (2018) 139–149.

[26] Q. Sun, M. Zhang, A.S. Mujumdar, Recent developments of artificial intelligence in drying of fresh food: a review, Crit. Rev. Food Sci. Nutr. 59 (2019) 2258–2275.

[27] J.M. Talavera, L.E. Tobón, J.A. Gómez, M.A. Culman, J.M. Aranda, D.T. Parra, L.A. Quiroz, A. Hoyos, L.E. Garreta, Review of iot applications in agro-industrial and environmental fields, Comput. Electron. Agric. 142 (2017) 283–297.

[28] M. Vahdati, K.G. HamlAbadi, A.M. Saghiri, Iot-based healthcare monitoring using blockchain, in: Applications of Blockchain in Healthcare, Springer, 2021, pp. 141–170.

[29] M. Vahdati, K.G. HamlAbadi, A.M. Saghiri, H. Rashidi, A self-organized framework for insurance based on Internet of things and blockchain, in: 2018 IEEE 6th International Conference on Future Internet of Things and Cloud (FiCloud), IEEE, 2018, pp. 169–175.

[30] A.M. Vaskovsky, M.S. Chvanova, M.B. Rebezov, Creation of digital twins of neural network technology of personalization of food products for diabetics, in: 2020 4th Scientific School on Dynamics of Complex Networks and Their Application in Intellectual Robotics (DCNAIR), IEEE, 2020, pp. 251–253.

[31] M. Vazquez-Briseno, C. Navarro-Cota, J.I. Nieto-Hipolito, E. Jimenez-Garcia, J. Sanchez-Lopez, A proposal for using the Internet of things concept to increase children's health awareness, in: CONIELECOMP 2012, 22nd International Conference on Electrical Communications and Computers, IEEE, 2012, pp. 168–172.

[32] P.E. Velmovitsky, F.M. Bublitz, L.X. Fadrique, P.P. Morita, Blockchain applications in health care and public health: increased transparency, J. Med. Inform. 9 (2021) e20713.

[33] P. Verboven, T. Defraeye, A.K. Datta, B. Nicolai, Digital twins of food process operations: the next step for food process models?, Curr. Opin. Food Sci. 35 (2020) 79–87.

[34] C.N. Verdouw, J. Wolfert, A. Beulens, A. Rialland, Virtualization of food supply chains with the Internet of things, J. Food Eng. 176 (2016) 128–136.

[35] WHO, Sugars and dental caries, https://www.who.int/news-room/fact-sheets/detail/sugars-and-dental-caries, 2017. (Accessed 18 October 2021).

[36] WHO, Healthy diet, https://www.who.int/news-room/fact-sheets/detail/healthy-diet, 2020. (Accessed 18 October 2021).

[37] WHO, Salt reduction, https://www.who.int/news-room/fact-sheets/detail/salt-reduction, 2020. (Accessed 18 October 2021).

[38] WHO, Fact sheets – malnutrition, https://www.who.int/news-room/fact-sheets/detail/malnutrition, 2021. (Accessed 18 October 2021).

[39] WHO, Nutrition, https://www.who.int/health-topics/nutrition, 2021. (Accessed 18 October 2021).

[40] WHO, Obesity and overweight, https://www.who.int/news-room/fact-sheets/detail/obesity-and-overweight, 2021. (Accessed 18 October 2021).

[41] L. Zhou, C. Zhang, F. Liu, Z. Qiu, Y. He, Application of deep learning in food: a review, Compr. Rev. Food Sci. Food Saf. 18 (2019) 1793–1811.

Digital twins for allergies

Kamran Gholizadeh HamlAbadi[a,b],
Monireh Vahdati[a,b], Ali Mohammad Saghiri[c], and
Kimia Gholizadeh[d]

[a]Young Researchers and Elite Club, Qazvin Branch, Islamic Azad University,
Qazvin, Iran, [b]Faculty of Computer and IT Engineering, Qazvin Branch,
Islamic Azad University, Qazvin, Iran, [c]Computer Engineering Department,
AmirKabir University of Technology, Tehran, Iran, [d]Department of
Computer and Electrical Engineering, Mazandaran University of Science and
Technology, Babol, Iran

16.1 Introduction

Allergies are a critical issue for everyone in today's advanced world, causing enormous costs for both individuals and governments. Allergies are a prevalent chronic illness. An allergy occurs when the body's immune system sees a substance as harmful and overreacts to it. The substances that cause allergic reactions are allergens. When an individual suffers from allergies, their immune system produces an antibody known as immunoglobulin E. (IgE). The symptom results are an allergic reaction. Fig. 16.1 depicts several types of allergens will be described as follows [1], [25]:

- **Drug (medicine):** The majority of adverse reactions to pharmaceuticals are more appropriately referred to be "adverse drug reactions." True drug allergies are uncommon and are the result of the immune system.
- **Food:** A food allergy occurs when the immune system of the body perceives a particular meal as hazardous and reacts with allergic symptoms.
- **Insect:** Insects that do not sting can potentially trigger allergic responses. Cockroaches, an insect-like dust mite, are the most prevalent.

325

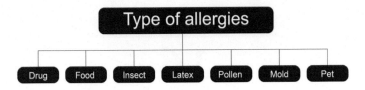

FIGURE 16.1   Type of allergies.

Allergies to insects such as bees, fine ants, hornets, yellow jackets, and wasps.
- **Latex:** An allergic reaction to natural rubber latex is called a latex allergy. Latex is found in natural rubber latex gloves, balloons, condoms, and other natural rubber items. Latex allergies can pose a major health concern.
- **Mold:** Mold and mildew are both types of fungi. Due to the fact that fungi may grow in a wide variety of environments, both indoors and outdoors, allergic reactions can develop at any time of year.
- **Pet:** Allergies to furry pets are quite frequent. It is critical to understand that there is no such thing as an allergy-free (hypoallergenic) dog or cat breed.
- **Pollen:** Pollen is a frequent allergen that promotes seasonal allergies. Although many people refer to pollen allergy as "hay fever," professionals prefer the term "seasonal allergic rhinitis".

The most frequent allergy symptoms include watery eyes, runny nose, sneezing, rash, or hives [1], [25]. Allergy is diagnosed in three stages such as personal and medical history, physical exams, and allergens test. Appropriate allergy therapy is individualized and dependent on medical history and the severity of symptoms. As a result, allergies are rising public health problems and the financial burden of health care, affecting many adults and children [1], [25].

There are very few studies on the influence of eHealth on asthma and allergies. Mobile health technology has immense promise and may become a realistic, cost-effective, and helpful tool in the future not only for allergy disorders, but for a wide variety of other medical fields [24]. Several novel informatic techniques have the potential to significantly enhance health care. New information representation approaches may help translate cutting-edge research into therapeutic practice [9]. In today's tech-driven society, patients express higher pleasure and demand for telemedicine services. Interim virtual visits may help overcrowded clinics, decrease travel costs, boost access to specialist treatment, and improve adherence to chronic allergy surveillance. Due to the outpatient nature of allergy immunology and the simplicity with which many components of a typical visit may be conducted through telemedicine, it is extremely desired to include telehealth training in fellowship programs [35].

FIGURE 16.2   Human DTs construction.

Studies in medicine, biotechnology, Artificial Intelligence (AI), computer science, and other domains have been conducted to reduce allergy threat. Numerous sciences have contributed to the diagnosis, prevention, management, and control of allergies, but they were insufficient. Among modern technologies, Digital Twins (DT) have the unique ability to reduce the number of theses for both patients and physicians. A DT is a digital representation of a real thing, either alive or nonliving. Through the establishment of a link between the physical and virtual worlds, data is conveyed fluidly, enabling the virtual entity to coexist with the actual entity [10]. In the process of construction of DT, using the IoT, real-time human data can be received through clinical sensors, wearable sensors, and more. The 5G/6G are also needed to facilitate communication among all sections from the real world to the digital world, and then data is modeled with the help of AI such as machine learning algorithms, making new knowledge of real humans, and then several graphical models are developed to organize the virtual entity, and Finally, an individual DT can be created [18]. Fig. 16.2 depicts DTs of human construction. To the best of our knowledge this paper is the first study about allergies and DT.

In this chapter, DT in allergies is created based on a formal DTs ecosystem for healthcare and well-being performed by [11]. To examine the existing model, anaphylaxis shock has been studied. The rest of this paper is organized as follows: Section 16.2 shows related works. Section 16.3 present an ecosystem of DTs for allergic disorders. Section 16.4 is described to the investigation of ecosystem in anaphylaxis shocks as a case study. Further discussion on the modeling findings and future approaches is given in Sections 16.5 and section 16.6 concludes this research.

16.2 Related works

Until now, no studies have been performed on the role of DT in allergies so we will review some carried out research on technologies which could play a vital role for creating DT.

16.2.1 Internet of things (IoT)

In modern society with the help of IoT, we can find some specific information about users [73], but a few studies have been conducted on the role of IoT in allergic diseases. In general, this section deals with previous studies on the role of the IoT in drug allergies and reducing air pollution.

In drug point of view for allergies, [28] proposes a solution for assisted living that addresses the issue of adverse medication responses and allergy identification in patients. Whereas [27] discusses how IoT is being used in a pharmaceutical system to examine drugs for adverse drug reactions, allergies, etc., and [29] shows the prospect of applying modern technologies to enhance the quality of medication delivery, to increase adherence to drug-adequate intake, and to decrease clinical errors caused by medication error, and drug interactions, while in [13], the IoT is applied to examine drugs in order to fulfill treatment, and to detect harmful side effects of allergies. For air quality perspective, Indoor air pollution has a significant negative impact on human health, particularly for persons who are allergic to certain allergens such as airborne allergens [59]. Asthana and Mishra [4] develop a real time monitoring the fundamental pollutants with the help of IoT and then it provides pollution level information directly to a smart device in real time. In [14], when the pollution level exceeds a specified threshold value, the proposed air pollution monitoring system can provide alerts to users. Pla et al. [59] discusse available mobile applications for pollen information that are powered by distributed intelligence systems with help of the IoT in smart cities. Li et al. [45] present the IoT sensor system's possible uses which includes epidemiological studies of asthma development and exacerbation, individualized asthma treatment, and environmental monitoring.

16.2.2 Machine learning (ML)

Developing ML learning algorithms could help to anticipate failures to accomplish clinically substantial satisfaction [42]. Consider the massive volumes of data created in the medical field, each patient gets a unique medical record that includes information on allergies, chronic diseases, and vaccines. If properly analyzed, healthcare can profit. ML may be a very useful solution not only to simplify analysis but also to reduce time [67]. References [17] and [64] propose cognitive recommender systems and cognitive IoT based on blockchain to encourage people consume healthy foods based on their physician evidence and health status. Furthermore, [2] provides a purely computational ML strategy for properly diagnosing food allergies and potentially identifying epigenetic targets for the illness using DNA methylation data. Kavya et al. [33] propose a paradigm for allergy diagnosis that is computer-aided and capable of addressing comorbidities. Mohabatkar et al. [53] present a computational technique for protein allergen prediction. Rong et al. [62] provide a deep learning-based solution for real-time detection and insight creation about one of Australia's most frequent chronic illnesses – pollen allergies. Omurca et al. [56] design an intelligent diagnostic assistant for anticipation of the type of an allergic disease across Turkey automatically by ML algorithms. Zewdie et al. [76] propose a method for the robust estimate of airborne Ambrosia pollen concentrations utilizing a variety of ML techniques, including deep learning and reinforcement learning. Harvey and Kumar [20] using ML, they construct prediction models to assess a dataset on child asthma health.

16.2.3 Blockchain technology

Blockchain technology, AI, and IoT will be used as the substructure of advanced applications [63] more specifically health systems [49], [72]. In terms of allergy, [55] focuses on cocreating a distributed ledger for patients' allergies in order to deal with the availability, integrity, and confidentiality of new allergy information for healthcare professionals. Likewise a blockchain-based system for the secure storage, administration, and access of measurement data is provided in [52]. Ngassam et al. [54] demonstrate how to create a Blockchain-based allergy card to address real-world challenges, namely registering, sharing, and tracing information regarding medication allergies.

16.2.4 Cloud and fog computing

Fog computing and cloud computing are two of the most important technologies that have significantly aided the growth of the healthcare industry [21]. Fog computing is the backbone of current healthcare systems; it not only reduces related treatment and medication costs, but also simplifies logistics [75]. Cloud-based services may be critical in the delivery of emergency care because they provide simple and fast access to patient data from almost any location and through almost any device [41]. In [71] a cloud-based conceptual framework is suggested which will benefit the healthcare sector as it implements IoT healthcare solutions. Cloud-computing and mobile-based client technologies may help keep asthma patients engaged, enhance control and treatment, as well as provide new insights for scientists and doctors [48].

16.2.5 5G and 6G wireless communication

The sixth generation (6G) wireless communications standard, now under development, will connect everything, provide full-dimensional wireless coverage, and integrate sensing, communication, computation, caching, and control [44]. However, 5G may limit the reliability of low-latency high-data-rate services, required for Augmented Reality (AR), as Mixed Reality (MR), and Virtual Reality (VR). No advanced IoT technologies that need communication, detection, control, and processing operations are supported by 5G. As a consequence, 6G is required to enable IoT technologies [50]. 6G will be more reliable and faster than 5G, and completely integrated with ML, IoT, and blockchain technologies [50], [47]. Aside from Super high definition (SHD and Extremely high definition (EHD) video capabilities, 6G wireless networks will support the Internet of Nano-Things, the Internet of Bodies, and continuous communications enabled by implanted nanodevices and nano sensors that consume very little power [50]. 6G communication has the potential to facilitate data transmission across a variety of healthcare organizations, including physicians, patients, labs, ambulances, hospitals, and smart houses [72]. These could facilitate transformation allergen data from people living cross-borders in real time.

16.2.6 AR/VR/Mix reality

AR, VR, and MR have been used in medical sectors such as medical education, surgical simulation, neurological rehabilitation, psychotherapy, and telemedicine. Related research shows that VR, AR, and MR lessen medical malpractice caused by unskilled operation and decrease the cost of medical education and training [26]. AR, VR, and MR are used to improve and optimize patients with allergic conditions with the help of technologies including use of digital diaries, wearable devices, remote monitoring of patients' physiologic information, use of electronic health record (EHR) with decision support, and analysis of Big Data [61].

16.2.7 Simulation techniques

Simulation models have demonstrated value in medical teaching, particularly for rare events. Allergy physicians frequently meet emergencies in their offices, which can cause considerable anxiety [34]. Manavi et al. [51] built models of both synthetic and natural allergens and analyzed the resulting aggregates to determine the effect of model resolution on antibody aggregation simulations. Barrett [5] modeled and simulated technique to aid in the development of ARS-1 (Intranasal Epinephrine) for adults and children with systemic allergies.

16.3 Ecosystem of the DT for allergy disease

In this section, we adapt the general framework for [11] and extract substantial components of DT for allergy diseases. They introduced a DT's ecosystem for health and well-being which is called *'Dtwins'*. This ecosystem is an outstanding reference model for medical fields in DT. This model can be well used for allergic diseases in DT and meets all its requirements. Therefore, in this chapter, based on this ecosystem, allergy disease is discussed in DT. The following is a brief description of this ecosystem.

Their proposed ecosystem provides a model for the DT system in preventive health care. In their model, preventive health is individualized based on DT. Real twin health data is collected via sensors, and then analyzed using AI to provide suitable recommendations [10] and feedback. DTs in their model can interact with real-world counterparts and with other DTs in real time [10]. Using AR, VR, and haptic technologies, this set of criteria enables the integration of disparate technologies into DTs and the creation of a personal healthcare system.

The proposed ecosystem for allergic diseases includes three stages which are allergy data source, AI interface, and multimodal interaction, which will be discussed in details.

16.3.1 Allergy data source

This layer contains a variety of resources related to health and public health services. To achieve an accurate picture of each day's allergens, data from several sources must be collected and intelligently processed without human interference. Table 16.1 summarizes the data sources. These sources include wearable sensors (which record all bodily activities), textual information (geographic location, weather conditions, air pollution, water pollution, and humidity levels), and each person's electronic health records (EHR), which are referred to as paraclinical records. Allergen data must be gathered from the sources of human data indicated below.

- **Medical records:** Every person has individuals' medical status such as cardiovascular disease, diabetes, cancer, chronic obstructive pulmonary disease, and mental disease, etc. [19].
- **Allergy history:** In order to accurate assessment of every person allergy status, the personal history information of allergy is significantly important [3].

- **Allergy symptoms:** A Most of the time people have variety of symptoms caused by allergies. The allergy symptoms include sneezing, nasal congestion, runny nose, post nasal drip [7], [74], anaphylaxis shocks [58] etc.
- **Allergies records:** Everyone has different allergies. It depends on their locations, circumstance, age, etc. The type of allergies illustrated in Fig. 16.1.
- **Allergies test:** Allergy testing can detect allergies to foods [46], drugs [57], etc. [38]. Having an allergist test and diagnosing would make people feel significantly better. An allergy test should be tailored to the patient's needs, as well as the cost-benefit ratio.
- **Vaccine records:** Some vaccines could cause allergies including influenza [31], tetanus [32], pneumococcal [30], hepatitis A and B [15], Coronavirus 2019 (Covid-19) [70].
- **Pet history:** Furry creatures kept as pets are a significant source of allergens [65].
- **Indoor environment:** Several indoor environmental risk factors were linked to allergies and asthma. Clinicians and public health workers may need to focus on indoor health risks [68].
- **Geographic history:** A correlation between urban/rural status and the prevalence of food allergies was identified in [16], therefore gathering history of residency may be important for determining and predicting allergies.
- **Family allergy history:** In comparison to having no family member who suffered from allergy, having one family member with symptoms increases the likelihood of developing certain allergies [40].
- **Family medical history:** Family medical history is a powerful predictor of diseases risk such as allergies. Adult sickness can be caused by inherited alterations in gene sequence or function. The common environment and heredity influence the risk of familial disease. Collecting and assessing useful family health histories is difficult [23].

Table 16.1 summarized the type of elements which have in human healthcare data source.

16.3.2 AI interface

Patients' data received from a variety of sources, such as IoT sensors and the paraclinical management system, must be analyzed in order to accurately summarize and extract each person's allergy-related features and characteristics. Using data fusion, the data is then classified into many categories, including numbers, sounds, images, texts, and continuous and discrete signals. The data is standardized using the ISO / IEEE 11073 standard [43], and data analysis is initiated using AI. ML techniques such as reinforcement learning, deep learning, and pattern recognition can be uti-

TABLE 16.1   Type of data which have in Human healthcare data source.

Data	Details
Medical records	Constitutional, Eyes Ear, Nose, Throat and Mouth problems, Cardiovascular, Genitourinary Musculoskeletal, Skin Problems, Neurological, Endocrine, Cancer, Psychiatric/Emotional, Hematologic Lymphatic
Allergy history	Constitutional, Eyes Mowing lawn, Walking on grass Sweeping, Dusting a vacuum cleaner, Moldy areas or articles, Contact to animals, Household cleaning, Strong odors, Air conditioning, Following rainfall, Trips, Smoke, Use of hair spray, Emotional upsets, Heavy physical exertion
Allergy symptoms	Sneezing, Nasal Congestion, Runny Nose, Post Nasal Drip, Itchy Nose, Eye Symptoms, Ear problems, Headaches, Migraine, Coughing, Wheezing, Cough or wheeze with exercise, Short of breath, Recurrent Infections, Hives, Swelling Eczema, Fatigue
Allergies records	Drug, Food, Insect, Latex, Mold, Pet, Pollen
Allergies test	Blood tests, Patch tests, Elimination diet, Challenge testing
Vaccine records	Influenza, Tetanus, Pneumococcal, Hepatitis A, Hepatitis B, Coronavirus 2019
Pet history	Cat, Dog, Bird, Others
Indoor environment	Type of home, Age of home, Describe neighborhood, Air conditioning, Air cleaner, Type of mattress, Age of Mattress
Geographic history	Previous Young Adults location, Childhood location, Place of birthday location
Family allergy history	Asthma, Nose allergies, Eye allergies, Hives, Eczema, Migraine, Recurrent Pneumonia, Immune Disorder, Drug allergies, Food allergies
Family medical history	Arthritis, Heart Disease, Hypertension, Stroke, Cancer, Diabetes, Epilepsy, Bleeding Disorder, Kidney Disease, Thyroid Disease, Mental Illness, Osteoporosis

lized in this domain to not only make intelligent decisions, but also to predict future outcomes. The data on allergy disease supplied in the data source can be used to forecast a variety of future problems associated with allergy disease, such as anticipating the adverse effects of drugs and foods and minimizing their dangers in patients. Additionally, this component can make personalized recommendations based on an individual's allergies. The recommender system is a successful way for inspiring and supporting real twins to achieve health and well-being goals through the modification of twin behavior and health habits.

16.3.3 Multimodal interaction

Through a number of methods, DT can communicate with other digital and real twins. Indeed, this layer enables the representation of DT via videos, ARs, VRs, holograms, haptics, and robotics. This form of display enables these twins to communicate with each other and with the real en-

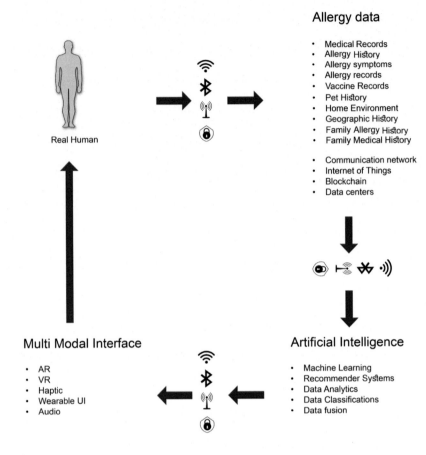

FIGURE 16.3   An ecosystem of the DT for allergy disease.

vironment. This structure is based on the preferences of real twins. This is critical in the case of allergic disorders. Because specialists can test various types of allergies on DT and then prescribe them to real twins if the necessary findings are obtained. This can aid in the control of many allergies and significantly lower the financial expenditures associated with allergy treatment. Fig. 16.3 depicts an ecosystem of the DT for allergy disease. Other critical aspects for bringing a great DT to life include the following:

- **Privacy:** Every person's allergy information is very important and very sensitive. Because each person may be allergic to a specific food, medicine, weather, etc. Therefore, maintaining the privacy profile of each patient is significantly crucial. DTs can determine what information can be shared with other digital and real twins. DTs can allow the sharing of data which directly affects allergies, such as weather, geographical location, allergy symptoms, food intake, or medication. But real twin consent is required to share this data. Sending physiological data to the clinical intelligence service is announced in a private manner and security measures are observed.

- **Communication:** The many components of the DT ecosystem for allergies are linked. Connecting disparate technologies, such as the IoT, requires a high-quality communication network, including 4G, 5G, or even 6G [36]. A way of communication that can send crucial sounds, information, and data. Cloud and fog computing technologies are also important because the predicted volume of data in allergic disease is considerable. These technologies can help achieve the utmost level of quality.
- **Feedback loops:** Interaction with real twins is largely used to provide feedback on drug, food, and other sensitivities in order to raise the subject's awareness of their current physical and emotional state. This link must frequently occur in real time in order to provide rapid feedback or motivation to the real twin. As a result, real twins can take suitable actions or follow the recommendations given to improve their well-being. This feedback is extremely helpful in diagnosing anaphylactic shocks, which will be discussed in more details in the following section as a case study.

16.4 Case study: anaphylaxis shocks

Anaphylaxis is a life-threatening allergic reaction in individuals who have been sensitized to a specific antigen by past exposure [37]. It is a serious, life threatening systemic hypersensitivity reaction that should be diagnosed as soon as possible. The subsequent stage will discuss the cause and symptoms of anaphylactic shocks.

Causes of anaphylaxis: Food allergy is the most common cause of anaphylaxis, followed by medications [60], and then animals and Hymenoptera (bees, wasps, and ants) bites and stings [58], [39].

Signs of anaphylaxis: Anaphylaxis's most common clinical symptoms are cutaneous, followed by respiratory, gastrointestinal, cardiovascular, and neurological problems [58] including urticaria, angioedema, respiratory distress, nausea, vomiting, diarrhea, and hypotension [60]. Every year, many people die as a result of the side effects of drugs and diet [69]. Anaphylactic shock prediction and diagnose may benefit greatly from the use of DTs. Using this technology, the elements that triggers anaphylaxis in the DT may be evaluated, and if the DT has a proper reaction to either drug or food, they are prescribed to the real twins. For great evaluation

TABLE 16.2   Services which are use in anaphylaxis shocks.

Service name	Descriptions
DS-Food-services	The Digital Shadow of food service could identify crucial materials of food which might lead to anaphylaxis shocks.
DS-Drug-services	The Digital Shadow of drug service could identify crucial materials of drug might which lead to anaphylaxis shocks.
DS-Animal-Allergy	The Digital Shadow of animal's allergy service could identify crucial materials of animal bites and stings which might lead to anaphylaxis shocks.

of anaphylaxis shocks from diets, drugs, and animals' bites and stings, in the first stage the DT of the drug, the DT of food, and the human DT for animals' bites and stings allergy should be constructed, and then the assessment of their symptoms in anaphylaxis shocks is evaluated in the DT and either suitable food and drugs is prescribed to a real human or alerts are sent to them in order to notify anaphylaxis shocks caused by animal's bites and stings. There is a life cycle in the food, medicine, and animal bites and stings, because food may spoil, the drug has an expiration date, and animal bites and stings have limited life cycle. On the other hand, we do not need to use a two-way communication between the real world and the digital world in this study to obtain information on these factors to investigate anaphylactic. Therefore, Digital Shadow (DS) can be used instead of DT to produce these three elements. In a DT, however, both virtual and physical entities must communicate with each other, in DS, a virtual model has one-way data flow with a physical model [12]. Life cycle is a key feature of DS / DT concepts that show an up-to-date view of a real estate in cyberspace. The DS can be seen as the preliminary stage of a DT. The DS should be interpreted as a simple assignment of status data by a measurement to a specific asset at a specific point in time t_n. From a scientific point of view, DS has no claim to knowledge production, but it forms the basis of statistical analysis between different samples of an asset or between steps in the process chain. The DS may not include all the physical features of an asset, but only those that relate to its primary purpose. The DS relies on data analysis (based on unsupervised learning) and a knowledge inference engine, allowing events to be identified as well as being able to decode them [6]. Our main goal in this case study is to identify the elements that are involved in anaphylaxis shock. [66] shows schematic visualization of both 'one-way' and 'bi-directional' data flow in order to distinguish DS from DT. We use this concept and present DT and DS, for three main elements of anaphylaxis shocks such as food (egg), drug (penicillin), and bite (bee sting) which are demonstrated in Fig. 16.4. Table 16.2 also illustrates description of three DS services which are used in Algorithm 1

We hope to develop DT in anaphylactic shocks; not only will it be able to mimic the many organs of people at risk of anaphylaxis shocks, but it

Algorithm 1: A DT of anaphylaxis shocks.

```
/* these data presents in Table 16.1                              */
```
Input: Allergan data is collected via sensors and paraclinical parameters
Output: Notify real human in order to predict and diagnose anaphylaxis shocks
 caused by food, drug, and animal bites and stings.
Data:
 DS-Food-services: Digital Shadow of food service
 DS-Drug-services: Digital Shadow of drug service
 DS-Animal-Allergy-Service: Digital Shadow of animals' bites **Anaphylaxis**
shocks: Drug, food, animal bits, and stings
 Anaphylaxis shocks symptoms: cardiovascular, angioedema, etc.

1 **Begin**
2 The sensors and paraclinical data management systems collect all allergy data;
3 Allergan data is gathered and stored in a data source;
4 The AI interface layer extracted relevant knowledge from the data source of Allergy;
5 Multimedia models are used to simulate the acquired knowledge.
6 DT is constructed.
```
   /* Evaluation anaphylaxis shocks cause by drug for example penicillin    */
```
7 Call DS-Drug-services;
8 **while** *test drug for anaphylaxis shocks in DT* **do**
9   Evaluate drug anaphylaxis shocks symptoms in DT
10   **if** *DTs positively react to anaphylaxis shocks* **then**
11   Prescribe drug to the real twins;
12   **else**
13   Repeat until positive feedbacks
```
          /* The feedback loop could be mannered by physician.       */
```

```
   /* Evaluation anaphylaxis shocks cause by food for example egg         */
```
14 Call DS-food-services;
15 **while** *test food for anaphylaxis shocks in DT* **do**
16   Evaluate food anaphylaxis shocks symptoms in DT;
17   **if** *DTs positively react to anaphylaxis shocks* **then**
18   Prescribe drug to the real twins;
19   **else**
20   Write "Anaphylaxis shock warning: 'Type of food' don't eat"

```
    /* Evaluation anaphylaxis shocks cause by animals' bites for example bee
       stings                                                              */
```
21 Call DS-Animal-Allergy-Service;
22 **while** *test animal allergy bites for anaphylaxis shocks in DT* **do**
23   Evaluate animal allergy bites anaphylaxis shocks symptoms in DT
24   **if** *DTs positively react to anaphylaxis shocks* **then**
25   write "Type of animals:' No Problem";
26   **else**
27   write "Anaphylaxis shock warning"

28 Update DT of human;
29 Recommend appropriate either drug or food to individuals;
30 Predict the future of Real human anaphylaxis shocks;
31 **End.**

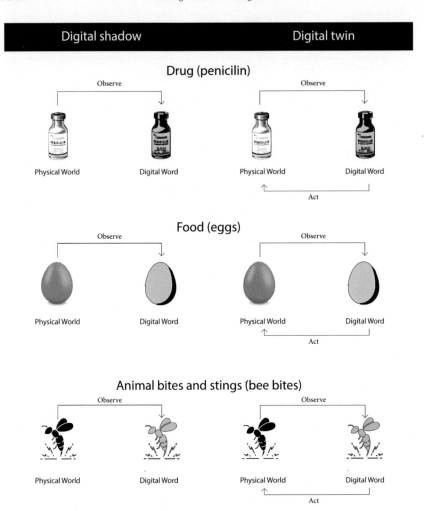

FIGURE 16.4   Data flow between physical world and digital world for both DT and DS for food (egg), drug (penicillin), and animals' bites (bees bites).

will also be able to screen, forecast, diagnose, and cure anaphylaxis shocks using ML. Fig. 16.5 depicts real-world human organs and human DT. The type of anaphylactic shocks illustrated in this image is for real individuals. Furthermore, the human DT will be created using the DT production procedure depicted in Fig. 16.2.

16.5 Discussion

Conventional allergy treatment has a number of difficulties, including a high degree of intricacy, unpredictability, and uncertainty. DT is capable of

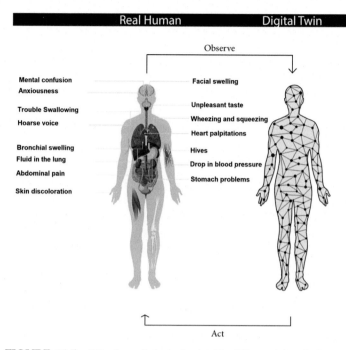

Real Human   Digital Twin

Observe

Mental confusion
Anxiousness

Facial swelling

Trouble Swallowing
Hoarse voice

Unpleasant taste

Wheezing and squeezing

Heart palpitations

Bronchial swelling
Fluid in the lung
Abdominal pain

Hives

Drop in blood pressure

Stomach problems

Skin discoloration

Act

FIGURE 16.5 DTs of anaphylaxis shocks. The Effects of Anaphylaxis on the Body illustrated in real humans which are reported in [22]. Moreover, the human DT will be made on the basis of the ecosystem of the DT for allergy disease.

resolving these issues. This section discusses the benefits, future dynamics, and development hurdles associated with the use of DT in allergic disorders in further details.

Benefits: Utilizing DT in allergies research has the following benefits:

- With the aid of DT, personalized allergies that include personalized medicine and treatments could be modeled.
- Allergy to drugs and resistance ratio can be evaluated with high accuracy in individuals.
- Detecting the optimal allergy pathway and optimal allergy monitoring in each person.
- With the help of DT, doctors could predict and diagnose the possibility of allergies in individuals.
- In recent years, allergies have resulted in huge financial and personal losses. Many deaths occurred as a result of medicine and food adverse effects. With the help of DT, it is possible to reduce the risks of drug, food, and insect sensitivity.

Future dynamics: As a possible future dynamic of research in this field, we might consider the possibility of some components of DT evolving to have amazing capacities for managing allergies and disorders. For instance, the advancement of AI toward artificial superintelligence and perfect cognitive systems may result in the discovery of a new allergy that humans are unable to explore due to the human mind's and abilities' limitations. Another example is that the evolution of the IoT to the nanoscale and molecular level may be utilized to monitor and manage any form of allergy.

Developmental hurdles: Among the primary development hurdles that we may encounter while transitioning from existing solutions to DT-based solutions are the following:

- Inadequate data for the construction of multiscale models and associated information. Allergy data are not centralized, and it may take multiple studies to get a proper model for each component of allergies.
- Because of the huge amount of data and massive number of participants, and technologies, security and privacy-preserving solutions are costly and complicated.
- We may require the creation of new professions to model allergy entities on the basis of a company comprised of a diverse group of professionals, including clinical immunologist, pathologists, nutritionist, and pharmacist.

Open Research: Several open research directions can be followed, which some of them are explained:

- To develop a fully functional DT for allergens, many stakeholders, including scientists and physicians, should collaborate. Throughout this collaboration, individuals from a variety of fields, including the food industry, medicine, and computing, aim to extract insights from DT in order to design effective allergy regimens.
- DT will introduce a revolutionary technique to treat allergies on an individual basis. Physicians can analyze various different treatment approaches for a patient's DT in order to determine the most appropriate one. A strategy that integrates all capabilities gained by DT and conventional treatment may be necessary.
- Although DT has a high potential for allergens, this is the first study of its kind undertaken by a minority on DT and allergies. Additional research can be conducted in the future.
- Some researchers could focus their future work on the application of DTs to various forms of allergies.
- Individuals of different ages suffer from a variety of allergies. As a result, future research will focus on a specific age range, such as children.
- Recommender systems could define a unique reward situation based on a user's health status. The reward system can be described in terms of cryptocurrency using blockchain technology and smart contracts.

- With the assistance of a cognitive system, the DT can learn and update itself in an unknown environment. As a result, the DT might become self-organized over time and adapt to new situations. Even so, this system can utilize the cognitive engines to build a cognitive recommendation system based on IoT and blockchain that operates completely autonomously.
- Although many studies have been carried out on anaphylactic shock testing for the Covid-19 vaccine [8], by using DT, physicians could test different types of Covid-19 vaccinations for variety of doses in a personalized way, and after evaluation of DT reactions, the suitable vaccine could be injected.

16.6 Conclusion

Many people die each year as a result of the informative effects of drugs. Using the potential of DT, the risks of allergy diseases can be reduced or eliminated. This article examines DT in Allergy. To the best of our knowledge, no studies have been conducted on DT technology and allergies. In this article, we used a public health ecosystem presented by [11]. In this ecosystem, user data about the allergy of patient is received through intelligent sensors and paraclinical information and collected in a data source. This data is then analyzed with the aid of AI and knowledge is acquired. This knowledge will be transformed into a model in three stages such as data source, AI interface, and Multimodal Interaction, and eventually, a DT will be created. To examine the existing model, Anaphylaxis shock has been studied as a case study. The authors of this article hope that this study will serve as a starting point for further research in the field of allergies and DTs.

Clearly the lessons learned

- In this chapter, we learned how digital twins are used to diagnosing, predict and treat allergies.
- This article focuses on Anaphylaxis shock as a case study. Apparently, this shock kills many people every year. With the help of a digital twin, the risk of this shock can be minimized.
- It is the starting point for a discussion about digital twins and allergies. With the help of digital twins, different allergies can be studied and minimized based on certain ages and geographical locations in the future.
- Digital Twin can simulate a human with all its features in real-time and historical data. Therefore, allergy tests can be performed using a per-

sonalized digital twin. If the test is successful, real human beings can be tested.

- Since every person has a digital twin, allergy testing is based on their individual characteristics and therefore more accurate than tests performed on laboratory animals.
- Digital twins can minimize the number of tests on laboratory animals as well as reduce animal mortality.

Acknowledgment

I wish to thank my parents Mr. Alireza Gholizadeh and Mrs. Maryam Eskandari for their love and encouragement, without whom I would never have enjoyed so many opportunities. The author would greatly appreciate Eng. Raika Monazzam for providing some graphical contents, and Eng. Rouzbeh Ghalandarzadeh for some reviews and grammatical checks.

References

[1] AAFA, Allergies and allergic reactions, https://www.aafa.org/allergies.aspx, 2021. (Accessed 18 October 2021).

[2] A. Alag, Machine learning approach yields epigenetic biomarkers of food allergy: a novel 13-gene signature to diagnose clinical reactivity, PLoS ONE 14 (2019) e0218253.

[3] B.M. Associates, Allergy and medical history, https://bergenmed.com/forms/ordered_allergy-medical-history_Date.pdf, 2021. (Accessed 3 November 2021).

[4] P. Asthana, S. Mishra, Iot enabled real time bolt based indoor air quality monitoring system, in: 2018 International Conference on Computational and Characterization Techniques in Engineering & Sciences (CCTES), IEEE, 2018, pp. 36–39.

[5] J. Barrett, Modeling and simulation strategy to support the development of ars-1 (intranasal epinephrine) for adult and pediatric subjects with systemic allergies, J. Allergy Clin. Immunol. 147 (2021) AB19.

[6] T. Bergs, S. Gierlings, T. Auerbach, A. Klink, D. Schraknepper, T. Augspurger, The concept of digital twin and digital shadow in manufacturing, Proc. CIRP 101 (2021) 81–84.

[7] L. Bielory, D.P. Skoner, M.S. Blaiss, B. Leatherman, M.S. Dykewicz, N. Smith, G. Ortiz, J.A. Hadley, N. Walstein, T.J. Craig, et al., Ocular and nasal allergy symptom burden in America: the allergies, immunotherapy, and rhinoconjunctivitis (airs) surveys, in: Allergy & Asthma Proceedings, 2014.

[8] C. COVID, R. Team, Allergic reactions including anaphylaxis after receipt of the first dose of pfizer-biontech Covid-19 vaccine—United States, December 14–23, 2020, Morb. Mort. Wkly. Rep. 70 (2020) 46.

[9] D. Dinakarpandian, Y. Lee, C. Dinakar, Applications of medical informatics in allergy/immunology, Ann. Allergy, Asthma, & Immun. 99 (2007) 2–10.

[10] A. El Saddik, Digital twins: the convergence of multimedia technologies, IEEE Multimed. 25 (2018) 87–92.

[11] A. El Saddik, H. Badawi, R.A.M. Velazquez, F. Laamarti, R.G. Diaz, N. Bagaria, J.S. Arteaga-Falconi, Dtwins: a digital twins ecosystem for health and well-being, IEEE COMSOC MMTC Commun. Front. 14 (2019) 39–43.

[12] I. Errandonea, S. Beltrán, S. Arrizabalaga, Digital twin for maintenance: a literature review, Comput. Ind. 123 (2020) 103316.

[13] K. Eswari, C. Priya, Drug identification and interaction checking using the Internet of things, An Industrial IoT Approach for Pharmaceutical Industry Growth 2 (2020) 87.

[14] M. Firdhous, B. Sudantha, P. Karunaratne, Iot enabled proactive indoor air quality monitoring system for sustainable health management, in: 2017 2nd International Conference on Computing and Communications Technologies (ICCCT), IEEE, 2017, pp. 216–221.

[15] W.H. Gerlich, D. Glebe, Development of an allergy immunotherapy leads to a new type of hepatitis b vaccine, EBioMedicine 11 (2016) 5–6.

[16] R.S. Gupta, E.E. Springston, B. Smith, M.R. Warrier, J. Pongracic, J.L. Holl, Geographic variability of childhood food allergy in the United States, Clin. Pediatr. 51 (2012) 856–861.

[17] K.G. HamlAbadi, A.M. Saghiri, M. Vahdati, M.D. TakhtFooladi, M.R. Meybodi, A framework for cognitive recommender systems in the Internet of things (iot), in: 2017 IEEE 4th International Conference on Knowledge-Based Engineering and Innovation (KBEI), IEEE, 2017, pp. 0971–0976.

[18] K.G. HamlAbadi, M. Vahdati, A.M. Saghiri, A. Forestiero, Digital twins in cancer: state-of-the-art and open research, in: 2021 IEEE/ACM Conference on Connected Health: Applications, Systems and Engineering Technologies (CHASE), 2021, pp. 199–204.

[19] Y. Haruyama, T. Yamazaki, M. Endo, R. Kato, M. Nagao, M. Umesawa, T. Sairenchi, G. Kobashi, Personal status of general health checkups and medical expenditure: a large-scale community-based retrospective cohort study, J. Epidemiol. 27 (2017) 209–214.

[20] J.L. Harvey, S.A. Kumar, Machine learning for predicting development of asthma in children, in: 2019 IEEE Symposium Series on Computational Intelligence (SSCI), IEEE, 2019, pp. 596–603.

[21] H.B. Hassen, N. Ayari, B. Hamdi, A home hospitalization system based on the Internet of things, fog computing and cloud computing, Inform. Med. Unlocked 20 (2020) 100368.

[22] Healthline, 15 effects of anaphylaxis on the body, https://bergenmed.com/forms/ordered_allergy-medical-history_Date.pdf, 2021. (Accessed 6 November 2021).

[23] V.C. Henrich, L.A. Orlando, B.H. Shirts, The growing medical relevance and value of family health history, Managing Health in the Genomic Era: A Guide to Family Health History and Disease Risk (2020) 1.

[24] S. Hofmaier, X. Huang, P.M. Matricardi, Telemedicine and mobile health technology in the diagnosis, monitoring and treatment of respiratory allergies, in: Implementing Precision Medicine in Best Practices of Chronic Airway Diseases, Elsevier, 2019, pp. 117–124.

[25] S. Holgate, M. Church, D. Broide, F. Martinez, Allergy, Elsevier, 2011.

[26] M.C. Hsieh, J. Lee, Preliminary study of vr and ar applications in medical and healthcare education, J. Nurs. Health Stud. 3 (2018) 1.

[27] A.J. Jara, A.F. Alcolea, M. Zamora, A.G. Skarmeta, M. Alsaedy, Drugs interaction checker based on iot, in: 2010 Internet of Things (IOT), IEEE, 2010, pp. 1–8.

[28] A.J. Jara, F.J. Belchi, A.F. Alcolea, J. Santa, M.A. Zamora-Izquierdo, A.F. Gómez-Skarmeta, A pharmaceutical intelligent information system to detect allergies and adverse drugs reactions based on Internet of things, in: 2010 8th IEEE International Conference on Pervasive Computing and Communications Workshops (PERCOM Workshops), IEEE, 2010, pp. 809–812.

[29] A.J. Jara, M.A. Zamora, A.F. Skarmeta, Drug identification and interaction checker based on iot to minimize adverse drug reactions and improve drug compliance, Pers. Ubiquitous Comput. 18 (2014) 5–17.

[30] W. Kamchaisatian, W. Wanwatsuntikul, J.W. Sleasman, N. Tangsinmankong, Validation of current joint American academy of allergy, asthma & immunology and American college of allergy, asthma and immunology guidelines for antibody response to the 23-valent pneumococcal vaccine using a population of hiv-infected children, J. Allergy Clin. Immunol. 118 (2006) 1336–1341.

[31] T. Kanaoka, K. Matsuoka, M. Shaker, Safe and effective intradermal influenza vaccine desensitization for delayed influenza vaccine allergy, Ann. Allergy, Asthma, & Immun. 120 (2018) 666–667.

[32] J.D. Kattan, G.N. Konstantinou, A.L. Cox, A. Nowak-Węgrzyn, G. Gimenez, H.A. Sampson, S.H. Sicherer, Anaphylaxis to diphtheria, tetanus, and pertussis vaccines among children with cow's milk allergy, J. Allergy Clin. Immunol. 128 (2011) 215–218.

[33] R. Kavya, J. Christopher, S. Panda, Y.B. Lazarus, Machine learning and xai approaches for allergy diagnosis, Biomed. Signal Process. Control 69 (2021) 102681.

[34] J.L. Kennedy, S.M. Jones, N. Porter, M.L. White, G. Gephardt, T. Hill, M. Cantrell, T.G. Nick, M. Melguizo, C. Smith, et al., High-fidelity hybrid simulation of allergic emergencies demonstrates improved preparedness for office emergencies in pediatric allergy clinics, J. Allergy Clin. Immunol., Practice 1 (2013) 608–617.

[35] A. Keswani, J.P. Brooks, P. Khoury, The future of telehealth in allergy and immunology training, J. Allergy Clin. Immunol., Practice 8 (2020) 2135–2141.

[36] L.U. Khan, W. Saad, D. Niyato, Z. Han, C.S. Hong, Digital-twin-enabled 6g: vision, architectural trends, and future directions, arXiv preprint, arXiv:2102.12169, 2021.

[37] L.U. Khan, W. Saad, D. Niyato, Z. Han, C.S. Hong, Digital-twin-enabled 6g: vision, architectural trends, and future directions, arXiv preprint, arXiv:2102.12169, 2021.

[38] J. Klier, D. Lindner, S. Reese, R.S. Mueller, H. Gehlen, Comparison of four different allergy tests in equine asthma affected horses and allergen inhalation provocation test, J. Equine Vet. Sci. 102 (2021) 103433.

[39] J.H. Klotz, S.A. Klotz, J.L. Pinnas, Animal bites and stings with anaphylactic potential, J. Emerg. Med. 36 (2009) 148–156.

[40] J.J. Koplin, K.J. Allen, L.C. Gurrin, R.L. Peters, A.J. Lowe, M.L. Tang, S.C. Dharmage, et al., The impact of family history of allergy on risk of food allergy: a population-based study of infants, Int. J. Environ. Res. Public Health 10 (2013) 5364–5377.

[41] V. Koufi, F. Malamateniou, G. Vassilacopoulos, Ubiquitous access to cloud emergency medical services, in: Proceedings of the 10th IEEE International Conference on Information Technology and Applications in Biomedicine, IEEE, 2010, pp. 1–4.

[42] K.N. Kunze, E.M. Polce, J. Rasio, S.J. Nho, Machine learning algorithms predict clinically significant improvements in satisfaction after hip arthroscopy, Arthroscopy 37 (2021) 1143–1151.

[43] F. Laamarti, H.F. Badawi, Y. Ding, F. Arafsha, B. Hafidh, A. El Saddik, An iso/IEEE 11073 standardized digital twin framework for health and well-being in smart cities, IEEE Access 8 (2020) 105950–105961.

[44] M. Lauridsen, P. Mogensen, T.B. Sorensen, Estimation of a 10 gb/s 5g receiver's performance and power evolution towards 2030, in: 2015 IEEE 82nd Vehicular Technology Conference (VTC2015-Fall), IEEE, 2015, pp. 1–5.

[45] B. Li, Q. Dong, R.S. Downen, N. Tran, J.H. Jackson, D. Pillai, M. Zaghloul, Z. Li, A wearable iot aldehyde sensor for pediatric asthma research and management, Sens. Actuators B, Chem. 287 (2019) 584–594.

[46] R.M. Lo, N. Purington, S.A. McGhee, M.B. Mathur, G.M. Shaw, A.R. Schroeder, Infant allergy testing and food allergy diagnoses before and after guidelines for early peanut introduction, J. Allergy Clin. Immunol., Practice 9 (2021) 302–310.

[47] Y. Lu, X. Zheng, 6g: a survey on technologies, scenarios, challenges, and the related issues, J. Ind. Inform. Integr. 100158 (2020).

[48] R.W. Lucas, J. Dees, R. Reynolds, B. Rhodes, R.W. Hendershot, Cloud-computing and smartphones: tools for improving asthma management and understanding environmental triggers, Ann. Allergy, Asthma, & Immun. 114 (2015) 431–432.

[49] M.F. Bublitz, A. Oetomo, S.K. Sahu, A. Kuang, X.L. Fadrique, E.P. Velmovitsky, M.R. Nobrega, P. Morita, Disruptive technologies for environment and health research: an overview of artificial intelligence, blockchain, and Internet of things, Int. J. Environ. Res. Public Health 16 (2019) 3847.

[50] M.N. Mahdi, A.R. Ahmad, Q.S. Qassim, H. Natiq, M.A. Subhi, M. Mahmoud, From 5g to 6g technology: meets energy, Internet-of-things and machine learning: a survey, Appl. Sci. 11 (2021) 8117.

[51] K. Manavi, B. Jacobson, B. Hoard, L. Tapia, Influence of model resolution on geometric simulations of antibody aggregation, Robotica 34 (2016) 1754–1776.

[52] R. Miron, M. Hulea, S. Folea, Food allergens monitoring system backed-up by blockchain technology, in: 2020 IEEE International Conference on Automation, Quality and Testing, Robotics (AQTR), IEEE, 2020, pp. 1–4.

[53] H. Mohabatkar, M. Mohammad Beigi, K. Abdolahi, S. Mohsenzadeh, Prediction of allergenic proteins by means of the concept of chou's pseudo amino acid composition and a machine learning approach, Med. Chem. 9 (2013) 133–137.

[54] R.G.N. Ngassam, R. Ologeanu-Taddei, J. Lartigau, I. Bourdon, A use case of blockchain in healthcare: allergy card, in: Blockchain and Distributed Ledger Technology Use Cases, Springer, 2020, pp. 69–94.

[55] R.G.N. Ngassam, R. Taddei, I. Bourdon, J. Lartigau, Digital service innovation enabled by the blockchain use in healthcare: the case of the allergic patients ledger, in: R&D management conference, 2019.

[56] S.İ. Omurca, E. Ekinci, B. Çakmak, S.G. Özkan, Using machine learning approaches for prediction of the types of asthmatic allergy across the Turkey, Data Sci. Appl. 2 (2019) 8–12.

[57] N.R. Pagani, M.A. Moverman, R.N. Puzzitiello, M.E. Menendez, C.L. Barnes, J.J. Kavolus, Preoperative allergy testing for patients reporting penicillin and cephalosporin allergies is cost-effective in preventing infection after total knee and hip arthroplasty, J. Arthroplast. 36 (2021) 700–704.

[58] H.J. Park, S.H. Kim, Factors associated with shock in anaphylaxis, Am. J. Emerg. Med. 30 (2012) 1674–1678.

[59] M.A.M. Pla, L.G. Lemus-Zúñiga, J.M. Montañana, J. Pons, A.A. Garza, A review of mobile apps for improving quality of life of asthmatic and people with allergies, Innovation in Medicine and Healthcare 2015 (2016) 51–64.

[60] P. Poowuttikul, D. Seth, Anaphylaxis in children and adolescents, Immunol. Allergy Clin. 41 (2021) 627–638.

[61] J.M. Portnoy, A. Pandya, M. Waller, T. Elliott, Telemedicine and emerging technologies for health care in allergy/immunology, J. Allergy Clin. Immunol. 145 (2020) 445–454.

[62] J. Rong, S. Michalska, S. Subramani, J. Du, H. Wang, Deep learning for pollen allergy surveillance from Twitter in Australia, BMC Med. Inform. Decis. Mak. 19 (2019) 1–13.

[63] A.M. Saghiri, K.G. HamlAbadi, M. Vahdati, The Internet of things, artificial intelligence, and blockchain: implementation perspectives, in: Advanced Applications of Blockchain Technology, Springer, 2020, pp. 15–54.

[64] A.M. Saghiri, M. Vahdati, K. Gholizadeh, M.R. Meybodi, M. Dehghan, H. Rashidi, A framework for cognitive Internet of things based on blockchain, in: 2018 4th International Conference on Web Research (ICWR), IEEE, 2018, pp. 138–143.

[65] A.M.M. Schoos, B.I. Nwaru, M.P. Borres, Component-resolved diagnostics in pet allergy: current perspectives and future directions, J. Allergy Clin. Immunol. 147 (2021) 1164–1173.

[66] S.M. Sepasgozar, Differentiating digital twin from digital shadow: elucidating a paradigm shift to expedite a smart, sustainable built environment, Buildings 11 (2021) 151.

[67] A. Smiti, When machine learning meets medical world: current status and future challenges, Comput. Sci. Rev. 37 (2020) 100280.

[68] E.R. Svendsen, M. Gonzales, A. Commodore, The role of the indoor environment: residential determinants of allergy, asthma and pulmonary function in children from a us-Mexico border community, Sci. Total Environ. 616 (2018) 1513–1523.

[69] C. Tacquard, D. Chassard, J.M. Malinovsky, M. Saucedo, C. Deneux-Tharaux, P.M. Mertes, Anaphylaxis-related mortality in the obstetrical setting: analysis of the French national confidential enquiry into maternal deaths from 2001 to 2012, Br. J. Anaesth. 123 (2019) e151–e153.

[70] P.J. Turner, H. Larson, È. Dubé, A. Fisher, Vaccine hesitancy: drivers and how the allergy community can help, J. Allergy Clin. Immunol., Practice 9 (2021) 3568–3574.

[71] S. Tyagi, A. Agarwal, P. Maheshwari, A conceptual framework for iot-based healthcare system using cloud computing, in: 2016 6th International Conference-Cloud System and Big Data Engineering (Confluence), IEEE, 2016, pp. 503–507.

[72] M. Vahdati, K.G. HamlAbadi, A.M. Saghiri, Iot-based healthcare monitoring using blockchain, in: Applications of Blockchain in Healthcare, Springer, 2021, pp. 141–170.

[73] M. Vahdati, K.G. HamlAbadi, A.M. Saghiri, H. Rashidi, A self-organized framework for insurance based on Internet of things and blockchain, in: 2018 IEEE 6th International Conference on Future Internet of Things and Cloud (FiCloud), IEEE, 2018, pp. 169–175.

[74] C.M. Visness, S.J. London, J.L. Daniels, J.S. Kaufman, K.B. Yeatts, A.M. Siega-Riz, A.H. Liu, A. Calatroni, D.C. Zeldin, Association of obesity with ige levels and allergy symptoms in children and adolescents: results from the national health and nutrition examination survey 2005–2006, J. Allergy Clin. Immunol. 123 (2009) 1163–1169.

[75] W. Wu, S. Pirbhulal, A.K. Sangaiah, S.C. Mukhopadhyay, G. Li, Optimization of signal quality over comfortability of textile electrodes for ecg monitoring in fog computing based medical applications, Future Gener. Comput. Syst. 86 (2018) 515–526.

[76] G.K. Zewdie, D.J. Lary, E. Levetin, G.F. Garuma, Applying deep neural networks and ensemble machine learning methods to forecast airborne ambrosia pollen, Int. J. Environ. Res. Public Health 16 (2019) 1992.

Index

Printed in the United States
by Baker & Taylor Publisher Services